AF699302

WITHDRAWN

ALLE·ZEIT·WACH
1842

Communications and Control Engineering Series

Editors: A. Fettweis · J. L. Massey · M. Thoma

Scientific Fundamentals of Robotics 5

M. Vukobratović
D. Stokić
N. Kirćanski

Non-Adaptive and Adaptive Control of Manipulation Robots

With 111 Figures

Springer-Verlag
Berlin Heidelberg New York Tokyo

D. Sc., Ph. D. MIOMIR VUKOBRATOVIĆ, corr. member of
Serbian Academy of Sciences and Arts

Ph. D. DRAGAN STOKIĆ
Ph. D. NENAD KIRĆANSKI
Institute »Mihailo Pupin«, Belgrade
Volgina 15, POB 15, Yugoslavia

TJ
211
.V845
1985

ISBN 3-540-13073-X Springer-Verlag Berlin Heidelberg New York Tokyo
ISBN 0-387-13073-X Springer-Verlag New York Heidelberg Berlin Tokyo

Library of Congress Cataloging in Publication Data.

Vukobratović, Miomir.
Non-adaptive and adaptive control of manipulation robots.
(Scientific fundamentals of robotics ; 5)
(Communications and control engineering series)
Includes bibliographies and index. 1. Robotics. 2. Manipulators (Mechanism)
I. Stokić, D. (Dragan) II. Kirćanski, N. (Nenad), 1953– . III. Title. IV. Series.
V. Series: Communications and control engineering series.
TJ211.V845 1985 629.8'92 85–17340
ISBN 0-387-13073-X (U.S.)

This work is subject to copyright. All rights are reserved, whether the whole or part of the material is concerned, specifically those of translation, reprinting, re-use of illustrations, broadcasting, reproduction by photocopying machine or similar means, and storage in data banks. Unter §54 of the German Copyright Law where copies are made for other than private use, a fee is payable to »Verwertungsgesellschaft Wort«, Munich.

© Springer-Verlag, Berlin, Heidelberg 1985
Printed in Germany

The use of registered names, trademarks, etc. in this publication does not imply, even in the absence of a specific statement, that such names are exampt from the relevant protective laws and regulations and therefore free for general use.

Offsetprinting: Mercedes-Druck, Berlin
Bookbinding: Lüderitz & Bauer, Berlin
2161/3020 5 4 3 2 1 0

Preface

The material presented in this monograph is a logical continuation of research results achieved in the control of manipulation robots. This is in a way, a synthesis of many-year research efforts of the associates of Robotics Department, Mihailo Pupin Institute, in the field of dynamic control of robotic systems. As in Vol. 2 of this Series, all results rely on the mathematical models of dynamics of active spatial mechanisms which offer the possibility for adequate dynamic control of manipulation robots.

Compared with Vol. 2, this monograph has three essential new characteristics, and a variety of new tasks arising in the control of robots which have been formulated and solved for the first time.

One of these novelties is nonadaptive control synthesized for the case of large variations in payload parameters, under the condition that the practical stability of the overall system is satisfied. Such a case of control synthesis meets the actual today's needs in industrial robot applications.

The second characteristic of the monograph is the efficient adaptive control algorithm based on decentralized control structure intended for tasks in which parameter variations cannot be specified in advance. To be objective, this is not the case in industrial robotics today. Thus, nonadaptive control with and without a particular parameter variation is supplemented by adaptive dynamic control algorithms which will certainly be applicable in the future industrial practice when parametric identification of workpieces will be required.

The third characteristic of this monograph is the presentation of a package for the synthesis and verification of control laws for manipulation robots with up to six degrees of freedom. Announced in Vol. 2 of the Series, this package is described in detail in the present volume. These details include: robot data entry and specification of control task, synthesis of trajectories, verification of the practical stability of the complete nonlinear robot model, simulation of the overall system, final selection of an adequate control law and microcomputer implementation. To this new quality of the monograph one should add the result relating to a detailed analysis of the numerical complexity of both nonadaptive and adaptive control. Among other points, this has also permitted formulation of the concept of a general controller for dynamic control of manipulation robots.

The monograph contains five chapters. Computer-assisted generation of dynamic robot models in analytical form is presented in Chapter 1. To permit efficient application of the models of robot mechanism dynamics to the synthesis of control algorithms, analytical representation of the variables of dynamic models and construction of real-time program code are described. It should be stated that the construction of real-time dynamics of manipulation robots is treated in detail in Vol. 4 of this Series, and only a brief description is repeated here to give autonomy to this text.

The task of nonadaptive control of manipulation robots with variable parameters is considered in Chapter 2. In contrast to Vol. 2, the problem solved in this monograph is the control robust to large payload variation - the most commonly used parameter in industrial manipulation. This chapter also provides a survey of nonadaptive control algorithms and an insight into various techniques employed in control synthesis, including the optimal regulator, "inverse problem" technique, force feedback, and decoupled and decentralized control. Algorithms for the synthesis of local controllers are described in detail. Particular attention is devoted to optimal servosystems, robust servoregulators, and regulators based on pole-placement method. A method for analyzing the practical stability of robots with variate parameters when only local controllers are applied is developed. Global control is also introduced either via force feedback or via on-line calculated dynamic robot model. Algorithms for the synthesis of both local and global control which are directly applicable to the computer-aided control synthesis as presented in Chapter 4, are described in detail.

Chapter 3 is devoted to the adaptive control of manipulators. Two concepts are considered, one based on the centralized and the other on the decentralized robot model. The application of self-tuning control strategy, algorithms based on hyperstability theory, and some quasi-gradient algorithms based on Lyapunov's theory are discussed within the first concept. However, the algorithms for decentralized control are much more important for practical application, because of the considerably lower numerical complexity.

Chapter 4 gives a description of the software package for robot control synthesis based on the theoretical results presented in previous chapters. All blocks of the software package, such as specification of kinematic and dynamic parameters of the mechanism, parameters of actuators, manipulation task, etc., are described in detail.

Chapter 5 is devoted to the implementation of control algorithms on today's low-cost microcomputers. The concept of a general-purpose controller, i.e., a controller capable of automatic generation of the model for a specified robot, is described. Examples of robot parameter identification by the control system are also presented. Numerous results obtained on a real manipulator are given at the end to illustrate the flexibility of control laws described in Chapters 2 and 3. They allow the readers to obtain an insight into the possibilities for realizing various types of control as well as into the advantages achievable by introducing dynamic control of robots.

This monograph is intended for all researchers in applied and theoretical robotics, robot designers, and for post graduate students of robotics. The background required for following the text includes a knowledge of the basic results from linear and nonlinear systems theory as well as a fundamental knowledge of mechanisms dynamics. Although the procedures from modern theory of large-scale nonlinear systems are used and developed in the monograph, the book has been conceived so as to allow easy following of the text and application of results to readers who are not very familiar with this theory.

The authors are grateful to Miss G. Aleksić for her help in preparing English version of this book. Our special appreciation goes to Miss V. Ćosić for her careful and excellent typing of the whole text.

May 1985
Belgrade, Yugoslavia

The Authors

Contents

Chapter 1
Computer-Assisted Generation of Robot Dynamic Models in Analytical Form

1.1 Introduction

The application of general control theory to complex mechanical systems, such as robots, aircrafts, complex hydraulic systems, etc., represents an extremely difficult problem because of the prominent nonlinearity and complexity of mathematical models of these systems. With industrial robots, which will be treated in this book, the application of such theory and the development of new control algorithms are unavoidable in order to achieve high positioning speed and accuracy. Several control concepts employing partially simplified or even exact mathematical models in on-line operation have so far been brought to an advanced level. In on-line control, the calculation of model equations must be repeated very often, preferably at sampling frequency which is not lower than 50 Hz, since the resonance frequency of most mechanical manipulators is about 10 Hz. However, the problem of forming the dynamic equations of manipulators in real time, i.e. after each 20ms or less, by means of today's microcomputers is rather difficult and complex.

In explaining a new path to solve this problem, which we have followed in this chapter, let us first mention some results in robot modelling achieved to date. It is possible to systematize these results according to different criteria. So, methods may be divided with respect to the laws of mechanics on the basis of which motion equations are formed. Taking this as a criterion, one may distinguish methods based on Lagrange's, Newton-Euler's, Appel's and other equations. The property of whether the method permits the solution of the direct or of the inverse problem of dynamics may represent another criterion. The direct problem of dynamics refers to determining the motion of the robot for known driving forces (torques), and the inverse problem of dynamics to determining the driving forces for known motion. Clearly, the methods allowing both problems of dynamics to be solved are of particular importance. The number of floating-point multiplications (divisions)/additions (subtractions) required to form the model is yet another criterion for comparing the methods. This criterion is certainly the most important one from the point of view of their on-line applicability.

The first result in the class of Lagrange's methods, which are used to solve either the direct or the inverse problem of dynamics, was reported by J.J.Uicker, [1]. Since the method referred mostly to certain classes of closed-loop mechanisms, M.Kahn [2] elaborated an algorithm for the modelling of open kinematic chains. The method was modified and preformulated by L.Woo and F.Freundenstein [3] through introducing screw-calculus, and by A.Yang [4]. N.Orlandea and T.Berenui gave a program implementation of this method [5] in the form of a program package for dynamic analysis of robots. Improved versions of this method in the sense of a reduced number of numerical operations were developed by S. Mahil [6] and M.Renaud [7]. A property common to these algorithms is that they do not employ recursive relations but closed-form expressions for the elements of model matrices. However, J.Hollerbach showed [8] that the number of multiplications/additions in these methods depended on n^4 (n - the number of degrees of freedom) and that over 5000 multiplications and approximately as many additions were required for manipulators with 3 degrees of freedom. The number of operations for manipulators with 6 degrees of freedom was more than 10 times larger. It is easy to show that these methods can hardly be implemented on today's microcomputers in real time. Namely, an average up-to-date 16-bit microcomputer (with an arithmetic processor or appropriate floating-point hardware support), can realize one multiplication and one addition in about 0.1ms. This, in turn means that about 200 floating-point multiplications and additions are implementable in 20 ms, and this is considerably less than the number of operations required in the above-mentioned methods.

M.Vukobratović and V.Potkonjak [9] developed an algorithm which employs recursive expressions for the construction of matrices of dynamic models. Although the number of operations was thus reduced by a number of times in comparison with Kahn-Uicker's method, even this method could hardly be implemented in real time, since it required several thousands of floating-point operations for a typical manipulator structure.

As may be seen from the preceding text, the algorithms based on Lagrange's equations, which permit either the direct or the inverse problem of dynamics to be solved, require too many computer efforts to be real-time implementable. This is why R.Waters and J.Hollerbach developed algorithms for solving only the inverse problem of dynamics on the basis of Uicker-Kahn's method. These procedures allow the determination of driving forces (torques) for a given known motion. The number of floating-point multiplications/additions was thus reduced to dependence on n^2 [10], or n [8]. These savings result from the fact that the inertia matrix of manipulator model is not explicitly calculated, i.e., the second derivatives of internal coordinates implicitly figure in manipulator equations. Therefore, such algorithms are not applicable to all control concepts (e.g. in [11]). In addition, regardless of the dependence of the number of multiplications on n, these methods were also shown to be hardly implementable in real time. So, in spite of the advanced preformulation of the algorithm, Hollerbach obtained that the number of floating-point multiplications amounted to 412n - 277, which gives about 1000 floating-point multiplications for manipulators with 3 degrees of freedom.

The application of Newton-Eulers' dynamic equations to modelling of joint-connected rigid bodies may be found in papers by T.Kane [12]. Kane derived dynamic equations not restricting his attention to robotic systems, but for a somewhat broader class including space vehicles (satellites) with several joint-connected rigid bodies. R.Huston made a concretization of Kane's results for robotic systems [13] and elaborated its program implementation. These algorithms permit either the direct or the inverse problem of dynamics to be solved using closed-form expressions for the elements of matrices of mathematical model. However, similarly to Kahn-Uicker's method, the number of operations required to form the model by these algorithms is very large. A much more efficient computer method was elaborated by M.Vukobratović and J. Stepanenko [14] through introducing recursive relations into the solution of either the direct or the inverse problem of dynamics. It was shown that the number of floating-point multiplications required to be performed in this method was $1.5n^3 + 18n^2 + 223.5n$. This means that even about 900 multiplications are required for a manipulator with 3 degrees of freedom. Walker and Orin [15] gave a new organization of the calculation of this method and thus reduced the number of operations to dependence on n^2. However, the number of operations was still not sufficiently reduced to make the method suitable for real time applications. A special case of Newton-Euler's method which allows the solution to the inverse problem of dynamics was treated by Luh, Walker and Paul [16]. By expressing the equations of motion of each link exclusively in local coordinate systems, they obtained a procedure much more efficient than preceding ones. According to this algorithm, the number of floating-point multiplications is 150n - 48, and this makes about 400 operations for manipulators with 3 degrees of freedom. In addition, as already noted for Hollerbach's method [8], this algorithm is not applicable to all control concepts, since it does not give the inertia matrix of the system but only driving forces and moments. Employing the preceding algorithm, J.Luh and S.Lin [17] developed, a procedure for the computation of driving forces by parallel calculation on a computer with several central processor units. The procedure was illustrated by the example of a six-degree-of-freedom Stanford manipulator controlled by 6 processor units. It was shown that the computation of the forces (torques) of actuators required about 20 floating-point multiplications and additions per one processor unit.

It follows from the preceding text that even real-time control of manipulators with no more than 3 degrees of freedom is implementable only by using several processor units in parallel operation. This is why an essential question arises as to whether the number of floating-point operations in previous methods is really minimal. Before answering the question, let us note a property common to all mentioned methods. The described methods are independent of the type of manipulator structure and encompass the general laws of kinematics and dynamics of a system of rigid bodies, which permit the construction of mathematical models or the calculation of driving forces. One can recognize that, for a given type of manipulator configuration, the actual analytical model comprises a considerably smaller number of numerical operations. Certain indications about the validity of this hypothesis were given by M.Aldon [18], M.Vukobratović and N.Kirćanski [19, 20] and M.Renaud [20]. For example, it was shown [19] that the construction of the model of an anthropomorphic manipulator with 3 degrees of freedom required no more than 44 floating-point multiplications and 23 additions, and this is over 10 times less than in any of the above-mentioned methods. Only 352 multiplications were shown to be required for a manipulator with 5 degrees of freedom [21]. However, none of the mentioned papers gives a solution to the problem of computer-aided generation of the analytical model for an arbitrary, given type of manipulator configuration. This problem is considered in detail in the volume 4 of the series "Scientific Fundamentals of Robotics", [22], and in the paper [26]. Here, it will be presented shortly.

In the algorithm for automatic generation of the analytical model, we will assume that the parameters of a manipulator (lengths, masses, inertias, etc.) are known and will treat them as constants. Joint coordinates (angles or linear displacements corresponding to rotational or linear degrees of freedom, respectively), as well as their derivatives will be treated as independent variables, i.e., as symbols. Moreover, for the construction of matrices of dynamic model (inertia matrix, matrices of Coriolis and centrifugal effects, of gravity vector), it will be shown in Paragraph 1.2. that only joint coordinates may be taken as independent variables. Accordingly, all quantities participating in the construction of the mathematical model will be treated as functions of joint coordinates. Lagrange's method (e.g., Uicker-Kahn's) or Newton-Euler's methods (Huston-Kane's algorithm) can be used to form the matrices of the dynamic model. However, the partial derivatives of transient matrices and the inertia tensors of links figure in Lagranges's methods. Of course, it would be much more convenient if only the main moments of inertia were used instead of inertia tensors, and the appropriate closed-form expressions instead of the derivatives of transient matrices. This is why Paragraph 1.2. contains a preformulation of these methods which gives very compact expressions for the elements of dynamic model matrices. Since the procedure for this preformulation was considered in detail in [22], only the obtained closed-form expressions will be given.

Formalism for manipulating the analytical expressions using a computer will be described in Paragraph 1.3. It will be shown that any analytical expression participating in the construction of the mathematical model of a manipulator can be represented by a polynomial or by a corresponding matrix. Since these matrices provide exact descriptions of the structure of analytical expressions, i.e. polynomials, they will be referred to as "polynomial matrices".

Further on, operations between polynomial matrices allowing the construction of all elements of the dynamic model matrices will be described. It will be shown that it is very simple to implement these operations on digital computers, and, accordingly, to construct the analytical models of manipulators by means of a computer. After that, the procedure for optimal calculation of the obtained analytical model in the sense of a minimal number of floating-point multiplications/additions will be presented. Previous results have provided a basis for constructing an "EXPERT PROGRAM" which generates a program for calculating the analytical model in a desired computer language (FORTRAN, assembler, etc.). This program-code can now be implemented on the microcomputer of the control system of the robot for which the model is formed.

Operation of the described program package will be illustrated in the next paragraph using the example of Stanford manipulator with 6 degrees of freedom. It will be shown that the generated analytical model comprises a very small number of numerical operations, which, in turn, allows very short calculation time to be achieved.

1.2 A Closed Form Dynamic Model of a Robotic Manipulator

To permit precise definition of the variables participating in the construction of the mathematical model of a robot, we will first give definitions of the main notions describing the physical model of a robot mechanism. We will give the definitions of a link, kinematic and dynamic variable, joint coordinate, etc.

We consider a mechanism comprising n rigid bodies (links) interconnected by rotational or prismatic joints, where one degree of freedom of motion exists between each two bodies. The first joint performs the motion of the first link with respect to the fixed coordinate system Oxyz attached to the support.

A link of a manipulator represents a rigid body whose parameters are determined by a set $C_i\left(K_i, D_i\right)$, where K_i denotes the set of kinematic and D_i the set of dynamic parameters. The index i denotes the serial number of the link in the chain, counting from the support to the end-effector.

Sets K_i and D_i can be defined in different ways. For example, in most Lagrange's methods the kinematic parameters are defined with respect to local coordinate frames assigned to manipulator links. In each frame one axis coincides with the joint axis and the other with the common normal between the axes of the two adjacent joints. This is Denavit--Hartenberg's formalism [23] which is very convenient for kinematic analysis. In contrast to this approach, in Newton-Euler's methods [14] local frames are most often attached to mass centers of the links, and the axes are directed along the main axes of inertia. The parameters defined with respect to such coordinate systems are more convenient for dynamic analysis. Following the second approach the sets K_i and D_i are defined as

$$K_i = \left(\tilde{e}_i,\ e_{\sim i+1},\ \tilde{r}_{ii},\ \tilde{r}_{i,i+1}\right)$$

$$D_i = \left(m_i,\ \tilde{J}_i\right)$$

$\tilde{e}_i$ denotes the unit vector of the axis of the ith joint,

$e_{\sim i+1}$ is the unit vector of the (i+1)-st axis in the ith frame,

$\tilde{r}_{ik}$ denotes the vector from the center of the kth joint to the mass center of the ith link,

m_i denotes the mass of the ith link,

$\tilde{J}_i = \mathrm{diag}\left(J_{i1},\ J_{i2},\ J_{i3}\right)$ is the diagonal 3×3 matrix comprising the main moments of inertia of the ith link.

The sign ~ denotes that the corresponding vector is expressed with respect to the local coordinate frame. Thus, $\tilde{e}_i$ and $\tilde{r}_{ik}$ are expressed with respect to the local coordinate frame assigned to the link i. As pointed out, ith frame is a Cartesian coordinate system attached to the mass center of the link, with the axes aligned along the main axes of inertia. Unit vectors of these axes will be denoted as $\vec{q}_{i1}$, $\vec{q}_{i2}$ and $\vec{q}_{i3}$. An example of a link is shown in Figure 1.1. The second link of an Elbow manipulator is chosen for illustration.

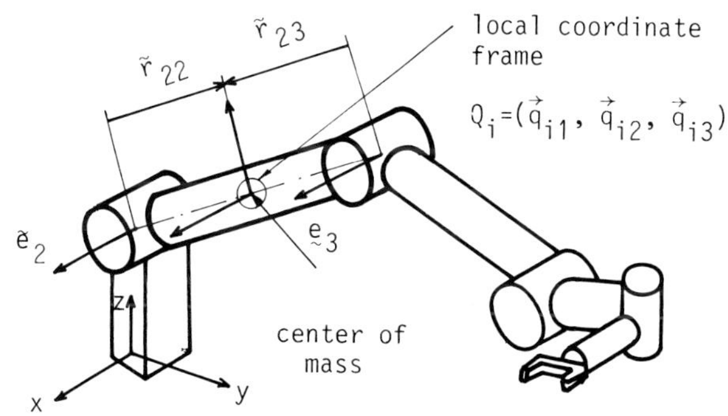

Fig. 1.1. The characteristic vectors for the second link of the Elbow manipulator

Relative position of the links is determined by joint coordinates $q_1,\dots,q_n$. In the case of a rotational joint, q_i is defined as the angle between the projections of vectors $\vec{r}_{i-1,i}$ and $\vec{r}_{ii}$ onto the plane perpendicular to the axis of the i-th joint e_i, Figure 1.2. In a sliding joint, q_i represents the linear displacement of the ith link with respect to the (i-1)-st link. The positive sense of rotation for revolute joints or displacement for prismatic joints is determined by the direction of the joint axes.

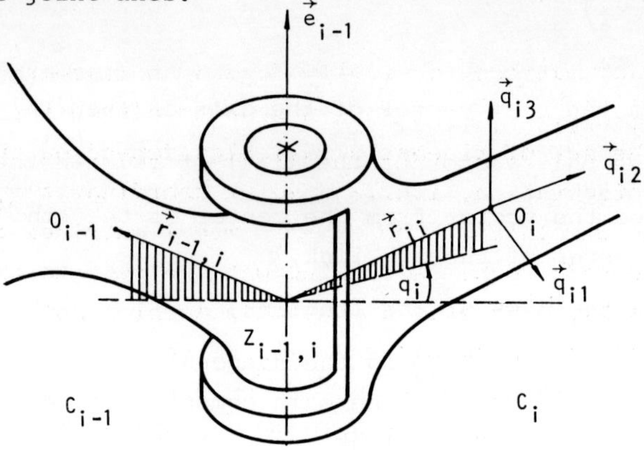

Fig. 1.2. Revolute joint

Dynamic model of a robot manipulator can be represented as [24]:

$$P = H(q, d)\ddot{q} + \dot{q}^T C(q, d)\dot{q} + g(q, d) \tag{1.2.1}$$

where H(q, d) denotes the n×n inertia matrix of the system, C(q, d) the n×n×n matrix of centrifugal and Coriolis effects, g(q, d) the n-dimensional gravitational vector, and $q = [q_1 \cdots q_n]^T$ - the joint coordinates vector. Vector d comprises the geometric and dynamic parameters of the mechanism (lengths, masses, inertias, etc.). The second term in (1.2.1) may be represented in the form

$$\dot{q}^T C(q, d)\dot{q} = \begin{bmatrix} \dot{q}^T C^1(q, d)\dot{q} \\ \vdots \\ \dot{q}^T C^n(q, d)\dot{q} \end{bmatrix} \tag{1.2.2}$$

where $C^i(q, d)$, i=1,...,n are the n×n matrices.

We have seen that the matrices of dynamic model depend on joint coordinates and parameters only. This is a very important property for obtaining an efficient algorithm for forming the analytical models of robots. It should be noted that a large number of robot modelling methods referred to in the preceding paragraph do not enable obtaining the model in form (1.2.1). For example, in some methods [9, 14] the model is obtained in form $P = H(q, d)\ddot{q} + h(q, \dot{q}, d)$ where joint velocities figure implicitly. Moreover, some methods provide only the solution of inverse dynamics problem, i.e., give the model in the form $P = P(q, \dot{q}, \ddot{q}, d)$, where all variables figure implicitly [8, 10, 17].

The dynamic model matrices obviously depend on kinematic variables $K_i = \left(Q_i, \vec{e}_i, \vec{e}_{i+1}, \vec{r}_{ii}, \vec{r}_{i,i+1}\right)$ and on dynamic parameters D_i. In addition, vectors from the set of kinematic variables depend on the instantaneous manipulator configuration, i.e., on joint coordinates $q_1,\ldots,q_n$. Determination of the elements of the dynamic model matrices therefore requires these vectors to be calculated with respect to the reference coordinate system Oxyz. This is a kinematic problem solvable in different ways. We will present a simple solution to this problem using the well known formula for finite rotations of vectors. Let us assume that the mechanism is in an initial configuration in which all joint angles are equal zero. Let the unit vectors of the link coordinate systems in this position be known, and let us denote them by $\vec{q}^{\,o}_{ij}$, $(j=1,2,3)$, $i=1,\ldots,n$. Now we perform rotation of all mechanism links about axis $\vec{e}_1$ by angle q_1. After that, we perform motion of all of mechanism links, except for the first, by angle q_2 about axis $\vec{e}_2$. This procedure is repeated to the (i-1)-th joint. In this spatial position the joint coordinates vector is $q = \left(q_1,\ldots,q_{i-1}, 0,\ldots 0\right)$. Let us now perform a rotation of the mechanism part which comprises links i, i+1,...,n by angle q_i about axis $\vec{e}_i$ (Fig. 1.3).

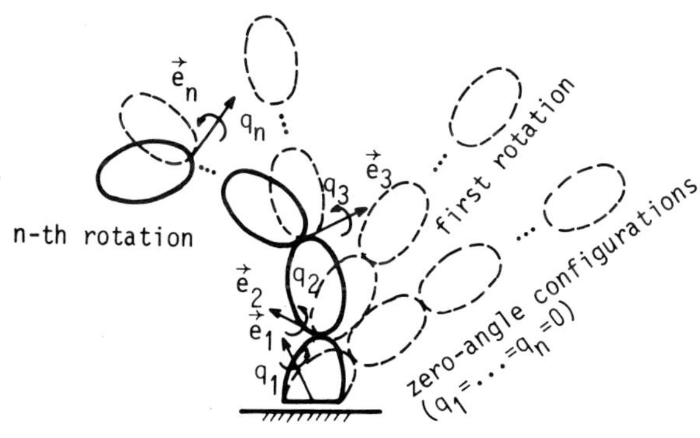

Fig. 1.3. Successive rotations

According to the finite rotation formula, we obtain that, after the rotation by angle q_i, the unit vectors of the ith link coordinate frame become

$$\vec{q}_{ij} = \vec{q}_{ij}^{(1)} \cos q_i + \vec{q}_{ij}^{(2)} \sin q_i + \vec{q}_{ij}^{(3)}, \qquad j=1,2,3 \tag{1.2.3}$$

with

$$\vec{q}_{ij}^{(1)} = \vec{e}_i \times \left(\vec{q}'_{ij} \times \vec{e}_i\right)$$

$$\vec{q}_{ij}^{(2)} = \vec{e}_i \times \vec{q}'_{ij}$$

$$\vec{q}_{ij}^{(3)} = \left(\vec{e}_i \cdot \vec{q}'_{ij}\right)\vec{e}_i$$

where $\vec{q}'_{ij}$ denotes the unit vector of the jth axis of the ith link coordinate frame before the rotation by angle q_i. If the ith joint is a prismatic one, $\vec{q}'_{ij} = \vec{q}_{ij}$ will obviously hold. Notice that the matrix $Q_i = \left[\vec{q}_{i1}\ \vec{q}_{i2}\ \vec{q}_{i3}\right]$ represents the 3×3 transformation matrix from the local coordinate system of the i-th link to the reference frame Oxyz. Accordingly, if $\underset{\sim}{e}_{i+1}$, $\tilde{r}_{ii}$ and $\tilde{r}_{i,i+1}$ denote vectors with respect to the local system of the ith link, in the fixed system these vectors will become

$$\vec{e}_{i+1} = Q_i\, \underset{\sim}{e}_{i+1}$$

$$\vec{r}_{ii} = Q_i\, \tilde{r}_{ii} \quad \text{and} \quad \vec{r}_{i,i+1} = Q_i\, \tilde{r}_{i,i+1} \tag{1.2.4}$$

If the i-th joint is a prismatic one, vector $\vec{r}_{ii}$ will become $\vec{r}_{ii} = Q_i\, \tilde{r}_{ii} + q_i \vec{e}_i$. We have thus formed all required vectors with respect to the reference system which belong to the set of kinematic variables.

Dynamic analysis of the mechanism in accordance with Newton-Euler's method comprises the calculation of the inertial forces and moments acting to the links during the motion of the mechanism. The inertial forces are obtained from Newton's law, and the moments from Euler's dynamic equations. Driving moments or forces are determined applying a kinetostatic procedure. This method obviously incorporates many laws of mechanics formulated in terms of recursive relations. The transformation of these recursive relations into nonrecursive yields the elements of dynamic model in a closed form [22]. This derivation will not be repeated here because it is rather complex; only the final expressions for the elements of the dynamic model matrices will be given.

Element (i, k) of the H(q, d) matrix is obtained in the form

$$H_{ik} = \sum_{j=\max(i,k)}^{n} \left[m_j <\vec{e}, \vec{r}>_{ji} \cdot <\vec{e}, \vec{r}>_{jk} + \vec{e}_i J_j \vec{e}_k \bar{\xi}_i \bar{\xi}_k \right] \tag{1.2.5}$$

with

$$\xi_i = \begin{cases} 0 & \text{when the } i\text{th joint is revolute} \\ 1 & \text{when the } i\text{th joint is a sliding one.} \end{cases}$$

Expression $<\vec{e}, \vec{r}>_{ij}$ represents the vector defined as

$$<\vec{e}, \vec{r}>_{ij} = \left(\vec{e}_j \times \vec{r}_{ij}\right)\bar{\xi}_j + \vec{e}_j \xi_j \tag{1.2.6}$$

with $\bar{\xi}_j = 1 - \xi_j$.

J_j represents the inertia matrix of the *j*th link with respect to the reference frame Oxyz. Having in mind that the Q_j matrix relates between the *j*th and the reference coordinate frame, one obtains

$$J_j = Q_j^T \tilde{J}_j Q_j.$$

Thus, the inertial part of expression (1.2.5) becomes

$$\vec{e}_i J_j \vec{e}_k = \sum_{\mu=1}^{3} \left(\vec{e}_i \cdot \vec{q}_{j\mu}\right)\left(\vec{e}_k \cdot \vec{q}_{j\mu}\right) J_{j\mu}$$

From the right hand side of expression (1.2.5) one can easily show that the inertia matrix is symmetric and positive definite.

Consider now $C^i(q, d)$ matrices describing centrifugal and Coriolis' effects. Element (k, ℓ) of the matrix C^i can be expressed as

$$C^i_{k\ell} = \sum_{j=\max(i,k)}^{n} \left\{ m_j <\vec{e}, \vec{r}>_{ji} \cdot \left(\vec{e}_\ell \times <\vec{e}, \vec{r}>_{jk}\right) + I^i_{k\ell} \right\} \bar{\xi}_\ell, \text{ for } k \geq \ell \tag{1.2.7}$$

Similarly to $H(q, d)$ matrix, the inertial part of the above expression may also be written as

$$I^i_{k\ell} = \frac{1}{2} \sum_{\mu=1}^{3} \left[\left(\vec{e}_i \cdot \vec{q}_{j\mu}\right)\vec{\varepsilon}_{\ell k} + \left(\vec{e}_k \cdot \vec{q}_{j\mu}\right)\vec{\varepsilon}_{i\ell} + \left(\vec{e}_\ell \cdot \vec{q}_{j\mu}\right)\vec{\varepsilon}_{ik} \right] \cdot \vec{q}_{j\mu} J_{j\mu} \bar{\xi}_k \bar{\xi}_i \tag{1.2.8}$$

It was shown in [22] that $C^i(q, d)$ matrix is symmetric and that $C^i_{k\ell} =$

$= -C^k_{i\ell}$ for $k,i > \ell$, and $C^i_{k\ell}=0$ for $i=k > \ell$ holds.

Gravitational vector g(q, d) is given by the expression

$$g(q,\ d) = -\sum_{j=i}^{n} m_j <\vec{e},\ \vec{r}>_{ji}\cdot\vec{g} \tag{1.2.9}$$

where $\vec{g}$ denotes the vector of gravitational acceleration.

At the end we may conclude that the matrices of dynamic model really depend only on internal coordinates and parameters. To form them, it is necessary to determine the kinematic variables as functions of internal coordinates, and then, by their scalar and vector product, determine the dynamic model matrices.

1.3 Analytical Representation of the Variables of Dynamic Model

In this paragraph we will give a solution to the problem of computer-aided construction of the analytical model of a robot. In order to achieve such a goal two main problems should be considered:

1) How to describe the dependence of an arbitrary model variable on joint coordinates?

2) How to realize the algebraic operations between such functions on a digital computer.

In the preceding paragraph it was shown that the elements of the dynamic model matrices are obtained by using scalar or vector products of vectors from the set of kinematic variables K_i. Dynamic parameters of mechanism D_i i.e., masses and moments of inertia of manipulator links, figure as the coefficients of proportionality. Let us denote by $\vec{k}_i$ any element from the set of kinematic variables of the ith member, i.e., let $\vec{k}_i$ represent one of vectors $\vec{q}_{ij}$ (j=1,2,3), $\vec{e}_i$, $\vec{r}_{ii}$ or $\vec{r}_{i,i+1}$. Let us assume for a moment that the joints from the first to the i-th joint are revolute, and that all joint coordinates are set to zero. Let us now perform a series of successive rotations around axes $\vec{e}_1,\dots,\vec{e}_i$ by angles $q_1,\dots,q_i$, respectively. After the first rotation, vector $\vec{k}_i$ may be expressed as

$$\vec{k}_i^{(1)} = \vec{k}_{i1}^{(1)}\cos q_1 + \vec{k}_{i2}^{(1)}\sin q_1 + \vec{k}_{i3}^{(1)} \tag{1.3.1}$$

where $\vec{k}_{ij}^{(1)}$ (j=1,2,3) are constants. This expression follows immediately from the finite rotation formula (1.2.3). The corresponding matrix form is

$$\vec{k}_i^{(1)} = \left[\vec{k}_{i1}^{(1)} \ \vec{k}_{i2}^{(1)} \ \vec{k}_{i3}^{(3)}\right]^T \begin{bmatrix} \cos q_1 \\ \sin q_1 \\ 1 \end{bmatrix} \tag{1.3.2}$$

or, simply

$$\vec{k}_i^{(1)} = K_i^{(1)} \mathrm{Rot}(q_1) \tag{1.3.3}$$

where $\mathrm{Rot}(q_1)$ represents the vector $[\cos q_1 \ \sin q_1 \ 1]^T$ and $K_i^{(1)}$ the 3×3 matrix on the right hand side of Equation (1.3.2). The symbol $\mathrm{Rot}(q_1)$ has already been used in the literature with a similar meaning [25]. Clearly, after a series of i successive rotations, vector $\vec{k}_i$ becomes

$$\vec{k}_i^{(i)} = K_i^{(i)} \ \mathrm{Rot}\left(q_1, \ldots, q_i\right) \tag{1.3.4}$$

where vector $\mathrm{Rot}\left(q_1, \ldots, q_i\right)$ contains the products of cosines and sines of angles $q_1, \ldots, q_i$, and $K_i^{(i)}$ the constants. Of course, if one of the degrees of freedom $1, \ldots, i$ is a sliding one, $\mathrm{Rot}\left(q_1, \ldots, q_i\right)$ will not contain cosines and sines of the corresponding joint coordinate. Thus, we conclude that the analytical structure of the kinematic variables can unifoldly be described by functions $\mathrm{Rot}\left(q_1, \ldots, q_i\right)$.

Let us now consider analytical structure of the elements of the dynamic model matrices. As already shown, these elements are obtained by employing scalar or vector products of kinematic vectors. Since the multiplication represents a nonlinear operation, it follows that a dynamic variable can not be presented in the same way as a kinematic variable (1.3.4). For example, a nonlinear operation may yield square cosines and sines which are not contained in vector $\mathrm{Rot}\left(q_1, \ldots, q_i\right)$. This is why we have to introduce a more general form than (1.3.4) is. An arbitrary scalar or vector variable $\underline{v}$ of the dynamic robot model can be presented by the following analytical expression

$$\underline{v} = \sum_{k=1}^{m} \underline{v}_k \left(\cos q_1\right)^{c_{1k}} \cdots \left(\cos q_n\right)^{c_{nk}} \left(\sin q_1\right)^{s_{1k}} \cdots \\ \cdots \left(\sin q_n\right)^{s_{nk}} q_1^{u_{1k}} \cdots q_n^{u_{nk}} \tag{1.3.5}$$

where $\underline{v}_k$ is a constant, and c_{ik}, s_{ik} and u_{ik} (i=1,...,n) are positive integers. The number of addends in the above sum, m, is not fixed, but depends on the analytical complexity of function $\underline{v}$. Let us introduce a vector of dimension N = 3n:

$$\left(x_1,\ldots,x_N\right)=\left(\cos q_1,\ldots,\cos q_n,\ \sin q_1,\ldots,\sin q_n,\ q_1,\ldots,q_n\right). \tag{1.3.6}$$

Accordingly, we introduced a new set of arguments $x_1,\ldots,x_N$ corresponding respectively to the functions on the right hand side of the Equation (1.3.5), i.e. $x_1 = \cos q_1$, $x_2 = \cos q_2,\ldots,x_N = q_n$. Similarly, let us introduce the vector of exponents appearing in (1.3.5) as

$$\left(\varepsilon_{1k},\ldots,\varepsilon_{Nk}\right)=\left(c_{1k},\ldots,c_{nk},\ s_{1k},\ldots,s_{nk},\ u_{1k},\ldots,u_{nk}\right) \tag{1.3.7}$$

where ε_{ik} (i=1,...,N) are positive integers or zeros. By substituting (1.3.6) and (1.3.7) in (1.3.5), we obtain

$$\underline{v} = \sum_{k=1}^{m} \underline{v}_k\, x_1^{\varepsilon_{1k}} \cdots x_N^{\varepsilon_{Nk}} \tag{1.3.8}$$

which represents a polynomial defined on the new set of arguments (1.3.6).

Now, we can clearly recognize the difference between the numerical methods, shortly mentioned in Paragraph 1.1., and the analytical method presented here. In a numerical method any kinematic or dynamic variable is considered as a real number (3 reals for a vector variable). For example, a distance vector at any time instant is identified by 3 real numbers, representing the 3 vector components with respect to a given reference frame. Here, this variable is presented by the corresponding polynomial. Although the variable is a symbolic one, we see from (1.3.8) that it can be described by the coefficients and exponents of the polynomial. Suppose that the coefficients $\underline{v}_k$ of the polynomial (1.3.8) represent the components of the vector $v^{(m)}$, i.e.

$$v^{(m)} = \begin{bmatrix} \underline{v}_1 \\ \vdots \\ \underline{v}_m \end{bmatrix} \tag{1.3.9}$$

and that the exponents $\varepsilon_{1k} \cdots \varepsilon_{Nk}$ form the integer matrix

$$E_v = \begin{bmatrix} \varepsilon_{11} & \cdots & \varepsilon_{N1} \\ \vdots & & \vdots \\ \varepsilon_{1m} & \cdots & \varepsilon_{Nm} \end{bmatrix} . \tag{1.3.10}$$

In such a way we obtain the pair

$$v^{(m)} = \left(\begin{bmatrix} \underline{v}_1 \\ \vdots \\ \underline{v}_m \end{bmatrix} , \begin{bmatrix} \varepsilon_{11} & \cdots & \varepsilon_{N1} \\ \vdots & & \vdots \\ \varepsilon_{m1} & \cdots & \varepsilon_{mN} \end{bmatrix} \right) \tag{1.3.11}$$

which uniquely defines the polynomial (1.3.8), i.e. the variable $\underline{v}$. The ith row of matrix (1.3.11)

$$\left(\left[\underline{v}_i \right], \left[\varepsilon_{i1} \cdots \varepsilon_{iN} \right] \right)$$

corresponds to the ith monomial

$$\underline{v}_i \; x_1^{\varepsilon_{i1}} \cdots x_N^{\varepsilon_{iN}}$$

of the polynomial (1.3.8).

The pair (1.3.11) will be referred to as a "polynomial matrix" in text to follow. The vector $v^{(m)}$ will be called the "vector of coefficients" and the matrix E_v - the "matrix of exponents".

We will now present a very important property of the matrix.

If v is an arbitrary kinematic variable of the mechanism, the corresponding matrix of exponents will contain elements which may be either 0 or 1. If v is a dynamic variable of the mechanism, the exponents will then be 0, 1 or 2.

In case of a kinematic variable, this statement is easy to prove, since no square cosines and sines figure in vector $\mathrm{Rot}\left(q_1,\ldots,q_n\right)$ which determines the structure of these variables. The second part of the theorem was proved in [22]. These properties are very important when one writes the computer program, since they show that the exponents of kinematic variables can be stored in a single memory location (one bit) and the exponents of dynamic variables within only 2 bits of a computer memory.

Let us now introduce the basic operations between polynomial matrices. Let $v_1 = \left(V_1^{(m_1)}, E_{v_1}\right)$ and $v_2 = \left(V_2^{(m_2)}, E_{v_2}\right)$ denote two arbitrary vector variables of the model. The following polynomial matrix corresponds to the sum of these two variables

$$S_{v_1+v_2}^{(m_1+m_2)} = \left(\begin{bmatrix} V^{(m_1)} \\ \hline V^{(m_2)} \end{bmatrix}, \begin{bmatrix} E_{v_1} \\ \hline E_{v_2} \end{bmatrix}\right) \tag{1.3.12}$$

Since the following polynomial corresponds to the vector product of variables v_1 and v_2

$$\vec{v}_1 \times \vec{v}_2 = \sum_{k=1}^{m_1} \sum_{\ell=1}^{m_2} \left(\vec{v}_{1k} \times \vec{v}_{2\ell}\right) x_1^{\varepsilon_{1k}^1+\varepsilon_{1\ell}^2} \cdots x_N^{\varepsilon_{Nk}^1+\varepsilon_{N\ell}^2} \tag{1.3.13}$$

where ε_{ik}^1 belongs to the matrix of exponents E_{v_1}, and ε_{ik}^2 to matrix E_{v_2}, the polynomial matrix of vector product has the form

$$S_{v_1 v_2}^{(m_1 m_2)} = \left(V^{(m_1)} \times V^{(m_2)}, E_{v_1} + E_{v_2}\right)$$

where $V^{(m_1)} \times V^{(m_2)}$ is the $3 \times (m_1 m_2)$ matrix whose rows are vectors $\vec{v}_{1k} \times \vec{v}_{2\ell}$ $(k = 1,\ldots,m_1,\ \ell = 1,\ldots,m_2)$, and $E_{v_1} + E_{v_2}$ the matrix whose rows have the form $\left[\varepsilon_{1k}^1 + \varepsilon_{1\ell}^2 \cdots \varepsilon_{Nk}^1 + \varepsilon_{N\ell}^2\right]$.

The scalar product of two polynomial matrices is defined in a similar way

$$S_{v_1 \cdot v_2}^{(m_1 m_2)} = \left(V^{(m_1)} \cdot V^{(m_2)}, E_{v_1} + E_{v_2}\right) \tag{1.3.14}$$

The operations follow directly from the corresponding polynomial forms. One can easily show that it is possible to define, for polynomial matrices, all operations of a vector space.

Let us now consider the issue of dimensions of a polynomial matrix. It can be seen from (1.3.6) that the arguments of the polynomial which corresponds to variable $\underline{v}$ are interdependent, since $x_i = \cos q_i$ and

$x_{n+i} = \sin q_i$ for $i = 1,\ldots,n$. As a result of trigonometric relation between arguments x_i and x_{n+i}, there appears identities such as

$$x_i^2 = 1 - x_{n+i}^2$$

$$x_i^4 = 1 - 2x_{n+i}^2 + x_{n+i}^4$$

and other. On the other hand, let us consider the polynomial

$$\underline{w} = \sum_{k=1}^{m+1} \underline{w}_k\, x_1^{\xi_{1k}} \cdots x_N^{\xi_{Nk}}$$

and let us assume that the following holds

$$\underline{w}_k = \underline{v}_k, \qquad (k=1,\ldots,m), \qquad \underline{w}_{m+1} = \underline{w}_m$$

$$\xi_{jk} = \varepsilon_{jk} \qquad (j=1,\ldots,N \quad \text{and} \quad k=1,\ldots,m-1)$$

$$\xi_{jm} = \varepsilon_{jm} \qquad (j=1,\ldots,i-1,\ i+1,\ldots,N), \quad {}_{im} = {}_{im} + 2$$

$$\xi_{j,m+1} = \varepsilon_{jm} \ (j=1,\ldots,n+i-1,\ n+i+1,\ldots,N), \ \xi_{n+i,m} = \varepsilon_{n+i,m} + 2$$

Having in mind that $x_i^2 + x_{n+1}^2 = 1$, it is easy to recognize that the polynomials of variables $\underline{v}$ and $\underline{w}$ are equivalent. So, they represent the same variable. Thus, we obtain

$$S_v^{(m)} = S_w^{(m+1)}$$

It follows directly therefrom that variable $\underline{v}$ can be represented by two different polynomials, i.e., by two different polynomial matrices. By induction, we conclude that for an arbitrary number $m_\ell > m$ there exists $\underline{v}_\ell$, and the following identity holds

$$\left(V_v^{(m)},\ E_v\right) = \left(V_{v_\ell}^{(m_\ell)},\ E_{v_\ell}\right)$$

i.e., an infinite number of matrices corresponds to any variable $\underline{v}$. These matrices will be referred to as equivalent analytical matrices. Since the dimensions of equivalent analytical matrices are different, it is necessary to determine the matrix of a minimal dimension

$$S_v^{(m)} = \min_{m_\ell}\left(V_{v_\ell}^{(m_\ell)},\ E_{v_\ell}\right) \tag{1.3.15}$$

The polynomial corresponding to this matrix is the simplest polynomial which describes variable $\underline{v}$ and which may be referred to as a minimal--complexity polynomial. Matrix (1.3.15) may be said to be unique for a given variable if it satisfies the following conditions:

1) None of the elements of the vector of coefficients is equal to zero.

2) There are not two identical rows in the matrix of exponents.

3) There are not two rows which can be compressed into one due to the trigonometric relation $x_\ell^2 + x_{n+\ell}^2 = 1$, for any $\ell \in \{1,\ldots,n\}$.

If the first condition is not satisfied, zero elements of the vector of coefficients as well as the corresponding rows of the matrix of exponents can be omitted in order to reduce dimension m. When the second condition is not satisfied, i.e., when two identical rows exist in the matrix of exponents, the corresponding elements in the vector of coefficients can be summed and the reduction in dimensions by 1 performed. The third condition is in some way clumsy for a formal and general definition (see Ref. [22]). Thus, we will explain it by a simple example. Consider a polynomial matrix with the ith and kth row

$$\left(\left[\underline{v}_i\right], \left[\varepsilon_{1i} \cdots \varepsilon_{Ni}\right]\right)$$

$$\left(\left[\underline{v}_k\right], \left[\varepsilon_{1k} \cdots \varepsilon_{Nk}\right]\right)$$

being

ith row = ([2.], [0 1 0 0 1 2 0 0 0])

kth rov = ([2.], [0 1 2 0 1 0 0 0 0]).

We see that v represents a scalar variable and that N = 3n = 9. This corresponds to a manipulator with 3 joints and 3 links. Denoting the joint coordinates by q_1, q_2 and q_3, the polynomial corresponding to the ith and kth row becomes

$$2.\cos q_2 \sin q_2 \sin^2 q_3 + 2.\cos q_2 \cos^2 q_3 \sin q_2 =$$

$$= 2.\cos q_2 \sin q_2.$$

It means that the ith and kth row can be reduced to one row:

([2.], [0 1 0 0 1 0 0 0 0])

due to the identity $\sin^2 q_3 + \cos^2 q_3 = 1$.

Minimization of the dimension of an analytical matrix, i.e., determination of the simplest polynomial describing a corresponding variable, can easily be automated on a computer. Each mathematical operation intended to give the elements of the dynamic model matrices, should be followed by the dimension reduction of the obtained analytical matrix.

Let us now consider the problem of calculating the obtained analytical matrices so as to achieve a minimal number of floating-point multiplications/additions. It follows from the preceding text that the analytical model reduces to a set of polynomials of N variables. Thus, the problem is how to calculate these polynomials so that the number of floating-point multiplications/additions is the smallest possible. This is a mathematical problem referred to as the problem of optimal factorization of polynomials with several arguments. Here, we will not present the algorithms for optimal factorization of polynomials, but only describe their essence by a simple example.

Consider polynomials P and Q of 3 independent variables x_1, x_2 and x_3, which are to be calculated with a minimal computational effort:

$$P = 0.1\, x_1\, x_2 + 0.7\, x_3 + 0.2\, x_1\, x_2\, x_3$$

$$Q = 0.8\, x_1\, x_2$$

Simple enumeration shows that the calculation of these polynomials requires 8 floating-point multiplications and 2 additions. Performing a partial factorization of the first polynomial:

$$P = x_1\, x_2 \left(0.1 + 0.2\, x_3\right) + 0.7\, x_3$$

and introducing the monomial $R = x_1 x_2$, the calculation reduces to

$$R = x_1\, x_2$$

$$P = R\left(0.1 + 0.2\, x_3\right) + 0.7\, x_3$$

$$Q = 0.8R$$

There are now 5 multiplications to be performed. Structure of the calculation of these polynomials may be represented graphically as follows

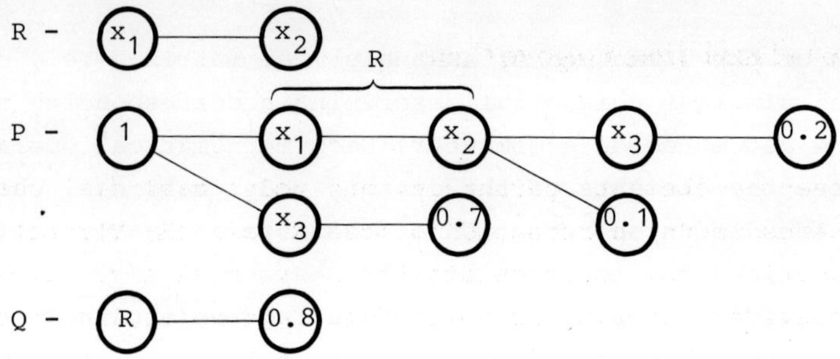

In the above structures, branching stands for the operation of addition, and the branches connecting two nodes for the operation of multiplication.

Now, consider only the structure representing polynomial P. To this structure we can assign two matrices which define it in a unifold way. The first matrix contains the elements corresponding to the elements in nodes, and the second matrix the number of branches originating from corresponding nodes. Thus, for polynomial P, we obtain the matrices

$$E = \begin{bmatrix} x_1 & x_2 & x_3 & c_2 \\ x_3 & c_3 & c_1 & 0 \end{bmatrix}, \qquad V = \begin{bmatrix} 1 & 2 & 1 & 0 \\ 1 & 0 & 0 & 0 \end{bmatrix}$$

where $c_1 = 0.1$, $c_2 = 0.2$ and $c_3 = 0.7$. These matrices describe the structure of the optimal calculation of polynomials. Instead of the symbolic elements (x_i and c_i) in the E matrix, it is more convenient to use the numeric elements. Renaming the constants c_i as

$$c_1 = x_4, \quad c_2 = x_5 \quad \text{and} \quad c_3 = x_6$$

and replacing the elements x_i by their subscripts "i", E matrix becomes

$$E = \begin{bmatrix} 1 & 2 & 3 & 5 \\ 3 & 6 & 4 & 0 \end{bmatrix} .$$

Thus, to each element of the dynamic model matrix, for which we have obtained the polynomial, we associate two matrices which describe how this element can be calculated with a minimal number of floating-point operations. This procedure can be performed by a computer, i.e., is fully automated.

1.4 Construction of Real-Time Program Code

As we have seen in the last paragraph, any polynomial h(x) can be described in terms of the corresponding matrix pair (E, V). Let us now state the question: how to calculate the polynomial h(x), given the E and V matrices. This problem is pointed-out and solved in Ref. [22]. Thus, we will here give only the final result.

Let $X_{i,j}$ denote the argument or the coefficient of polynomial h(x) which corresponds to the (i, j)-th element of matrix E. We suppose that E and V matrices have dimensions $n_r \times n_c$. In addition, let us introduce two auxiliary vectors p and q of dimension n_r. The polynomial can be calculated according to the following algorithm:

<u>1. step</u>: $q(i) = X_{i,n_c}, \quad i=1,\ldots,n_r$

<u>2. step</u>: $p(i) = X_{i,n_c-1}\left(q\left(k_{i-1}+1\right)+ \cdots +q\left(k_i\right)\right), \quad i=1,\ldots,n_r$

where $k_i = \sum_{j=1}^{i} V\left(j,\, n_c-1\right), \quad k_o = 1$

⋮

<u>(2ℓ)-th step</u>: $p(i) = X_{i,n_c-2\ell+1}\left(q\left(k_{i-1}+1\right)+ \cdots +q\left(k_i\right)\right), \quad i=1,\ldots,n_r$

where $k_i = \sum_{j=1}^{i} V\left(j, n_c-2\ell+1\right), \quad k_o = 1$

<u>(2ℓ+1)-th step</u>: $q(i) = X_{i,n_c-2\ell}\left(p\left(k_{i-1}+1\right)+ \cdots +p\left(k_i\right)\right), \quad i=1,\ldots,n_r$

where $k_i = \sum_{j=1}^{i} V\left(j, n_c-2\ell\right), \quad k_o = 1$

⋮

<u>(n_c+1)-th step</u>:

$$h(x) = \begin{cases} \sum_{i=1}^{n_r} p(i) & n_c \text{ - odd} \\ \sum_{i=1}^{n_r} q(i) & n_c \text{ - even} \end{cases}$$

As may be seen, the algorithm contains recursive relations each of which comprises one multiplication and a few additions. It is easy to see that these relations describe the calculation of polynomials according to the graphical structures presented in the preceding paragraph. It may also be seen that the analytical matrices, i.e., the polynomials, could be calculated by a computer if these relations were mapped into a series of program statements. It will be shown that even the problem of mapping can be efficiently solved by means of a computer program. The program which does this will be referred to as "expert-program". This is a very important program because it permits fully automated construction of the mathematical model of robot in analytical form. The output of expert-program is also a PROGRAM which is written in a higher or lower-level programming language by the computer itself.

According to the presented algorithm, the basic relation to be generated by expert-program is of the type

$$p_i = x_j\left(q_{i1} + \cdots + q_{in}\right)$$

where i,j,n are given. For example, let i=1, j=2 and n=3. We will present the expert-program output statement corresponding to this expression in FORTRAN and ASSEMBLER for a microcomputer (INTEL 8086 with coprocessor 8087).

1. FORTRAN code

```
P1 = X2*(QI1 + QI2 + QI3)
```

2. ASSEMBLER 8086

```
FLD  QI1  ;  QI1→ST
FADD QI2  ;  ST*QI2→ST
FADD QI3  ;  ST+QI3→ST
FMUL X2   ;  ST*X2→ST
FST  P1   ;  ST→P1
```

where ST denotes the first location of STACK-memory, and → the data transfer.

The generated programs can immediately be translated by appropriate COMPILERs and used for real-time operation, simulation or control synthesis. Notice that, in a general case, the programming language in which expert-program is written does not have to be the same as that of the generated program code. The general block-diagram for the overall program system, including analytical model construction and optimization, is given in Fig. 1.3. The operation of expert program will be illustrated in the next section using the example of STANFORD manipulator of (Fig. 1.4).

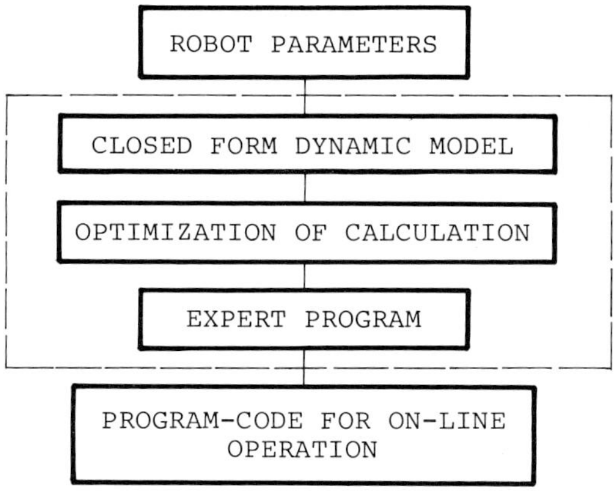

Fig. 1.3. General block-diagram for on-line model generation

1.5 Example

Although the results presented in this book refer to a general n-joint manipulator, manipulator shown in Fig. 1.4. will be used as a reference in illustrating the developed algorithms. This manipulator (Stanford arm) has 6 joints and links in addition to the gripper. Link 1 rotates about vertical axis on the platform, while link 6 is the hand [17, 25]. There are 6 orthonormal coordinate systems $\left(\vec{q}_{i1}, \vec{q}_{i2}, \vec{q}_{i3}\right)$ assigned to mass centers of the links. Fig. 1.4 also shows the unit vectors of joint axes $\vec{e}_i$. Numerical data about geometric parameters, masses and moments of inertia may be found in [25], so they will not be repeated here.

In the text to follow we will consider only the first 3 joints for the sake of clear illustration of the described method. The 6-joint manipulator is considered in Ref. [26]. First, we have to define the manipulator configuration when all joint coordinates are equal to zero (zero-angle configuration). Let us accept the configuration presented in

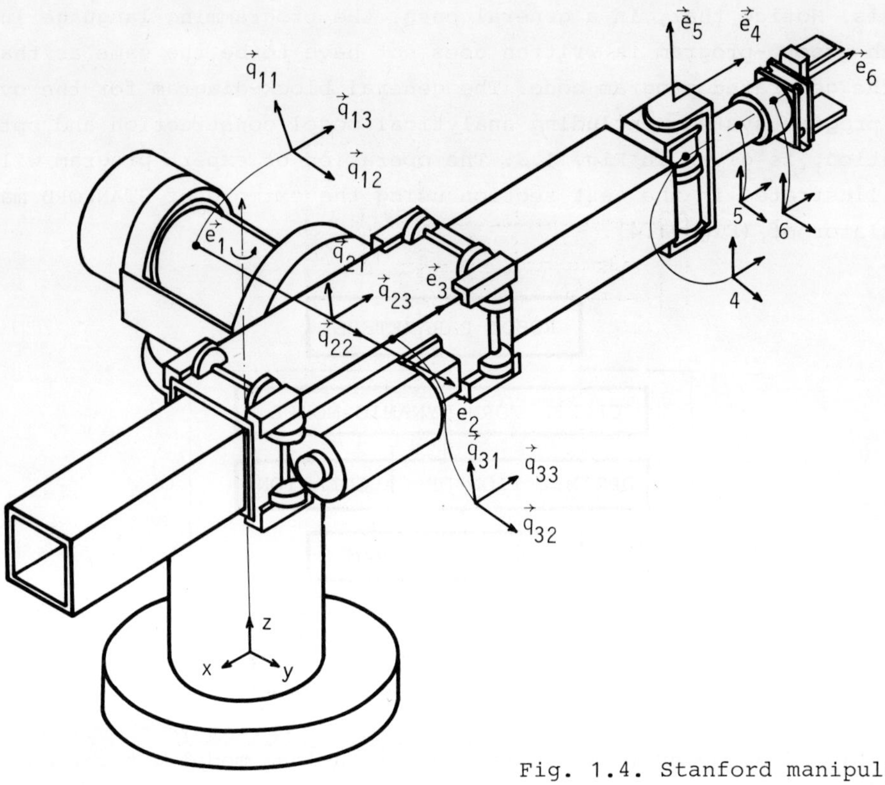

Fig. 1.4. Stanford manipulator

Fig. 1.5. The positive sense of rotation for revolute joints and displacement for the prismatic joint is determined by the direction of the axes $\vec{e}_i$. We can also accept that, at the zero-angle configuration the axes of local coordinate frames are paralled to the axes of the reference frame. The characteristic vectors at this configuration are given in Table 1.1.

Now, we have to determine the characteristic vectors at any desired manipulator configuration $q = \left[q_1 \; q_2 \; q_3\right]^T$. Applying rotations about $\vec{e}_1$ and $\vec{e}_2$ successively, and the translation along $\vec{e}_3$ for displacement q_3, we can derive the analytical expressions for $\vec{e}_i$, $\vec{r}_{ii}$ and $\vec{r}_{i,i+1}$ (i = 1,2,3) as functions of q_1, q_2 and q_3. In the case of a simple robot structure, as in this example, we can obtain these expressions "by hand", i.e. without the aid of a computer. But, in a general case, it is much more convenient to apply the polynomial-matrix algebra, described in preceding sections. Such an algebra is fully "numeric", and convenient for programming. The polynomial matrices corresponding to

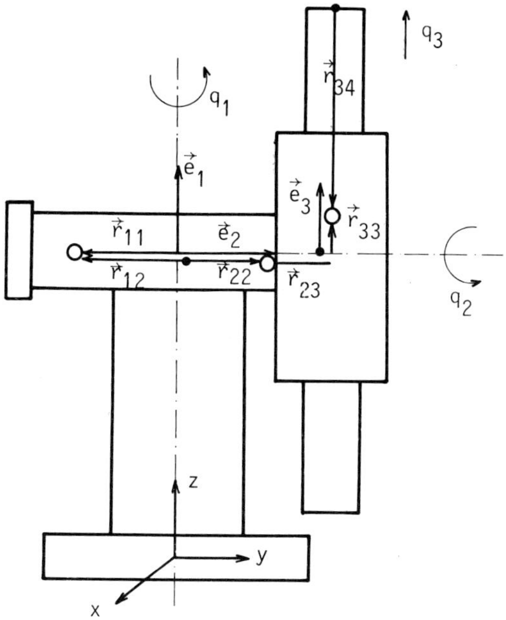

Fig. 1.5. Stanford arm when all joint coordinates are zero

i	$\vec{e}_i$	$\vec{r}_{ii}$	$\vec{r}_{i,i+1}$
1	(0, 0, 1)	(0, -0.1105, 0)	(0, -0.1105, 0)
2	(0, 1, 0)	(0, 0.0475, 0)	(0, -0.1054, 0)
3	(0, 0, 1)	(0, 0, 0)	(0, 0.323, 0)

Table 1.1. Characteristic vectors

the characteristic vectors are given in Table 1.2. The unit vectors $\vec{q}_{ij}$ of local coordinate frames are given in Table 1.3. For clarity, let us explain the relation between the polynomial matrix of $\vec{r}_{33}$ and the analytical expression for $\vec{r}_{33}$. The first row of the polynomial matrix ([0. 0. 1.], [010 000 001]) corresponds to the monomial

$$\begin{bmatrix} 0. \\ 0. \\ 1. \end{bmatrix} x_2 x_9$$

i	$\vec{e}_i$
1	$([0. \quad 0. \quad 1.],\ [0\,0\,0 \quad 0\,0\,0 \quad 0\,0\,0])$
2	$\left(\begin{bmatrix} 0. & 1. & 0. \\ -1. & 0. & 0. \end{bmatrix} \begin{bmatrix} 1\,0\,0 & 0\,0\,0 & 0\,0\,0 \\ 0\,0\,0 & 1\,0\,0 & 0\,0\,0 \end{bmatrix}\right)$
3	$\left(\begin{bmatrix} 0. & 0. & 1. \\ 1. & 0. & 0. \\ 0. & 1. & 0. \end{bmatrix}, \begin{bmatrix} 0\,1\,0 & 0\,0\,0 & 0\,0\,0 \\ 1\,0\,0 & 0\,1\,0 & 0\,0\,0 \\ 0\,0\,0 & 1\,1\,0 & 0\,0\,0 \end{bmatrix}\right)$
	$\vec{r}_{ii}$
1	$\left(\begin{bmatrix} 0. & -0.1105 & 0. \\ 0.1105 & 0. & 0. \end{bmatrix}, \begin{bmatrix} 1\,0\,0 & 0\,0\,0 & 0\,0\,0 \\ 0\,0\,0 & 1\,0\,0 & 0\,0\,0 \end{bmatrix}\right)$
2	$\left(\begin{bmatrix} 0. & 0.475 & 0. \\ -0.475 & 0. & 0. \end{bmatrix} \begin{bmatrix} 1\,0\,0 & 0\,0\,0 & 0\,0\,0 \\ 0\,0\,0 & 1\,0\,0 & 0\,0\,0 \end{bmatrix}\right)$
3	$\left(\begin{bmatrix} 0. & 0. & 1. \\ 1. & 0. & 0. \\ 0. & 1. & 0. \end{bmatrix}, \begin{bmatrix} 0\,1\,0 & 0\,0\,0 & 0\,0\,1 \\ 1\,0\,0 & 0\,1\,0 & 0\,0\,1 \\ 0\,0\,0 & 1\,1\,0 & 0\,0\,1 \end{bmatrix}\right)$
	$\vec{r}_{i,i+1}$
1	$\left(\begin{bmatrix} 0. & -0.1105 & 0. \\ 0.1105 & 0. & 0. \end{bmatrix}, \begin{bmatrix} 1\,0\,0 & 0\,0\,0 & 0\,0\,0 \\ 0\,0\,0 & 1\,0\,0 & 0\,0\,0 \end{bmatrix}\right)$
2	$\left(\begin{bmatrix} 0. & -1.054 & 0. \\ 1.054 & 0. & 0. \end{bmatrix}, \begin{bmatrix} 1\,0\,0 & 0\,0\,0 & 0\,0\,0 \\ 0\,0\,0 & 1\,0\,0 & 0\,0\,0 \end{bmatrix}\right)$

Table 1.2. Polynomial matrices of $\vec{e}_i$, $\vec{r}_{ii}$ and $\vec{r}_{i,i+1}$

where $x_2 = \cos q_2$ and $x_9 = q_3$. The second row

$([1. \quad 0. \quad 0.],\ [1\,0\,0 \quad 0\,1\,0 \quad 0\,0\,1]$

and the third row

$$([0. \quad 1. \quad 0.],\ [0\ 0\ 0 \quad 1\ 1\ 0 \quad 0\ 0\ 1])$$

correspond to

$$\begin{bmatrix} 1. \\ 0. \\ 0. \end{bmatrix} x_1 x_5 x_9 \quad \text{and} \quad \begin{bmatrix} 0. \\ 1. \\ 0. \end{bmatrix} x_4 x_5 x_9$$

respectively, with $x_1 = \cos q_1$, $x_4 = \sin q_1$ and $x_5 = \sin q_2$.

i	j	$\vec{q}_{ij}$
1	1	$\left(\begin{bmatrix} 1. & 0. & 0. \\ 0. & 1. & 0. \end{bmatrix}, \begin{bmatrix} 1\,0\,0 & 0\,0\,0 & 0\,0\,0 \\ 0\,0\,0 & 1\,0\,0 & 0\,0\,0 \end{bmatrix} \right)$
	2	$\left(\begin{bmatrix} 0. & 1. & 0. \\ -1. & 0. & 0. \end{bmatrix}, \begin{bmatrix} 1\,0\,0 & 0\,0\,0 & 0\,0\,0 \\ 0\,0\,0 & 1\,0\,0 & 0\,0\,0 \end{bmatrix} \right)$
	3	$([0. \quad 0. \quad 1.],\ [0\,0\,0 \quad 0\,0\,0 \quad 0\,0\,0])$
2	1	$\left(\begin{bmatrix} 1. & 0. & 0. \\ 0. & 1. & 0. \\ 0. & 0. & -1. \end{bmatrix}, \begin{bmatrix} 1\,1\,0 & 0\,0\,0 & 0\,0\,0 \\ 0\,1\,0 & 1\,0\,0 & 0\,0\,0 \\ 0\,0\,0 & 0\,1\,0 & 0\,0\,0 \end{bmatrix} \right)$
	2	$\left(\begin{bmatrix} 0. & 1. & 0. \\ -1. & 0. & 0. \end{bmatrix}, \begin{bmatrix} 1\,0\,0 & 0\,0\,0 & 0\,0\,0 \\ 0\,0\,0 & 1\,0\,0 & 0\,0\,0 \end{bmatrix} \right)$
	3	$\left(\begin{bmatrix} 0. & 0. & 1. \\ 1. & 0. & 0. \\ 0. & 1. & 0. \end{bmatrix}, \begin{bmatrix} 0\,1\,0 & 0\,0\,0 & 0\,0\,0 \\ 1\,0\,0 & 0\,1\,0 & 0\,0\,0 \\ 0\,0\,0 & 1\,1\,0 & 0\,0\,0 \end{bmatrix} \right)$

Table 1.3. Polynomial matrices of $\vec{q}_{ij}$

This follows from the definition of polynomial matrices (1.3.11). By adding the monomials, we obtain the analytical expression of $\vec{r}_{33}$ as

$$\vec{r}_{33} = \begin{bmatrix} q_3 \cos q_1 \sin q_2 \\ q_3 \sin q_1 \sin q_2 \\ q_3 \cos q_2 \end{bmatrix}$$

which we can easily prove "by hand". It is evident from Table 1.2. that none of the elements of any matrix of exponents exceeds 1.

In order to obtain the analytical expressions for inertia-matrix elements H_{ik}, we have to apply the expression (1.2.5). Obviously, this expression is burdened with a rather large number of scalar and vector products and vector additions. For this reason, it is much more profitable to apply the algebra of polynomial matrices and use the facilities of a digital computer. The obtained polynomial matrices are presented in Table 1.4. Likewise, the elements of $C^i(q, d)$ matrices (1.2.7) describing centrifugal and Coriolis' effects, and the terms of gravitational vector (1.2.9), are given in Tables 1.5. and 1.6. The elements which are not listed in Tables 1.5. and 1.6. can be obtained from the properties of dynamic model matrices. In view of the importance of the properties, we will repeat them once again:

1) Matrices $H(q, d)$ and $C^i(q, d)$, for any superscript $i \in \{1,\dots,n\}$, are symmetric.

2) Given $C^i(q, d)$ and $C^k(q, d)$ matrices, then, for any $k \geqslant \ell$ and $i \geqslant \ell$, the relation $C^i_{k\ell} = -C^k_{i\ell}$ holds.

Using one of these elements, for example H_{11}, we will illustrate the algorithm for computer-aided optimal calculation of the model matrices described in Paragraph 1.3. As already stated, this algorithm is intended to optimize the number of floating-point multiplications/additions which have to be performed to calculate these elements. The following analytical expression corresponds to element H_{11}:

$$H_{11} = 11.251 + 2.618 \sin^2 q_2 + 0.106 \cos^2 q_2 + 4.250(\sin^2 q_2)q_3^2$$

or in polynomial form

$$H_{11} = 11.251 + 2.618\ x_5^2 + 0.106\ x_2^2 + 4.250\ x_5^2 x_9^2$$

Denoting the coefficients of the polynomial by

i	j	H_{ij}
1	1	$\left(\begin{bmatrix} 11.251 \\ 2.618 \\ 0.106 \\ 4.250 \end{bmatrix}, \begin{bmatrix} 0\ 0\ 0 & 0\ 0\ 0 & 0\ 0\ 0 \\ 0\ 0\ 0 & 0\ 2\ 0 & 0\ 0\ 0 \\ 0\ 2\ 0 & 0\ 0\ 0 & 0\ 0\ 0 \\ 0\ 0\ 0 & 0\ 2\ 0 & 0\ 0\ 2 \end{bmatrix}\right)$
2	1	([-6.498], [0 1 0 0 0 0 0 0 1])
2	2	$\left(\begin{bmatrix} 2.528 \\ 4.250 \end{bmatrix}, \begin{bmatrix} 0\ 0\ 0 & 0\ 0\ 0 & 0\ 0\ 0 \\ 0\ 0\ 0 & 0\ 0\ 0 & 0\ 0\ 2 \end{bmatrix}\right)$
3	1	([-6.498], [0 0 0 0 1 0 0 0 0])
3	2	([0.], [0 0 0 0 0 0 0 0 0])
3	3	([4.25], [0 0 0 0 0 0 0 0 0])

Table 1.4. H_{ij} polynomial matrices

$x_{10} = 11.251$, $x_{11} = 2.618$, $x_{12} = 0.106$ and $x_{13} = 4.250$

we obtain the form

$$H_{11} = x_{10} + x_{11}x_5^2 + x_{12}x_2^2 + x_{13}x_3^2x_9^2 \qquad (1.5.1)$$

whose arguments are now x_i, $i=1,\dots,13$.

We see that the calculation of this polynomial requires 8 multiplications and 3 additions. Using the algorithm desribed in Paragraph 1.3, we can get the matrices which describe the algorithm of optimal calculation of H_{11}, i.e. the calculation of H_{11} with a minimal number of floating-point multiplications/additions. The E-matrix, whose elements correspond to the indices of arguments in (1.5.1), is obtained as

$$E = \begin{bmatrix} 5 & 5 & 9 & 9 & 13 \\ 2 & 2 & 11 & 0 & 0 \\ 10 & 0 & 12 & 0 & 0 \end{bmatrix}.$$

i	k	ℓ	$C^i_{k\ell}$
1	2	2	([6.498] , [0 0 0 0 1 0 0 0 1])
	3	2	([-6.498] , [0 1 0 0 0 0 0 0 0])
	3	3	([0.] , [0 0 0 0 0 0 0 0 0])
2	1	1	$\left(\begin{bmatrix}-2.513\\-4.250\end{bmatrix}, \begin{bmatrix}0&1&0&0&1&0&0&0&0\\0&1&0&0&1&0&0&0&2\end{bmatrix}\right)$
	3	3	([0.] , [0 0 0 0 0 0 0 0 0])
3	1	1	([-4.250] , [0 0 0 0 2 0 0 0 1])
	2	1	([-0.0027], [0 0 0 0 1 0 0 0 0])
	2	2	([-4.250] , [0 0 0 0 0 0 0 0 1])

Table 1.5. $C^i_{k\ell}$ polynomial matrices

i	g_i
1	([0.] , [0 0 0 0 0 0 0 0 0])
2	([-41.693] , [0 0 0 0 1 0 0 0 1])
3	([41.693] , [0 1 0 0 0 0 0 0 0])

Table 1.6. g_i polynomial matrices

The V-matrix, whose elements represent the numbers of "branches" originating from nodes of the corresponding graph, is obtained as

$$V = \begin{bmatrix} 1 & 2 & 1 & 1 & 0 \\ 1 & 1 & 0 & 0 & 0 \\ 0 & 0 & 0 & 0 & 0 \end{bmatrix} .$$

On the basis of these matrices we can drow the graph of the optimal calculation of H_{11} (Fig. 1.6).

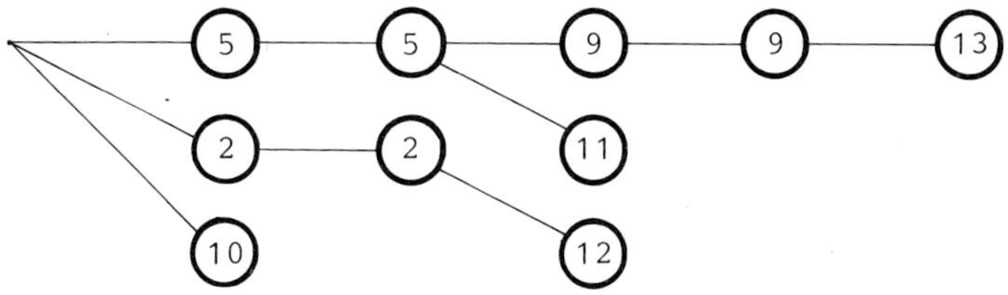

Fig. 1.6. Graph of the optimal calculation of H_{11} element

The following polynomial corresponds to this graph

$$H_{11} = x_5^2\left(x_9^2 x_{13}+x_{11}\right) + x_2^2 x_{12} + x_{10}$$

which is identical to the original polynomial for H_{11}. Enumeration shows that it is necessary to perform 6 multiplications, and this number is reduced by 25% in comparison with the number of operations in the previous polynomial which is not optimized.

Using the algorithm given in Paragraph 1.4, the computer also generates the program code for calculating model matrices in some programming language. For H_{11} element the following FORTRAN program is obtained

```
P1=X9* 0.42500E+01
Q1=X9*P1
P2=X2* 0.10600E+00
P1=X5*(Q1+0.26185E+01)
Q2=X2*P2
Q1=X5*P1
H(1,1)=Q1+Q2+0.11251E+02
```

Once again, let us emphasize that this program code is obtained by a computer, not "by hand", i.e., by a fully automated procedure. The program which generates this code (see Paragraph 1.4) is called EXPERT-PROGRAM.

Finally, for proper illustration of the operation of the developed program system, we will present the program code for calculating all elements of any dynamic model matrix for the 3 degree-of-freedom Stanford manipulator (Fig. 1.7).

```
      X7=Q(1)
      X8=Q(2)
      X2=COS(X8)
      X5=SIN(X8)
      X9=Q(3)
C
      Q1=X2* 0.41693E+02
      G(3)=Q1
      P1=X5*(-0.41693E+02)
      Q1=X9*P1
      G(2)=Q1
      G1)=0.
      Q1=X9*(-0.42500E+01)
      C(3,2,2)=Q1
      Q1=X5*(-0.27495E-02)
      C(3,2,1)=Q1
      Q1=X5*(-0.42500E+01)
      P1=X5*Q1
      Q1=X9*P1
      C(3,1,1)=Q1
      C(2,3,3)=0.
      P1=X9*(-0.42500E+01)
      Q1=X9*P1
      P1=X2*(Q1 -0.25125E+01)
      Q1=X5*P1
      C(2,1,1)=Q1
      C(1,3,3)=0.
      Q1=X2*(-0.64983E+01)
      C(1,3,2)=Q1
      P1=X5*0.64983E+01
      Q1=X9*P1
      C(1,2,2)=Q1
      H(3,3)=0.42500E+01
      H(3,2)=0.
      Q1=X5*(-0.64983E+01)
      H(3,1)=Q1
      P1=X9*0.42500E+01
      Q1=X9*P1
      H(2,2)=Q1+0.25280E+01
      P1=X2*(-0.64983E+01)
      Q1=X9*P1
      H(2,1)=Q1
      P1=X9*0.42500E+01
      Q1=X9*P1
      P2=X2*0.10600E+00
      P1=X5*(Q1+0.26185E+01)
      Q2=X2*P2
      Q1=X5*P1
      H(1,1)=Q1+Q2+0.11251E+02
```

Fig. 1.7. Computer-generated FORTRAN program for the calculation of dynamic model matrices

As may be seen by simple enumeration, the construction of the complete model of this manipulator requires 26 floating-point multiplications and 5 additions to be carried out. The time needed to calculate all elements of the matrices of dynamic model by PDP 11/70 minicomputer is

0.46ms. This time is about 40 times shorter than 20ms, which may be taken as a reference for real-time operation. Consequently, it is possible to construct this model in real time even by microcomputers which operate 40 times slowlier than PDP 11/70. For example, an INTEL 8086/ /8087 - based microcomputer needs about 4ms for the same task.

The same example but for 6 degree-of-freedom Stanford manipulator is given in Ref. [26]. In that case, more than 300 floating-point multiplications is required. It consumes about 5ms on PDP 11/70, which is still convenient for real-time implementation.

At the end, it should be pointed out that for practical application in most cases it is necessary to take into account the dynamics of the first 3 links only. The last 3 links (see links 4, 5 and 6 in Fig. 1.4) are oftenly small compared with the first 3 links, and determine the orientation of the end-effector. Thus, their dynamics is oftenly negligeable comparing to the dynamics of the first 3 links. As we have seen from the previous example, the dominant dynamic effects can be calculated in a few miliseconds on a single up-date 16-bit microcomputer.

References

[1] Uicker J.J., "Dynamic Force Analysis of Spatial Linkages" ASME Journal of Applied Mechanics, pp. 418-424, June, 1976.

[2] Kahn M.E., "The near Minimum Time Control of Open Loop Articulated Kinematic Chains", Ph.D. Thesis, Stanford University, MEMO AIM, 106, 1969.

[3] Voo L.S., Freudenstein F., "Dynamic Analysis of Mechanisms Using Screw Coordinates", ASME Journal of Engineering for Industry, February, 1971.

[4] Yang A.T., "Inertia Force Analysis of Spatial Mechanisms", ASME Journal of Engineering for Industry, February, 1971.

[5] Orlandea N., Berenyi T., "Dynamic Continuous Path Synthesis of Industrial Robots Using Adams Computer Program", ASME Journal of Mechanical Design, No. 5, 1981.

[6] Mahil S.S., "On the Application of Lagrange's Method to the Description of Dynamic Systems", IEEE. Trans. on SMC, Vol. 12, No 6, pp. 877-890, 1982.

[7] Renaud M., "Contribution à L'etude de la Modelisation et de la Commande des Systèms Mècaniques Articulès", Thèse de Docteur--Ingènieur, Toulouse, France, 1975.

[8] Hollerbach J.M., "A Recursive Formulation of Lagrangian Manipulator Dynamics", IEEE Trans. on SMC, Vol. 10, No 11, pp. 730-736, 1980.

[9] Vukobratović M., Potkonjak V., "Contribution to Automatic Forming of Active Chain Models via Lagrangian Form", Journal of Applied Mechanics, No 1, 1979.

[10] Waters R.C., "Mechanical Arm Control", A.I. Memo 549, MIT Artificial Intelligence Laboratory, 1979.

[11] Saridis G.N., Lee C.S.G., "An Approximation Theory of Optimal Control for Trainable Manipulators", IEEE Trans. on SMC., Vol. 9, No 3, pp. 152-160, 1979.

[12] Kane T., Dynamics, New York, Holt, Rinehart and Winston, 1968.

[13] Huston R.L., Kelly F.A., "The Development of Equations of Motion of Single Arm Robots", IEEE Trans. on SMC., Vol. 12, No 3, pp. 259-266, 1982.

[14] Vukobratović M., Stepanenko Yu., "Mathematical Models of General Anthropomorphic Systems", Mathematical Biosciences, Vol. 17, pp. 191-242, 1973.

[15] Walker M.W., Orin D.E., "Efficient Dynamic Computer Simulation of Robotic Mechanisms", Trans. of ASME Journal of Dynamic Systems, Measurement and Control, Vol. 104, No 3, pp. 205-211, 1982.

[16] Luh J.Y.S., Walker M.W., Paul R.P.C., "On-Line Computational Scheme for Mechanical Manipulators", Trans. of ASME Journal of Dynamic Systems, Measurement and Control, Vol. 102, No 2, pp. 69-76, 1980.

[17] Luh J.Y.S., Lin C.S., "Scheduling of Parallel Computation for a Computer-Controlled Mechanical Manipulator", IEEE Trans. on SMC, Vol. 12, No 2, pp. 214-234, 1982.

[18] Aldon M.J., Liègeois A., "Génération et Programmation Automatiques des Equations de Lagrange des Robots et Manipulateurs", Rapport. de Recherche, INRIA.

[19] Vukobratović M., Kirćanski N., "New Method for Real-Time Manipulator Dynamic Model Forming on Microcomputers", Proceeding of First Yugoslav-Soviet Symp. on Applied Robotics, Moscow, February, 1983.

[20] Kirćanski N., "Computer-Aided Procedure of Forming of Robot Motion Equations in Analytical Forms, Proc. of VI IFToMM Congres, New Delhi, 1983.

[21] Renaud N., "An Efficient Iterative Analytical Procedure for Obtaining a Robot Manipulator Dynamic Model", Proc. of First International Symp. of Robotics Research, Bretton Woods, New Hampshire, USA, 1983.

[22] Vukobratović M., Kirćanski N., Scientific Fundamentals of Robotics 4, Real-Time Dynamics of Manipulation Robots, Springer-Verlag, 1985.

[23] Denavit J., Hartenberg R.S., "A Kinematic Notation for Lower Pair Mechanisms Based on Matrices", Journal of Applied Mechanics, pp. 215-221, June, 1955.

[24] Vukobratović M., Potkonjak V., Scientific Fundamentals of Robotics 1, Dynamics of Manipulation Robots, Springer-Verlag, 1982.

[25] Paul P.R., Robot Manipulators: Mathematics, Programming, and control, MIT Press, 1981.

[26] Vukobratović M., N.Kirćanski, "Computer-Assisted Generation of Robot Dynamic Models in an Analytical Form", Journal of Applied Mathematics and Mathematical Applications - Acta Applicandea Mathematicae, Vol. 3, pp. 49-70, 1985.

Chapter 2
Non-Adaptive Control of Manipulation Robots with Variable Parameters

2.1 Introduction

In this chapter we present the synthesis of non-adaptive control of manipulation robots with variable parameters. This book deals with the lowest hierarchical level (the so-called executive level) of manipulation control. The tactical control level which performs distribution of a movement to the motions of each degree of freedom of the robot has been considered in detail in the third book of this series [1].

The executive control level has already been treated in the second book of this series [2]. However, in that book we have treated the control synthesis for manipulation robots with constant and well-known parameters; we have assumed that all parameters of the manipulator are precisely defined. This assumption has allowed us to introduce nominal programmed control which has to be synthesized using the exact and complete model of robot dynamics. This programmed control realizes the desired motion of the robot under ideal conditions when no perturbation acts upon the system. Then, the model of deviation from the nominal trajectory and control has been considered and decentralized control synthesized. Nominal, programmed control compensates for coupling among the subsystems at the nominal level; thus, coupling among subsystems is reduced at the second control level.

In many cases in robotic practice the assumption that the parameters of the system are precisely known is valid. However, in a general case this need not be true, since some parameters of the robot (e.g. payload, clearances, frictions) may be uncertain and variable. On the other hand, calculation of nominal programmed control using a centralized model may be time consuming and cumbersome, because, the models of large-scale mechanical systems may be highly nonlinear and complex. Thus, we have assumed that the calculation of nominal control should be performed off-line, and calculated nominal control and trajectories have to be stored. However, this approach is acceptable in cases when the desired motion is well known in advance, when system parameters are precisely defined and constant and when there are only a few trajectories to be realized (which have to be repeated many times, as it is the common case in industrial robotics). However, if these assumptions are not satisfied the synthesis of nominal programmed control using complete, centralized model is not recommendable.

In this book we shall consider control synthesis for the robotic system with variable parameters. Our aim is to find simple non-adaptive and robust decentralized control capable of withstanding all expected variations of parameters. However, if it is impossible to find such a simple robust control it is necessary to introduce adaptive control.

In this chapter we shall consider the synthesis of non-adaptive control of manipulation robots with variable parameters, and in the next, Chapter 3, we shall consider the synthesis of adaptive control.

In Chapter 1 we have considered the mathematical model of the dynamics of the mechanical part of the robot, i.e. the mathematical model of the robot mechanism. In Paragraph 2.2 we consider the mathematical model of the complete robotic system when actuators dynamics is also taken into account, and we define the control task imposed on the executive control level. In Appendix 2.A we present how we can set mathematical models of actuators and discuss various complexities of these models from the standpoint of control synthesis. In Paragraph 2.3 we present various methods for solving the control tasks which have been considered by other authors, i.e. we shall make a brief survey of non-adaptive control methods for manipulation robots.

As explained above, the centralized nominal control is not useful for the case involving parameter variation. We consider the system as a set of subsystems, each of them being associated to one degree of freedom of the mechanism. For each subsystem we synthesize a local controller which stabilizes the decoupled (free) subsystem (Paragraph 2.4). Then, we directly examine the stability of the whole (coupled) system for the given set of allowable parameters values (Paragraph 2.5). This allows us to find the set of parameter values for which the chosen local controllers are satisfactory, or, vice-versa, we can find the local controllers which stabilize the whole system for the given set of allowable parameter values. If the set of allowable parameter values is given, it may happen that we cannot find unique local regulators which correspond to the system for all the values of the given region of parameters. In this case we have to introduce either global control (Paragraph 2.6) or adaptive control (Chapter 3).

In Volume 2 [2] we have analyzed asymptotic stability of the system around nominal trajectory. Now, when only local controllers are applied, we must directly analyze the practical stability of the whole system around the nominal trajectory. In this chapter we shall apply methods for practical stability analysis of large scale mechanical systems in finite regions around the set of nominal trajectories and for the given region of parameter variation. In Appendix 2.B we shall explain in detail the method for practical stability analysis.

Thus, in this chapter we present the synthesis of robust non-adaptive control which can be used as the executive level of the robot control. The control synthesis for a particular manipulation robot will be presented in Paragraph 2.7.

2.2 Mathematical Model of Manipulation Robots and Control Task Definition

In this paragraph we shall consider the mathematical model of the robotic manipulator system and define the task which has to be accomplished by robot control system.

2.2.1. Mathematical model

The robotic system consists of the mechanical part of the system and actuators. Mechanical part S^M of the robotic system has n degrees of freedom. Let us assume that each degree of freedom is powered by one actuator S^i, $i = 1,2,\ldots,n$. D.C. electro-motors, hydraulic servo-actuators, etc. can serve as actuators for robotic systems.

The mathematical model of the mechanical part of the robotic system has been considered in detail in the previous chapter. As stated in Paragraph 1.2 the mathematical model of the mechanical part S^M can be written in the general form [3, 4]:

$$S^M: P_i = H_i(q, d)\ddot{q} + \dot{q}^T C_i(q, d)\dot{q} + g_i(q, d) \equiv$$

$$\equiv H_i(q, d)\ddot{q} + h_i(q, \dot{q}, d) \qquad (2.2.1)$$

where $P_i \in R^1$ is the torque or force acting in the *i*th joint of the mechanism (*i*th d.o.f.); $H_i: R^n \times R^{\ell} \to R^{1\times n}$ is the vector of inertia; $C_i: R^n \times R^{\ell} \to R^{n\times n}$ is the matrix of centrifugal and Coriolis effects, $g_i: R^n \times R^{\ell} \to R^1$ is the gravity moment (force), h_i is defined as: $h_i(q, \dot{q}, d) \equiv \dot{q}^T C_i(q, d)\dot{q} + g_i(q, d)$, so that $h_i: R^n \times R^n \times R^{\ell} \to R^1$ are centrifugal, Coriolis and gravitational forces; $q \in R^n$ is the vector of robot angles (or displacements), $q=(q_1, q_2, \ldots, q_n)^T$, q_i corresponds to the *i*th d.o.f.; $d \in R^{\ell}$ is the vector of parameters of the mechanism (for example, payload, friction coefficients, etc.) which has to belong to the finite region of allowable parameter values D, i.e., $d \in D \subset R^{\ell}$; I denotes set $I=\{i: i=1,2,\ldots,n\}$. As we have mentioned in Paragraph 2.1, we shall consider the case when all parameters of the mechanism are not known in advance, or when some of them vary but within the given allowable boundaries (defined by the finite region D). It should be noted, that masses and moments of inertia of actuators must be also taken into account in the model (2.2.1). However the inertia terms are also included in the actuators models so we must substract the terms J_R^i from H_{ii}, where J_R^i is the moment of inertia (mass) of the rotor of the actuator which drives the *i*th joint. The mathematical models of actuators are nonlinear, in the general case. However, satisfactory results can be achieved if we suppose that the models of actuators are linear by state and nonlinear by input. Thus, in order to simplify the control synthesis, we shall adopt the mathematrical models of actuators in the following form:

$$S^i: \dot{x}^i = A^i(\theta^i)x^i + b^i(\theta^i)N(u^i) + f^i(\theta^i)M_i, \quad \forall i \in I \qquad (2.2.2)$$

where $x^i \in R^{n_i}$ is the vector of the state coordinates of the *i*th actuator S^i; n_i is the order of the mathematical model of actuator S^i; $A^i: R^{\ell_i} \to R^{n_i \times n_i}$ is the subsystem matrix; $b^i: R^{\ell_i} \to R^{n_i}$ is the input distribution vector, $f^i: R^{\ell_i} \to R^{n_i}$ is the load distribution vector; $u^i \in R^1$ is the input to actuator S^i; $M_i \in R^1$ is the load acting upon the *i*th actuator (for example, with D.C. motors M_i represents the load acting upon the output shaft of the reducer gear); $N(u^i)$ represents a nonlinear function of input u^i of amplitude saturation type:

$$N(u^i) = \begin{cases} -u_m^i & \text{for} \quad u^i < -u_m^i \\ u^i & \text{for} \quad |u^i| < u_m^i \\ u_m^i & \text{for} \quad u^i > u_m^i \end{cases} \qquad (2.2.3)$$

where u_m^i is the upper bound on input amplitude; $\theta^i \in R^{\ell i}$ is the vector of parameters of the ith actuator (viscous friction coefficient, moment and electromotor constant etc.).

We shall suppose that the parameters θ^i of the actuators are precisely identified and that they may vary very slowly. We also assume that the parameters of actuators must belong to some finite sets of parameters Θ^i, $\theta^i \in \Theta^i \subset R^{\ell i}$.

The order n_i of the actuator model (2.2.2) depends on the complexity of its adopted mathematical model. It has been shown [3] that for D.C. motors and hydraulic actuators (which are most commonly implemented with robotic systems) it is quite acceptable from the standpoint of control synthesis to adopt the third-order models. In Appendix 2.A we shall consider various complexities of actuators models (for D.C. motors and hydraulic actuators). The state vector x^i of the actuator model (2.2.2) depends on the type of actuator and the adopted model. For example, for D.C. motors we can adopt the third-order model with state vector x^i in the form $x^i = (l^i, \dot{l}^i, i_R^i)^T$ where l^i is the output shaft rotation and i_R^i is the rotor current; for hydraulic actuators we can also adopt the third-order model, but the state vector can be adopted as $x^i = (l^i, \dot{l}^i, p^i)^T$, where l^i is now the piston travel and p^i is the fluid pressure drop in the cylinder. The models of D.C. motors and hydraulic actuators have been presented in [2].

Now, we have to establish the connection between the model of the mechanical part of the robot S^M (2.2.1) and the models of actuators S^i (2.2.2). The connection between the motion of the actuator l^i and the angular or linear displacement q_i of the corresponding d.o.f. is, in the general case, nonlinear, i.e. we can write*)

$$q_i = \tilde{g}^i(l^i) \tag{2.2.4}$$

where $\tilde{g}^i$: $R^1 \to R^1$ is some nonlinear function of l^i. For example, if a rotational d.o.f. of the mechanism is powered by an actuator with a

---

*) It should be mentioned that in the general case one actuator might drive more than one mechanical d.o.f.; for example, two mechanical coordinates might be a complex function of displacements of two actuators: $(q_i, q_j) = g(l^i, l^j)$. For example ASEA manipulators have two actuators driving simultaneously two joints of the mechanism. However, we shall restrict ourselves to the case given by (2.2.4) where mechanical coordinate q_i depends only on l^i.

linear displacement (e.g. hydraulic linear actuator), $\tilde{g}^i$ is then given by cosines formula. If a rotational joint is powered by an actuator with rotational motion the connection (2.2.4) might be trivial: $q_i = l^i$; the same might hold if a linear d.o.f. of the mechanism is powered by an actuator with a linear displacement. Generally, the connection between the mechanism joint coordinate q_i and the state vector x^i of the corresponding actuator S^i can be written in the form:

$$q_i = G^i(x^i) \tag{2.2.5}$$

where $G^i\colon R^{n_i} \to R^1$.

Similarly, we have to establish the relation between the load M_i acting upon the actuator S^i and torque or force P_i developed in the ith d.o.f. of the mechanism. Analogously to relation (2.2.4), we can write that, in the general case, the relation between P_i and M_i is given by nonlinear mapping:

$$P_i = w^i(M_i, q_i) \tag{2.2.6}$$

where $w^i\colon R^1 \times R^1 \to R^1$ is some nonlinear function of M_i and q_i. As explained above, when the rotational joint of the mechanism is driven by the linear actuator the relation between the load acting upon the actuator M_i and the torque developed in mechanism joint P_i might be given by cosines formula or some other relation which depends on the joint position. For example, if a rotational joint is driven by a D.C. motor or vane hydraulic actuator the relation (2.2.6) might be trivial: $P_i = M_i$, but even in this case the relation (2.2.6) might be complex if some special clutches are implemented. For simplicity, we shall restrict our considerations to the case when relation (2.2.6) may be written in the form:

$$P_i = z^i(q_i) \cdot M_i \tag{2.2.7}$$

where $z^i\colon R^1 \to R^1$ is some nonlinear function of q_i.

The schematic presentation of the robotic manipulator model is given in Fig. 2.1.

Now, we can obtain the complete model of the robotic system S in the centralized form. In order to form the centralized model of the system, we shall unite the model S^M of mechanical part of the system and models

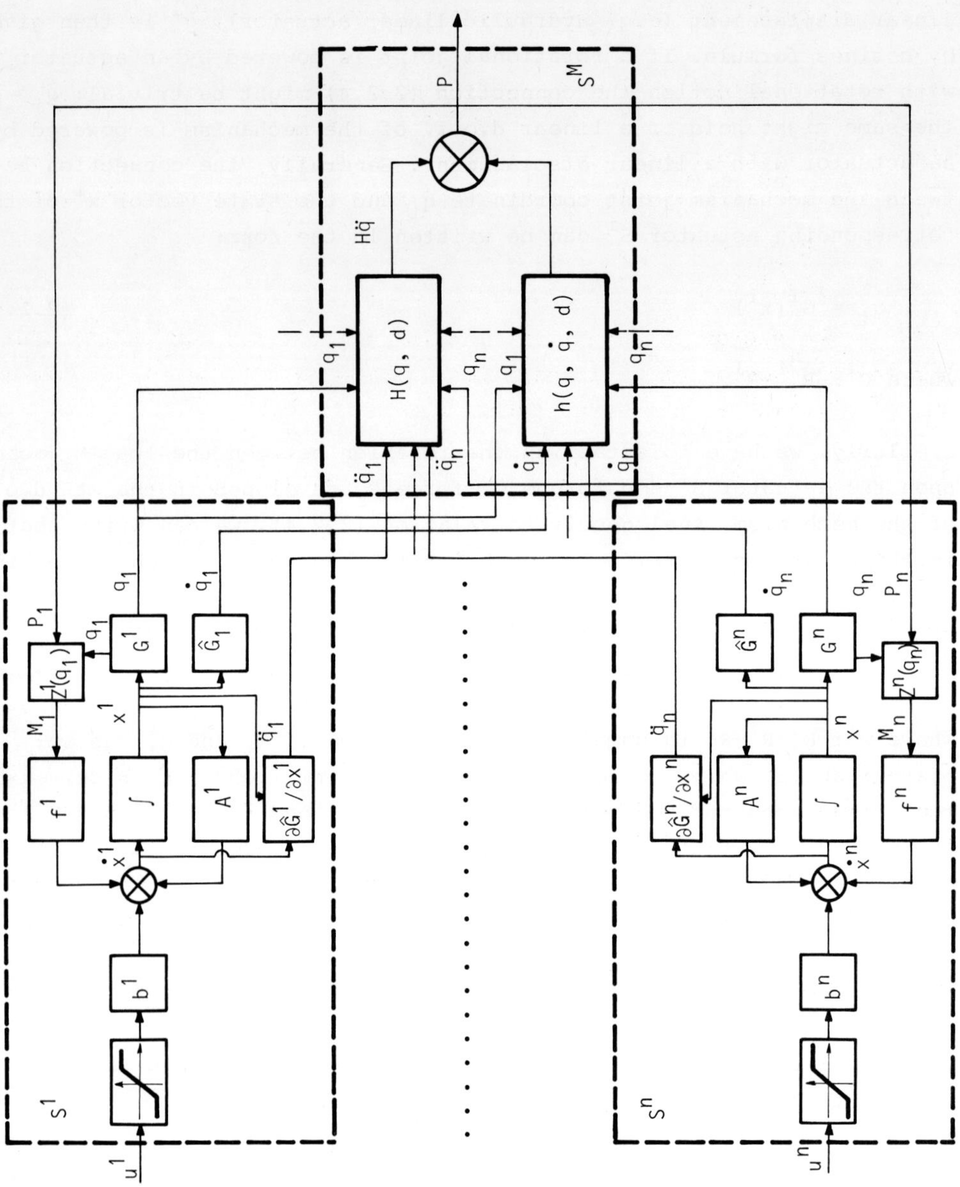

Fig. 2.1. Scheme of the model of the robotic manipulator

of actuators S^i. We get the centralized model in the form:

$$S: \quad \dot{x} = A_D(x, \theta, d) + B_D(x, \theta, d)N(u) \tag{2.2.8}$$

where $x \in R^N$ is the state vector of the robotic system, $x(t) = (x^{1T}(t), x^{2T}(t), \dots, x^{nT}(t))^T$; N is the order of the complete robotic system:

$$N = \sum_{i=1}^{n} n_i$$

$u \in R^n$ is the vector of inputs, $u(t) = (u^1(t), u^2(t), \dots, u^n(t))^T$; $N(u)$ denotes the vector of nonlinearities: $N(u) = (N(u^1), N(u^2), \dots, N(u^2))^T$; A_D: $R^N \times R^L \times R^{\ell} \to R^N$ is the vector given by:

$$A_D(x, \theta, d) = \left[I_N - F(\theta)z'(G(x)) \cdot H(G(x), d)\tilde{G}(x)\right]^{-1} \cdot \left[A(\theta) + \right.$$
$$\left. + F(\theta)z'h(G(x), \hat{G}(x), d)\right] \tag{2.2.9}$$

and B_D: $R^N \times R^L \times R^{\ell} \to R^{N \times n}$ is the input distribution vector given by:

$$B_D(x, \theta, d) = \left[I_N - F(\theta)z'(G(x)) \cdot H(G(x), d)\tilde{G}(x)\right]^{-1} B(\theta) \tag{2.2.10}$$

In (2.2.9) and (2.2.10) the following notation is used: A: $R^L \to R^{N \times N}$ denotes matrix $A(\theta) = \text{diag}(A^i(\theta^i))$, B: $R^L \to R^{N \times n}$ denotes matrix $B(\theta) = \text{diag}(b^i(\theta^i))$, F: $R^L \to R^{N \times n}$ denotes matrix $F(\theta) = \text{diag}(f^i(\theta^i))$, $H(q, d)$ is $n \times n$ inertia matrix defined by $H(q, d) = [H_i(q, d)]$, and $h(q, \dot{q}, d)$ is the $n \times 1$ vector of centrifugal, Coriolis forces defined by $h(q, \dot{q}, d) = [h_i(q, \dot{q}, d)]$, z': $R^n \to R^{n \times n}$ denotes matrix $z(q) = \text{diag}(z^i(q_i)^{-1})$; $\theta \in \Theta \subset R^L$ denotes the vector of all actuators' parameters $\theta = (\theta^{1T}, \theta^{2T}, \dots, \theta^{nT})^T$, Θ is the finite set defined by $\Theta = \Theta^1 \times \Theta^2 \times \dots \times \Theta^n$, L is the total number of parameters:

$$L = \sum_{i=1}^{n} \ell_i$$

Obviously, according to (2.2.5) the vector of mechanical coordinates can be expressed by the state vector x in the following way:

$$q = G(x) = (G^1(x^1), G^2(x^2), \dots, G^n(x^n))^T \tag{2.2.11}$$

Starting from (2.2.11) we get:

$$\dot{q} = \frac{\partial G}{\partial x}\dot{x} = \hat{G}(x) = (\hat{G}^1(x^1), \hat{G}^2(x^2), \dots, \hat{G}^n(x^n))^T \tag{2.2.12}$$

where by $\hat{G}^i$: $R^n \to R^1$ we denote $\hat{G}^i = \partial G^i/\partial x^i \cdot \dot{x}^i = \partial \tilde{g}^i/\partial l^i \cdot \dot{l}^i$, since $\tilde{g}^i$ is a function of l^i only, and not of other state variables in x^i. Similarly, we get:

$$\ddot{q} = \frac{\partial \hat{G}(x)}{\partial x} \dot{x} = \left(\frac{\partial \hat{G}^1}{\partial x^1} \dot{x}^1, \frac{\partial \hat{G}^2}{\partial x^2} \dot{x}^2, \ldots, \frac{\partial \hat{G}^n}{\partial x^n} \dot{x}^n\right)^T = \tilde{G}(x) \cdot \dot{x} \qquad (2.2.13)$$

where $\tilde{G}$: $R^N \to R^{n \times N}$ denotes matrix $\tilde{G} = \mathrm{diag}(\frac{\partial \hat{G}^i}{\partial x^i})$. Expressions (2.2.12) and (2.2.13) are also used in (2.2.9) and (2.2.10).

Thus, by (2.2.8) the model of robotic manipulator S is expressed in the centralized form. This model corresponds to the situation when the robot is not in mechanical contact with its environment, i.e. when there is no interaction between the robot and environment, so no external forces are acting upon the robot. In this book, we shall restrict ourselves to this case. The problem of interaction between the robot and environment has been treated in [2].

2.2.2. Control task definition

Now, we have to define the control task for manipulation robots. We shall synthesize the control which has to accomplish this control task.

As we have said in the Introduction, we shall consider the lowest - executive level of robot control system. Higher control levels are out of the scope of this book. Higher control levels define the task that the robotic system has to accomplish [5].

A robotic task is naturally specified in terms of its terminal device (manipulator tip) coordinates with respect to world reference frame. These coordinates we shall call external coordinates. The external coordinates S(t) consist of hand position, described by a position vector p(t) of the tip with respect to absolute Cartesian coordinates, and gripper orientation $\psi(t)$, which can be specified in various manners [1]. Here, we shall not discuss the ways of specifying external coordinates: this problem will be briefly presented in Chapter 4, but it has been considered in detail in [1]. The relation between the external coordinates $S(t) = [p^T(t), \psi^T(t)]$ of the manipulator and its joint coordinates q(t) is given by:

$$S(t) = f_S(q(t)) \qquad (2.2.14)$$

where f_s: $R^n \rightarrow R^m$ is a vector function; m is the order of vector S(t) of external coordinates $S(t) \in R^m$. If m=n the manipulator is nonredundant and the relation (2.2.14) is unique: for the given q(t) the unique value of S(t) can be obtained and vice versa, if the inverse vector function is nonsingular. If m<n the robot is redundant and for the given S(t) we cannot find a unique solution for q(t), so we have to introduce some additional criteria to get the unique solution.

Starting from (2.2.14) the control system has to calculate q(t) which corresponds to the given S(t). This problem is solved at the tactical control level which has been considered in [1]. So, the tactical level calculates q(t) and sends it to the executive control level. The executive control level has to generate commands for joint actuators which should ensure realization of joint coordinates, q(t) so the desired external coordinates S(t) will be accomplished.

Thus, our task is to synthesize the control for the executive level which should ensure that the internal coordinates get the values specified by the tactical control level. However, several problems arise concerning the specification of control task at the executive level [6]. The first problem is: must the robot follow some specified path in space, i.e. is it necessary only to place the manipulator terminal device in some specified positions in space $S(t_j)$ or there is some path S(t) in space which has to be followed by the robot? The second problem, closely related to the first one, is: is the work space free from obstacles? A further problem arising is: must the robot follow some specified velocity profile along the given path, i.e. must the velocity of the robot gripper follow some specified trajectory? If the robot gripper has to follow some specified trajectory (and its velocity has to follow some trajectory) the question is: what are the allowable deviations of the real robot state from the specified nominal trajectory? The next question which could be set here is: are all parameters of the manipulator d, θ sufficiently precisely known in advance, or are some of them variable or unknown (for example payload is often variable and cannot be known in advance but its variation must be within certain limits). Depending on how we answer all these questions, we can define various control tasks at the executive level.

Most of the robots that are installed today in industry do not follow any specified path in the space, nor are they capable of withstanding parameter variations. With such robots the control system has to ensure that the robot terminal device passes through some specified points $S(t_i)$ in space. In that case the control at the executive level is reduced to the positional control: the specified points in external coordinates $S(t_i)$ are transformed into corresponding internal (joint) coordinates $q(t_i)$ and then the robot position is controlled from point to point in internal coordinates. Of course, in this case there is no coordination among the motions of robot joints, so the path of the robot hand between the specified points $S(t_i)$ is quite arbitrary.

This is acceptable in many simple tasks in industry when the work space is free. However, in the general case this approach cannot be adopted due to the presence of various obstacles in the work space, and since in many tasks it is necessary to follow the specified trajectory in space; even more, in many complex industrial tasks it is necessary to ensure desired velocity and acceleration of the robot terminal device along a desired trajectory (for example, welding or transfer of liquids). Thus, in modern robotics the requirements upon robotic manipulators performances become more and more severe.

The most general task in robotic practice can be regarded as a task of practical stabilization of the robot around the specified trajectory. Thus, in this book we shall consider such a control task at the executive control level. We suppose that a tactical control level supplies the lowest control level with desired nominal trajectory $x^o(t)$, $\forall t \in T$ for all state coordinates of the robot; here, T denotes the time interval in which the desired trajectory should be realized $T=\{t: t \in (0, \tau)\}$; τ is a specified time period. Naturally, nominal trajectories of state coordinates $x^o(t) = (x^{o1T}(t), x^{o2T}(t), \dots, x^{onT}(t))^T$ correspond to desired trajectories of joint coordinates $q^o(t)$ and joint velocities and accelerations $\dot{q}^o(t)$, $\ddot{q}^o(t)$ and, accordingly, $x^o(t)$ corresponds to desired trajectories $S^o(t)$, velocities $\dot{S}^o(t)$ and accelerations $\ddot{S}^o(t)$ of external coordinates of the robot. Namely, we can state that the nominal trajectory is imposed in external coordinates $S^o(t)$, $\dot{S}^o(t)$, $\ddot{S}^o(t)$ and tactical and executive control levels have to realize it*).

However, in the general case it is not possible to ensure realizations of nominal trajectory without any error in position or velocity or

---

*) It should be mentioned that it is not indispensable to divide the control system into two levels: tactical and executive, as it is assumed in this text. Some authors try directly so control external coordinates (see Paragraph 2.3).

acceleration; on the other hand, some tolerances around trajectory are always allowed. Let us denote by $X_s(t)\subset R^m$ the set of alowable deviations of real external coordinates from nominal trajectory $S^o(t)$; analogously, let us denote by $X_s^v(t)$ and $X_s^a(t)$ the set of allowable tolerances of external velocities and accelerations of the robot, respectively. On the basis of (2.2.14), relation between the external and internal coordinates, we can map the set $X_s(t)$, in the space of external coordinates, into the corresponding set $Q(t)\subset R^n$, in the space of internal (joint) coordinates:

$$\underline{F}_s^{-1}(X_s(t)) \rightarrow Q(t), \qquad \forall t\in T \tag{2.2.15}$$

where mapping $\underline{F}_s^{-1}$: $R^m \rightarrow R^n$ is assumed to be nonsingular for all points in the set $X_s(t)$, $\forall t\in T$. Similarly, using (2.2.14) we can map the sets of allowable deviations of external velocities and external accelerations $X_s^v(t)$, $X_s^a(t)$ into corresponding sets of allowable deviations of joint velocities and joint accelerations $Q^v(t)$, $Q^a(t)$, $\forall t\in T$. In this way we get the regions of allowable deviations of internal coordinates, velocities and accelerations $Q(t)$, $Q^v(t)$, $Q^a(t)$, $\forall t\in T$ around nominal trajectories of joint coordinates $q^o(t)$, velocities $\dot{q}^o(t)$ and accelerations $\ddot{q}^o(t)$.

Using (2.2.5), relation between the joint coordinates of the robot q_i and state coordinates x^i of the corresponding actuator, we can map the regions $Q(t)$, $Q^v(t)$, $Q^a(t)$, $\forall t\in T$ into the regions $X^{t(i)}(t)\subset R^{n_i}$ $\forall i\in I$ in the subsystems state spaces:

$$\underline{G}^{-1}(Q(t),\ Q^v(t),\ Q^a(t)) \rightarrow X^t(t), \qquad \forall t\in T \tag{2.2.16}$$

where mapping*) $\underline{G}^{-1}$: $R^n \times R^n \times R^n \rightarrow R^N$ is nonsingular in all points of regions $Q(t)$, $Q^v(t)$, $Q^a(t)$, $\forall t\in T$; $X^t(t)\subset R^N$ denotes the region of allowable deviations of state coordinates $x(t)$ around nominal trajectory $x^o(t)$, $\forall t\in T$, which might be taken in "decoupled" form:

$$X^t(t) = X^{t(1)}(t) \times X^{t(2)}(t) \times \cdots \times X^{t(n)}(t), \quad \forall t\in T \tag{2.2.17}$$

It should be mentioned that the region $X^t(t)$ is not only defined by the

---

*) Here we assummed that $n_i=3$, $\forall i\in I$, which is commonly the case. If we choose $n_i=2$, it is sufficient to define regions $Q(t)$, $Q^v(t)$. If $n_i>3$, additional restrictions upon the subsystems state coordinates have to be imposed.

allowable deviations of external coordinates about nominal trajectory due to tolerances in tracking the given trajectory; this region is also constrained by the presence of different obstacles in the work space and by various constraints upon the manipulator kinematics. The sets $Q^v(t)$, $Q^a(t)$ are also constrained by the possibilities of the actuators (for more details on imposing the region $X^t(t)$, see Paragraph 4.2).

By these restrictions, the region $X^t(t)$ around $x^o(t)$ is defined to which the system state must belong during the tracking of $x^o(t)$. Perturbations that are acting upon the robotic system can be reduced, in principle, to perturbations of initial conditions type. However, initial error of the robot state can be assumed to be constrained, i.e. we can assume that initial condition $x(0)$ must belong to some finite region (set) $\in X^I$ in the state space $X^I \subset R^N$. Thus, our control task at the executive level is to synthesize the control which will ensure that for $\forall x(0) \in X^I$ it must be satisfied $x(t) \in X^t(t)$, $\forall t \in T$, where $x^o(0) \in X^I$, $x^o(t) \in X^t(t)$, $\forall t \in T$, and $X^I \subseteq X^t(0)$ must be satisfied (see Fig. 2.2).

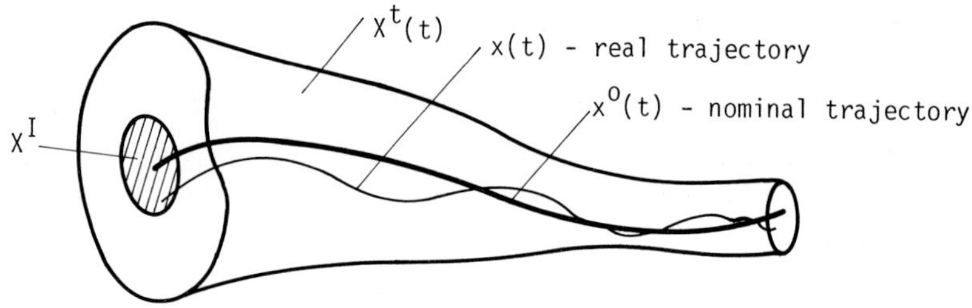

Fig. 2.2. Regions of practical stability of robotic system

This control task can be viewed as a task of practical stabilization of robotic system in the finite time interval, around the given nominal trajectory $x^o(t)$. Similar control has already been considered in our previous book [2]*). However, previously we have assumed that all parameters of the robots are well identified and known in advance. Now, we shall consider the case when all parameters of the robots d, θ cannot be defined in advance or some of them are variable. However, we shall divide the parameters of the robot into two groups: d parameters of the mechanical part of the robot are assumed to be variable and unknown, and we shall assume that their variation is relatively fast;

*) Actually, in [2] we have also introduced the region X^F of allowable deviations around terminal point on the trajectory $x^o(\tau)$. However, for the reason of simplicity, we have omitted this region X^F since this can be defined by a proper definition of $X^t(\tau_s)$.

θ parameters of actuators are assumed to vary very slowly. Thus, we shall assume that variation of actuator parameters θ may be neglected and their identification can be performed off-line. So we can synthesize the control assuming that the models of actuators are constant and precisely defined. During robot operation, the control system must occasionally check the parameters of actuators θ to identify their values, and, if they significantly vary, either synthesize new feedback gains or indicate variation of the parameters (see Paragraph 5.4).

On the other hand, some of the parameters of the mechanical part of the robot, may vary significantly (for example, payload). Thus during the synthesis of control we have to take care about the parameter variations. However, in the general case, variations of such parameters are not known in advance. So, we have to synthesize the control which is capable of withstanding all expected variations of parameters. As we have already noted, we assume that $d \in D$, i.e. parameter values must belong to some constrained set D of the allowable values of parameters. Thus, our control task at the executive level can be stated as follows (practical stabilization of robotic system):

(2.2.18) Definition of control task: The control $u(t, x)$ has to be synthesized which will ensure that $\forall x(0) \in X^I$ and $\forall d \in D$ implie $x(t, x(0), d, u(t, x(t)) \in X^t(t)$, $\forall t \in T$*), where X^I and X^t are finite regions in the state space around prescribed nominal trajectory $x^o(t)$.

Such a definition of control task is sufficiently broad to cover all possible control tasks in robotics (at the lowest control level). Typical industrial tasks such as arc welding, paint spraying, parts assembly, etc. can be included in the above definition. If we have to deal with some simple control tasks it is easy to "reduce" the above statement to suit the desired task; for example, if we have to deal with positional control only, i.e. if the robot does not have to follow some specified path, instead of the nominal trajectory $x^o(t)$ we can put only the given points $q^o(t_i)$ (corresponding to $S^o(t_i)$) and reduce the time-varying region $X^t(t)$ to set of time-constant regions $X^t(t_i)$ around the given points $q^o(t_i)$ which define allowable tolerances in positioning the robot around the points $S^o(t_i)$; the velocities and accelerations of robot are not important in such a case, so the regions $X^t(t_i)$ are restricted with respect to velocity and accelerations by actuators

*) For simplicity, in the text to follow we shall denote real state of the robotic system by $x(t)$, instead of $x(t, x(0), d, u(t, x(t)))$.

(2.2.20) Definition of extended control task: The control $u(t, x)$ has to be synthesized which will ensure that $\forall x(0)\in X_j^F$ and $\forall d\in D_j$ imply $x(t)\in X_j^t(t)$, $\forall t\in T_j$, $\forall j\in J$, where the regions X_j^F and $X_j^t(t)$ are finite regions in the state space around the corresponding prescribed nominal trajectory $x_j^o(t)$, $\forall t\in T$, $\forall j\in J$ (i.e. $x_j^o(0)\in X_j^I$ and $x_j^o(t)\in X_j^t(t)$, $\forall t\in T_j$).

In the general case regions X_j^I and $X_j^t(t)$ can be arbitrarily chosen; but, in accordance with what we have said above, we shall restrict to the case when these regions are defined analogously to (2.2.18). So, we can assume that the regions X_j^I and $X_j^t(t)$ are given by:

$$X_j^I = X_j^{I(1)}\times X_j^{I(2)}\times \cdots \times X_j^{I(n)}, \quad X_j^{I(i)}\subset R^{n_i}, \quad \forall i\in I$$

$$X_j^t(t) = X_j^{t(1)}(t)\times X_j^{t(2)}(t)\times \cdots \times X_j^{t(n)}(t), \quad X_j^{t(i)}(t)\subset R^{n_i}, \quad \forall i\in I$$

$$X_j^{I(i)} = \left\{x^i(0):\ ||\Delta x_j^i(0)|| \leqslant \bar{X}_j^{I(i)}\right\} \tag{2.2.21}$$

$$X_j^{t(i)} = \left\{x^i(t):\ ||\Delta x_j^i(t)|| \leqslant \bar{X}_j^{t(i)}\exp(-\alpha_j^{(i)}t)\right\}, \quad \forall t\in T_j$$

where $\bar{X}_j^{t(i)} \geqslant \bar{X}_j^{I(i)} > 0$, $\alpha_j^{(i)} \geqslant 0$, $\forall i\in I$, $\forall j\in J$. Here, $\bar{X}_j^{t(i)}$, $\bar{X}_j^{I(i)}$, $\alpha_j^{(i)}$ are real-valued positive numbers and $\Delta x_j^i(t)$ denotes the deviation of the *i*th subsystem S^i state vector from the *j*th nominal trajectory, $\Delta x_j^i(t) = x^i(t) - x_j^{oi}(t)$, $\Delta x_j^i(t)\in R^{n_i}$, $\forall t\in T_j$, $\forall j\in J$, $\forall i\in I$. Obviously, $x_j^o(t) = (x_j^{o1T}(t), x_j^{o2T}(t),\ldots,x_j^{onT}(t))^T$.

Although our aim is to synthesize control which will accomodate the extended control task (2.2.20), for the sake of simplicity we shall present control synthesis and stability analysis for the simpler task (2.2.18), and the extension of the results to the case (2.2.20) is trivial. We shall show how we can accommodate task (2.2.20) on an example (see Paragraph 2.7).

2.3 Survey of Non-Adaptive Control Algorithms

In this Paragraph we shall briefly survey various approaches to the synthesis of nonadaptive control for robotic systems. We shall concentrate mainly upon the problems of control synthesis of the executive control level. But, as already mentioned, in some approaches the tactical and executive control levels are undivided so we cannot speak solely about the executive control level but we must also touch some problems concerning trajectories planning and their generation.

The main problem with control of robotic systems is: to what extent it is necessary to include dynamic terms in the control law? Namely, as it has been explained in Chapter 1, the mathematical model of the robotic mechanism S^M (2.2.1) in general case is very complex system of nonlinear differential equations. The problem of on-line computation of robot dynamic model is very difficult. In Chapter 5 we shall discuss computation complexity of the various control laws and dynamic models. Various methods for computing the input torques P by (2.2.1) require various numbers of additions and multiplications to be performed each time the driving torques P are needed (every sampling period). However, the analytical models of robot dynamics, which have been proposed by Vukobratović and Kirćanski (Chapter 1), lead to the fastest computation of robot dynamic models achieved so far. Unfortunately, even in this case computation of driving torques is too complex (for some types of robot structures) to be computed by temporary sixteen-bit microprocessors. There have been made numerous attempts to use simplified models of the robot dynamics in which some terms have been neglected in order to simplify on-line computation of control. We shall mention some of them while discussing various control laws. Usually, centrifugal and Coriolis terms in (2.2.1) have been neglected since these terms are insignificant while the robot approaches its goal position and orientation (since the joint velocities are low during that time). However, if we attempt to ensure precise tracking of fast trajectories, these velocity dependent terms have to be compensated for, too. To implement on-line computation of dynamic model of the robot we have to use complex multiprocessor system. Thus, trade-off between the complexity of control equipment and the performance of the robot has to be made: the more complex law we use the more complex is the controller hardware but the performance of the robot would be better.

One of the dilemmas arising in the synthesis of control for robotic system is whether or not we have to minimize some criterion. Actually, we have defined our control task (2.2.18) (or (2.2.20)) without introducing any numerical criterion. Our control task is functionally imposed. However, in principle we can add some criterion which has to be minimized by the control (which also has to accomplish the given control task (2.2.20))*)

---

*) Obviously, we assume that the criterion (2.3.1) is set so as to be consistent with the control task (2.2.20).

$$J = \int_0^{t_s} \underset{\sim}{L}\,(x(t),\ u(t))dt + \underset{\sim}{g}(x(t_s)) \tag{2.3.1}$$

where time interval t_s is not fixed (but is shorter than τ); $\underset{\sim}{L}$: $R^N \times R^n \rightarrow R^1$ is a scalar, continuous and differentiable function of state and control and $\underset{\sim}{g}$: $R^N \rightarrow R^1$ is the function of the terminal state.

There were a few attempts to solve the control task together with minimization of some criterion (2.3.1). However, most of the authors tried to simplify the problem by synthesizing suboptimal control or avoiding the introduction of any criterion; or, they simplified the model of the robot in order to achieve some acceptable solution. Here, we shall briefly present some of these approaches to control synthesis for robotic systems. We have tried to choose the most representative contributions to the dynamic control synthesis. For clarity of presentation, we have, conditionally, divided all approaches to six directions, as can be seen in Fig. 2.3. Let us note that we have restricted ourselves to dynamic control of manipulation robots, i.e. to such control laws that take into account dynamic effects of the robotic systems. Thus, various kinematic approaches, that are based execusively on kinematic models of the robots are not included in this survey.

2.3.1. Optimal control synthesis

To synthesize the control which will satisfy the given control task (2.2.20) and, at the same time, minimize the criterion (2.3.1) is a very difficult problem. In [2] we have discussed this problem and underlined several reasons for this: (a) complexity of the highly nonlinear model of the robotic system (2.2.8), (b) the choice of the criterion (2.3.1) is often conditional and heuristic, (c) the constraints imposed by the control task (through region $X^t(t)$) are usually severe, so it is difficult to take them into account during optimization procedure, etc. Now, when we have assumed that the parameters of the robot are unknown and variable, the problem of optimization becomes more difficult. However, we shall briefly present a few attempts to minimize some criterion in the control synthesis.

First, let us consider the case when initial $x(0)$ and final $x^o(t_s)$ states of the robot are given (but not the complete trajectory $x^o(t)$). We have to synthesize the trajectory $x^o(t)$ which minimizes some criterion (2.3.1). The search for an optimal solution to the problem stated

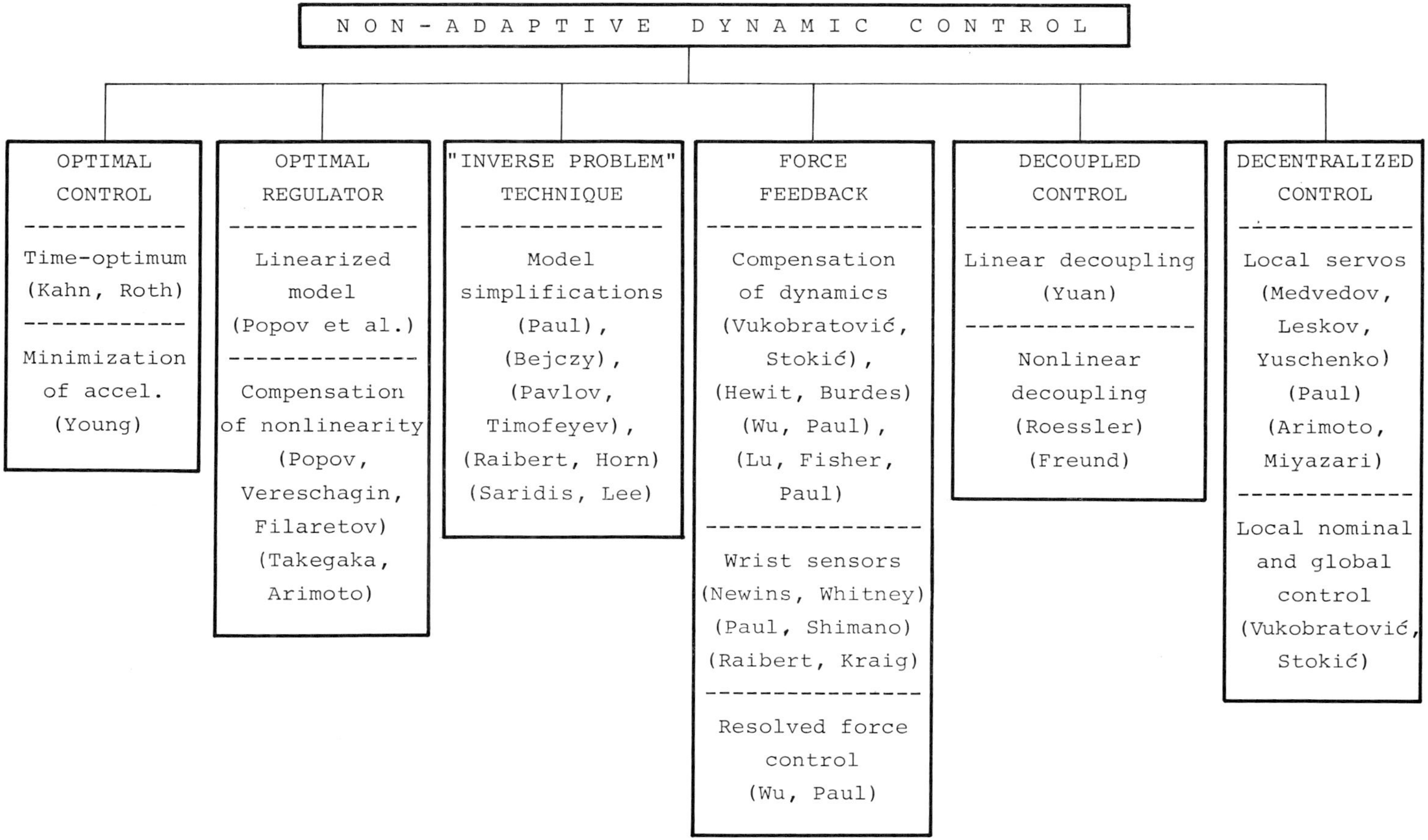

Fig. 2.3. Survey of control strategies

is determined by the minimum principle of Pontryagin.

Kahn and Roth [11] have considered time-optimal control problem: given the robotic system defined by (2.2.8) with the initial state $x(0)$, and the terminal constraint $x^o(t_s)$; find the control $u(t)$ which transfers the state of the system from $x(0)$ to $x^o(t_s)$ in the minimum time, i.e. such that the functional J, defined by (2.3.1), where: $\underset{\sim}{L}(x(t), u(t)) = 1$, $\underset{\sim}{g}(x(t_s)) = 0.$, is a minimum. Obviously, the control is constrained by (2.2.3). However, it can be shown that the analytical solution to this problem cannot be found even for a particular manipulation robot with three revolute degrees of freedom which has been considered in [11]. Owing to the nonlinearity of the model, only a numerical approach is obtainable which yields the optimal control only as a function of time and does not account for any unexpected disturbances which may act on the system or parameter variations; in addition, the computations must be repeated for each new set of initial and final conditions.

In order to obtain a feedback control an approximation to the optimal control has been proposed [11]. The suboptimal control is obtained by approximating the nonlinear system by a linear system for which the optimal control can be found analytically. The mathematical model (2.2.8) is linearized around the terminal point and the decoupling of the model is performed by neglecting the coupling and compensating only gravitational moments in the terminal point. Thus, the robot's model decouples into n double integrator systems for which analytical solution to the above stated optimization problem is well-known (it has been assumed that $n_i = 2$, $x^i = (q_i, \dot{q}_i)^T$, $\forall i \in I$). A switching surface accounting for the gravity loads on the robot is then obtained and an approximation is made for the effects of the angular velocity terms in the nonlinear system. It has been shown that for a particular robot suboptimal control results in response times and trajectories which are reasonably close to the optimal solutions. However, for more complex manipulation robots and longer distance between the terminal points this solution might be too suboptimal. On the other hand, the suboptimal control for this problem is bang-bang which is hardly acceptable for robotic practice, due to overload of actuators. Also, in this case trajectory of the robot gripper is not predictable which might lead to collision with obstacles; so, this approach can be applied in very limited manipulator applications.

A similar approach has been proposed by Young [12]. Instead of time-

-optimal control, Young adopted criterion (2.3.1) with

$$L(\underset{\sim}{x}(t),\ u(t)) = 0.5\cdot\ddot{q}^T M\ \ddot{q}, \quad g(x(t_s)) = 0.$$

where $M \in R^{n\times n}$ is a positive definite matrix, i.e. he wanted to minimize the integral of a quadratic function in acceleration. Similarly to the above mentioned approach [11], here again the robotic system is considered as a set of n double integrator plants (by ignoring the dependance of $\ddot{q}$ on $q(t)$ and $\dot{q}(t)$). Optimal trajectories $q^{oi}(t)$, $\dot{q}^{oi}(t)$, $\ddot{q}^{oi}(t)$, which drive the double integrators from given initial state $x^{oi}(0) = (q^{oi}(0), \dot{q}^{oi}(0))^T$ to desired state $x^{oi}(t_s) = (q^{oi}(t_s), \dot{q}^{oi}(t_s))^T$ and minimize (2.2.3), are well known. Now, from (2.2.1) the driving torques $P^o(t)$ are calculated that would drive the system along desired trajectories:

$$P^o(t) = H(q^o(t),\ d^o)\ddot{q}^o(t) + h(q^o(t),\ \dot{q}^o(t),\ d^o) \tag{2.3.2}$$

where d^o are well-defined parameters. Using (2.2.2) we can calculate "optimal" control. Obviously, this results in open loop control (explicit functions of time), and this control requires on-line calculation of the complete model. Of course, the control (2.3.2) compensates for all nonlinearities of the robot, and the robot´s model is formally reduced to n double integrators. However, the obtained control is far from being optimal for the complete nonlinear model of the robotic system. On the other hand, calculation of the complete dynamics might be too complex. Parameter variations cannot be compensated by this control law.

We have shown two attempts to synthesize optimal control for robotic systems. Numerous problems that are encountered, in optimization procedures result in enormous simplifications, which make the obtained control far from optimal. The obtained control laws are too complex even for simple manipulator structures for which they were developed [11, 12]. For more complex robotic structure their on-line imprementation might demand a rather complex multiprocessor system. On the other hand, the control laws are explicit functions of time so they are not robust and cannot withstand parameter variations.

2.3.2. Optimal regulator

The approaches described above were intended for positional control of robots. The problem of tracking some prescribed path in work space is more complex, so approaches by optimal synthesis are not likely to yield an acceptable solution.

An alternative aproach is to calculate (off-line) nominal trajectory by some optimal [1, 13] or suboptimal procedure [1, 14], and then, at the executive level, to track desired path*). In this bock, we shall pay attention to various attempts to synthesize control for tracking the given nominal trajectory.

The control for tracking the given nominal trajectory $x^o(t)$, $\forall t \in T$ (executive control level only) can be synthesized by minimizing some criterion (2.3.1). The most commonly applied criterion is the standard quadratic criterion where

$$\underset{\sim}{L}(x(t),\ u(t)) = \frac{1}{2}\,[\Delta x^T(t)Q(t)\Delta x(t) + \Delta u^T(t)\underline{R}(t)\Delta u(t)],$$

$$\underset{\sim}{g}(x(t_s)) = \frac{1}{2}\,\Delta x^T(t_s)Q_T\Delta x(t_s) \tag{2.3.3}$$

where $Q(t)$: $T \to R^{N\times N}$, $Q_T \in R^{N\times N}$ are positive semidefinite matrices and $\underline{R}(t)$: $T \to R^{n\times n}$ is a positive definite matrix. Here $\Delta x \in R^N$ is the deviation of the state vector from the nominal trajectory $\Delta x(t) = x(t) - x^o(t)$, $\Delta u(t) \in R^n$ is the deviation of the control vector from the nominal programed control $u^o(t)$, corresponding to $x^o(t)$ (see [2]) $\Delta u(t) = u(t) - u^o(t)$. Interval t_s, and the weighting matrices $Q(t)$, $\underline{R}(t)$, Q_T are chosen to satisfy the conditions of practical stability of the robot ((2.2.18) or (2.2.20)).

If we try to minimize criterion (2.3.1), (2.3.3) using the nonlinear model of the robot (2.2.8) we shall encounter numerous problems; we can find only a numerical solution to this problem resulting in an open-loop control law whose implementation is practically impossible [2]. A common approach in solving this problem is to consider some approximative model of the robot. The approximation that leads to an elegant well-known solution of control is to linearize the model (2.2.8) around the nominal trajectory $\Delta x^o(t)$, $\forall t \in T$.

Let us consider the model of deviations of the robot´s state from the nominal trajectory $x^o(t)$ [2].

---

*) Actually, this approach has been assumed in Paragraph 2.2.2 (when we have spoken about nominal trajectory), but our definitions of control tasks are wide enough to include cases considered in Paragraph 2.3.1, when only terminal points are given.

Starting from the robot's model (2.2.8) and assuming that nominal (programmed) control $u^o(t)$ satisfies:

$$\dot{x}^o(t) = A_D(x^o(t), \theta^o, d^o) + B_D(x^o(t), \theta^o, d^o)u^o(t) \qquad (2.3.4)$$

where θ^o, d^o denote some nominal values of parameters $\theta^o \in \Theta$, $d^o \in D$, we get the model of deviation in the form:

$$S: \Delta\dot{x} = \bar{A}^o(\Delta x, \theta, d, t) + \bar{B}^o(\Delta x, \theta, d, t)N(t, \Delta u), \qquad (2.3.5)$$

where $\bar{A}^o(\Delta x, \theta, d, t): R^N \times \Theta \times D \times T \to R^N$ is the vector function of order N,

$$\bar{A}^o(\Delta x, \theta, d, t) = A_D(x, \theta, d) - A_D(x^o(t), \theta^o, d^o) + \left[B_D(x, \theta, d) - B_D(x^o(t), \theta^o, d^o)\right]u^o(t),$$

and $\bar{B}^o(\Delta x, \theta, d, t): R^N \times \Theta \times D \times T \to R^{N\times n}$ is the matrix function of dimensions $N\times n$ given by

$$\bar{B}^o(\Delta x, \theta, d, t) = B_D(\Delta x, \theta, d, t) = B_D(\Delta x + x^o(t), \theta, d)$$

The nonlinearity of the amplitude saturation type of inputs $\Delta u(t)$ is given by $N(t, \Delta u) = (N(t, \Delta u^1), N(t, \Delta u^2), \ldots, N(t, \Delta u^n))^T$, where

$$N(t, \Delta u^i) = \begin{cases} -u_m^i - u^{oi}(t) & \text{for } \Delta u^i \leqslant -u_m^i - u^{oi}(t) \\ \Delta u^i & \text{for } -u_m^i \leqslant u^i + u^{oi}(t) \leqslant u_m^i \\ u_m^i - u^{oi}(t) & \text{for } \Delta u^i \geqslant u_m^i - u^{oi}(t) \end{cases} \qquad (2.3.6)$$

and $\Delta u = (\Delta u^1, \Delta u^2, \ldots, \Delta u^n)^T$.

Now, as we said above, we want to consider the linearized model of the robot around the nominal trajectory. Linearization of the deviation model (2.3.5) can be performed analytically [15], or by some numerical procedures for model linearization on a digital computer [16], or by means of various identification methods [2]. The linearized model is obtained in the form:

$$\Delta\dot{x} = \tilde{A}^o(\theta^*, d^*, t)\Delta x + \tilde{B}^o(\theta^*, d^*, t)N(t, \Delta u), \qquad (2.3.7)$$

where $\tilde{A}^o(\theta^*, d^*, t): T \rightarrow R^{N \times N}$ is an N×N matrix the elements of which are continuous functions of time $\tilde{A}^o = [\tilde{a}^o_{ij}]: \tilde{a}^o_{ij} = \frac{\partial \bar{A}^o_i}{\partial \Delta x_j}\Big|_{\Delta x=0,\ \theta=\theta^*, d=d^*}$, i=1,2,...,N, j=1,2,...,N, $\tilde{B}^o(\theta^*, d^*, t): T \rightarrow R^{N \times n}$ is an N×n matrix whose elements are piecewise continious in the interval T, $\tilde{B}^o(\theta^*, d^*, t) = \bar{B}^o(0, \theta^*, d^*, t)$. θ^* and d^* denote some particular values of parameters $\theta^* \in R^L$, $d^* \in D$ which might coincide with θ^o, d^o.

Popov and co-authors [15] were the first to synthesize an optimal linear regulator for robotic manipulators. It is well known that the optimal control minimizing the criterion (2.3.1), (2.3.3), under the constraint (2.3.7), is obtained in the form:

$$\Delta u(t) = -\underline{R}^{-1}(t)\, \tilde{B}^{oT}(\theta^*, d^*, t) K(\theta^*, d^*, t) \Delta x(t) =$$

$$= D(\theta^*, d^*, t) \Delta x(t) \tag{2.3.8}$$

where $K(t): T \rightarrow R^{N \times N}$ is a positive definite symmetrical matrix of dimensions N×N, which is the solution of the differential matrix equation of Riccati type:

$$-\dot{K}(\theta^*, d^*, t) = K(\theta^*, d^*, t)\tilde{A}^o(\theta^*, d^*, t) +$$

$$+ \tilde{A}^{oT}(\theta^*, d^*, t)K(\theta^*, d^*, t) + Q(t) -$$

$$- K(\theta^*, d^*, t)\tilde{B}^o(\theta^*, d^*, t)\underline{R}^{-1}(t)\tilde{B}^{oT}(\theta^*, d^*, t) \cdot$$

$$\cdot K(\theta^*, d^*, t),\ K(\theta^*, d^*, \tau) = Q_T \tag{2.3.9}$$

$D(t): T \rightarrow R^{n \times N}$ is the n×N matrix of feedback gains. In (2.3.8) it is assumed that the matrix pair $[\tilde{A}^o, \tilde{B}^o]$ is controllable in the interval $(0, t_s)$ and that all state coordinates of the robot S are measured, and we have neglected constraints upon inputs' amplitudes.

Obviously, control (2.3.8) requires time-varying gains in the feedback loops. The implementation of such control is very difficult and demands large memory capacity of the control computer to store time-variable gains, since on-line calculation of $K(\theta^*, d^*, t)$ is difficult to implement.

In order to simplify implementation of control law, a linear time-invariant model of deviation from the nominal trajectory is obtained by "time-averaging" the model (2.3.5). So, we get the approximative model of the robot's dynamics in the form:

$$\Delta\dot{x} = \bar{\tilde{A}}^{o}(\theta^{*}, d^{*})\Delta x + \bar{\tilde{B}}^{o}(\theta^{*}, d^{*})N(t, \Delta u) \tag{2.3.10}$$

where $\bar{\tilde{A}}^{o}(\theta^{*}, d^{*})\in R^{N\times N}$ is an N×N matrix, and $\bar{\tilde{B}}^{o}(\theta^{*}, d^{*})\in R^{N\times n}$ is an N×n matrix. Instead of criterion (2.3.3) we consider criterion (2.3.1) with:

$$\begin{aligned} \underset{\sim}{L}(x(t), u(t)) &= \frac{1}{2}\left[\Delta x^{T}(t)Q\Delta x(t) + \Delta u^{T}(t)\underline{R}\Delta u\right] \\ \underset{\sim}{g}(x(t_{s})) &= 0 \end{aligned} \tag{2.3.11}$$

where $Q\in R^{N\times N}$ is a positive semidefinite N×N matrix and $\underline{R}\in R^{n\times n}$ is a positive definite n×n matrix, both time-invariant. We also assume that $t_{s}\to\infty$. It is well known that the control is now obtained in the form:

$$\Delta u(t) = -\underline{R}^{-1}\bar{\tilde{B}}^{oT}(\theta^{*}, d^{*})K(\theta^{*}, d^{*})\Delta x(t) = D(\theta^{*}, d^{*})\Delta x(t) \tag{2.3.12}$$

where $K(\theta^{*}, d^{*})\in R^{N\times N}$ is the solution of the matrix equation of Riccati type:

$$\begin{aligned} &K(\theta^{*}, d^{*})\bar{\tilde{A}}^{o}(\theta^{*}, d^{*}) + \bar{\tilde{A}}^{oT}(\theta^{*}, d^{*})K(\theta^{*}, d^{*}) - \\ &\quad - K(\theta^{*}, d^{*})\bar{\tilde{B}}^{o}(\theta^{*}, d^{*})\underline{R}^{-1}\bar{\tilde{B}}^{oT}(\theta^{*}, d^{*})K(\theta^{*}, d^{*}) + Q = 0 \end{aligned} \tag{2.3.13}$$

and $D(\theta^{*}, d^{*})\in R^{n\times N}$ is the n×N matrix of constant feedback gains. The choice of the weighting matrices Q, $\underline{R}$ has to ensure that the closed-loop linear model of the robot is practically stable.

Three main problems concerning the optimal linear regulator (2.3.12) arise:

(a) It is evident that the linear regulator requires a complex control structure with many feedback loops (N×n). The control scheme is presented in Fig. 2.4. For example, for the manipulator with n = 6 d.o.f., if we take models of actuators of the third order $n_{i} = 3$, $\forall i\in I$, it follows that we have to implement 6×18=108 feedback loops. Thus, such a control scheme is too complex and unreliable, and is very difficult to maintain.

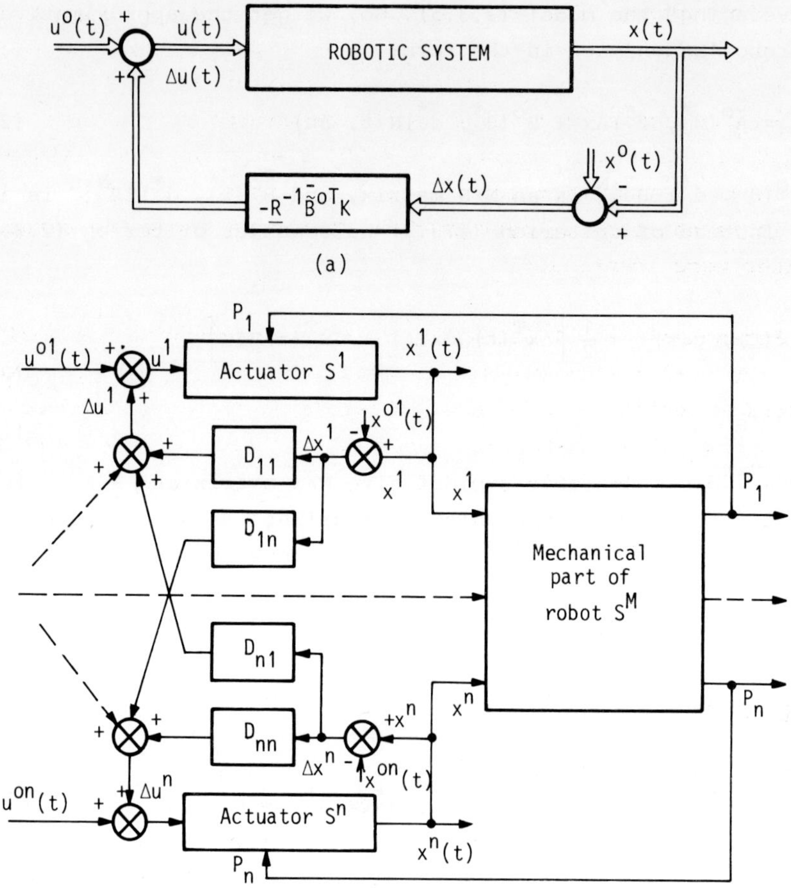

Fig. 2.4. Control scheme of linear optimal regulator
a) global scheme
b) control structure

(b) The second problem concerning the optimal regulator is that it guarantees the practical stability of the linearized model of the robot only. Since robot models are usually highly nonlinear, it is very questionable whether such linear control can accomplish the stated control task for an actual nonlinear model of the robot. In [2] we have shown how we can analyze stability of the nonlinear model of the robot when the linear regulator is applied.

(c) The third problem is the fact that the linear regulator is not robust enough to withstand all parameter variations. We have consistently denoted that the linearized model ((2.3.7) or (2.3.10)) and gains of optimal regulator (2.3.8) or (2.3.11)) depend on the chosen parameter values θ^*, d^*. This means that to each set of parameter values θ^*, d^* there corresponds a gain matrix $D(\theta^*, d^*)$, and it may be difficult to find unique feedback gains D which can withstand all parameter variations $\forall d \in D$. Even more, synthesis of D for given θ^*, d^* is relatively complex for complex robotic manipulators, so it is not easy to perform this synthesis after the parameter values were identified.

Popov, Vereschagin and Filaretov [17] have compensated for nonlinearities of the real robots models, in order to make the control law more adequate for nonlinear models of robots (i.e. to solve the above mentioned problem (b)). To achieve this, they observed the model of the robot (2.3.5) in the form*):

$$\Delta\dot{x} = \bar{\tilde{A}}^o \Delta x + f(\Delta x, \theta, d) + \bar{\tilde{B}}^o N(t, \Delta u) \tag{2.3.14}$$

where $f: R^N \times \Theta \times D \to R^N$ is a vector function of order N, given by:

$$f(\Delta x, \theta, d) = \bar{A}^o(\Delta x, \theta, d, t) - \bar{\tilde{A}}^o(\theta^*, d^*)\Delta x.$$

In (2.3.14) it has been assumed that $\bar{\tilde{B}}^o = \bar{B}^o(\Delta x, \theta, d, t)$; this assumption is valid in some cases (see Appendix 2.A). We have also assumed that f is time-invariant; this is the case when only the terminal position is set, i.e instead of $x^o(t)$, $\forall t \in T$, $x^o(\tau)$ is given only, and (2.3.14) becomes a model of deviation around given state $x^o(\tau)$. The control minimizing criterion (2.3.1), (2.3.11), with constraint given by (2.3.14), is achieved in the form:

$$\Delta u(t) = -\underline{R}^{-1}\bar{\tilde{B}}^{oT}K\Delta x(t) - \underline{R}^{-1}\bar{\tilde{B}}^{oT}(\bar{\tilde{A}}^o - K\bar{\tilde{B}}^o\underline{R}^{-1}\bar{\tilde{B}}^{oT})^{-1}K\, f(\Delta x, \theta, d) \tag{2.3.15}$$

i.e. we can state that control is in the form:

$$\Delta u = \Delta u_{lin} + \Delta u_N \tag{2.3.16}$$

where Δu_{lin} is the linear and Δu_N the nonlinear part of the control law. The matrix K is obtained from (2.3.13) Actually, we should substitute $\bar{\tilde{A}}^o + \partial f/\partial x$ for $\bar{\tilde{A}}^o$ in (2.3.13), but we can assume that in the region

---

*) For the sake of simplicity we have denoted $\bar{\tilde{A}}^o(\theta^*, d^*)$, $\bar{\tilde{B}}^o(\theta^*, d^*)$ by $\bar{\tilde{A}}^o$, $\bar{\tilde{B}}^o$ etc.

$x^t(t)$ around $x^o(\tau)$, the eigenvalues of the matrix $\bar{\tilde{A}}^o + \partial f/\partial x$ do not differ a lot from eigenvalues of the matrix $\tilde{A}^o$ [17]. Thus, the matrix K can be precalculated.

The control scheme is presented in Fig. 2.5. Obviously it differs from the classical optimal regulator (presented in Fig. 2.4) due to the addition of nonlinear part Δu_N whose role is to compensate for nonlinear effects of robot model. However, the main problem arising here is that such control is very complex. In addition to complex control structure of optimal linear regulator, on-line calculation of the nonlinear term f is demanded here. This means that the complete model of the robot dynamics has to be calculated on-line what might be a very difficult task (see next Paragraph). On the other hand, this nonlinear model depends on parameters d. We have assumed that these parameters might be variable and unknown. Thus, the robustness of such control scheme to parameter variations is questionable.

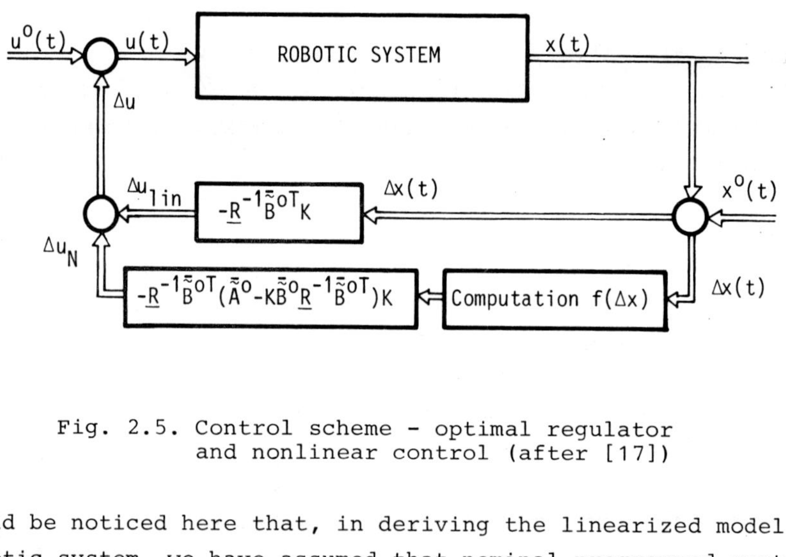

Fig. 2.5. Control scheme - optimal regulator and nonlinear control (after [17])

It should be noticed here that, in deriving the linearized model of the robotic system, we have assumed that nominal programmed control is introduced. However, the linearized model (and optimal regulator (2.3.13), or (2.3.15)) can be applied when nominal programmed control is not introduced; in this case it should be examined whether the optimal regulator can ensure tracking of nominal trajectory, or stability around the terminal point.

A similar idea of compensating for nonlinear effects of robot model by direct on-line calculation of model has also been exploited by several other authors.

For example, Takegaki and Arimoto [18] have applied positional control in the form*):

$$u(t) = G(q^o(\tau)) - K_1(q(t)-q^o(\tau)) - K_2\dot{q}(t) \tag{2.3.17}$$

where $G(q^o(\tau)$ is an $n\times 1$ vector of gravitational forces (moments) in a desired position $q^o(\tau)$), $K_1 \in R^{n\times n}$ is an $n\times n$ matrix of positional feedback gains, $K_2 \in R^{n\times n}$ is an $n\times n$ matrix of damping gains, and $x(t) = (q^T(t), \dot{q}^T(t))^T$. It has been shown that the control (2.3.17) is optimal with respect to some criterion (2.3.1). In this case gravitational forces in the terminal point only are to be calculated on-line, but this is just positional control; the tracking of nominal trajectory demands calculation of some more complex terms in the mechanical model of the robot. On the other hand, it has been shown [18] that this control allows us to avoid the inverse of Jacobian matrix, i.e. the calculation of joint coordinates of the desired position $q^o(\tau)$. Namely, as we have explained in the previous Paragraph a desired position (or trajectory) is usually imposed in external coordinates $S^o(\tau)$; so, if we want to realize the control (2.3.17), we have to solve eq. (2.2.14) for $q(\tau)$, i.e. to calculate the inverse of (2.2.14). It is well known that by differentating (2.2.14) by q we get:

$$\dot{S}(t) = J(q(t))\dot{q}(t) \tag{2.3.18}$$

where $J = [\partial f/\partial q]$, $J(q)$: $R^n \to R^{m\times n}$ is an $m\times n$ Jacobian matrix. The solution of inverse problem of (2.3.18) might be very complex i.e. the computation of inverse matrix $J^{-1}(q)$ might be too time consuming. So it is recommendable to avoid this calculation. It has been shown [18] that instead of (2.3.17) we can use the control

$$\Delta u = G(S(\tau)) - J^T(K_1(S(t)-S^o(\tau)) + K_2\dot{S}(t)) \tag{2.3.19}$$

where S(t) is the vector of actual external coordinates of robotic manipulator. Thus, in control law (2.3.19) we have to calculate on-line the Jacobian matrix and not the inverse of this matrix, which is, by far, an easier problem. However, this control with constant gains K_1

*) Here, only the model of mechanical part of the robot has been assumed. However, it is easy to extend the results to the general model which includes actuators models.

and K_2 cannot be acceptable for trajectory tracking and when parameter variations have to be taken into account.

Up to now we have presented a few attempts to synthesize control at the executive level by minimization of some criterion (either to calculate the optimal trajectory or just to calculate the optimal control for tracking some given nominal trajectory). We have shown that high nonlinearity of the robot model leads to the conclusion that it is impossible to calculate the optimal control for the exact nonlinear model of the system. Thus, we have either to calculate optimal control using some approximative (linearized) model of the robot or, by adding some nonlinear term, to compensate for nonlinearities in the robot models. In both cases the control is far from being optimal.

2.3.3. "Inverse problem" technique

We have shown that it is impossible to get an acceptable and simple control by standard optimization techniques. Actually, it is necessary to take care about dynamics of the complete robotic system in order to achieve good tracking of a desired trajectory. This means that we have to include the complete dynamic model of the robot in the control synthesis. Then, it is impossible to get a closed-loop solution by minimizing certain criteria. This was the reason why most researchers have attempted to synthesize dynamic control of the robots by some suboptimal approaches. They tried to directly include the mathematical model of robot dynamics in the control scheme.

Paul [19] has investigated the "inverse problem" technique, which was called the "computed torque" technique by Bejczy [20]. A similar approaches has been taken by Pavlov and Timofeyev [21]. Their approaches include on-line computation of the complete model of robot dynamics, i.e. they involve computation of driving torques by eq. (2.2.1) using the measured values of internal coordinates q and velocities $\dot{q}$ of the robot and the computed values of internal accelerations $\ddot{q}^0(t)$. Namely, if the desired trajectory is computed we can obtain $q^0(t)$, $\dot{q}^0(t)$, $\ddot{q}^0(t)$, $\forall t \in T$. It has been shown [21] that the robot is asymptotically stable around the nominal trajectory if the driving torques are computed by:

$$P(t) = H(q, d)\cdot\left[\ddot{q}^0(t)+K_1(q(t)-q^0(t))+K_2(\dot{q}(t)-\dot{q}^0(t))\right]+h(q, \dot{q}, d) \tag{2.3.20}$$

where $K_1 \in R^{n\times n}$ is an n×n matrix of position feedback gains and $K_2 \in R^{n\times n}$ is an n×n matrix of velocity feedback gains; K_1 and K_2 must be chosen in such a way, that a trivial solution of the equation

$$\ddot{e} = K_1 e + K_2 \dot{e}$$

is asymptotically stable, where $e \in R^n$. It can easily be shown that the driving torques (2.3.20) also ensure the practical stability of the robot (2.2.18). However only driving torques are computed in (2.3.20). It is necessary to include also the models of actuators (2.2.2), i.e. to calculate inputs u^i that correspond to computed driving torques (2.3.20).

The control scheme is presented in Fig. 2.6. Obviously, this scheme combines a closed-loop controller with "nominal" control signals computed on the basis of the equations of motion. In this scheme compensation is provided for time varying gravitational, centrifugal and Coriolis forces; the feedback gains are adjusted according to the changes in matrix H(q, d) of moments of inertia; an acceleration feedforward term is also included to compensate for changes along nominal trajectory.

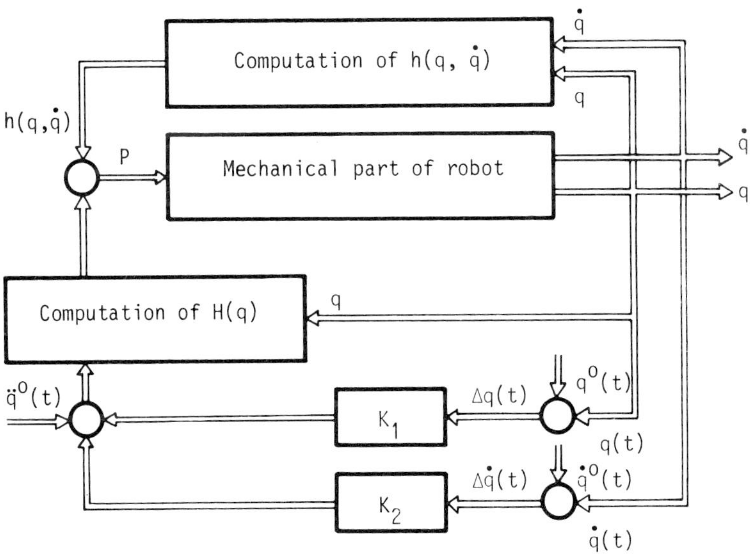

Fig. 2.6. Control scheme of "inverse problem" technique (after [19 - 21])

However, these approaches suffer from several disadvantages. The main problem is that in (2.3.20) computation of the complete dynamic model of robot is required. For complex robot structures this requirement may be very difficult to satisfy. This is the reason why some authors have tried to implement an approximative model of robot dynamics. They have omitted cross-coupling terms of moment of inertia in the matrix $H(q, d)$ and they have omitted centrifugal and Coriolis forces. This means, that the model is reduced to diagonal terms in inertia matrix $H(q, d)$ and to gravity terms, i.e. the control (2.3.20) is reduced to:

$$P_i(t) = H_{ii}(q, d)\left[\ddot{q}^{oi}(t)+K_1(q(t)-q^o(t))+K_2(\dot{q}(t)- - \dot{q}^o(t))\right]+g_i(q(t), d) \tag{2.3.21}$$

where $g_i: R^n \times D \to R^1$ is gravity force (moment) in the *i*th joint. The control scheme is now reduced to the one presented in Fig. 2.7. By this,

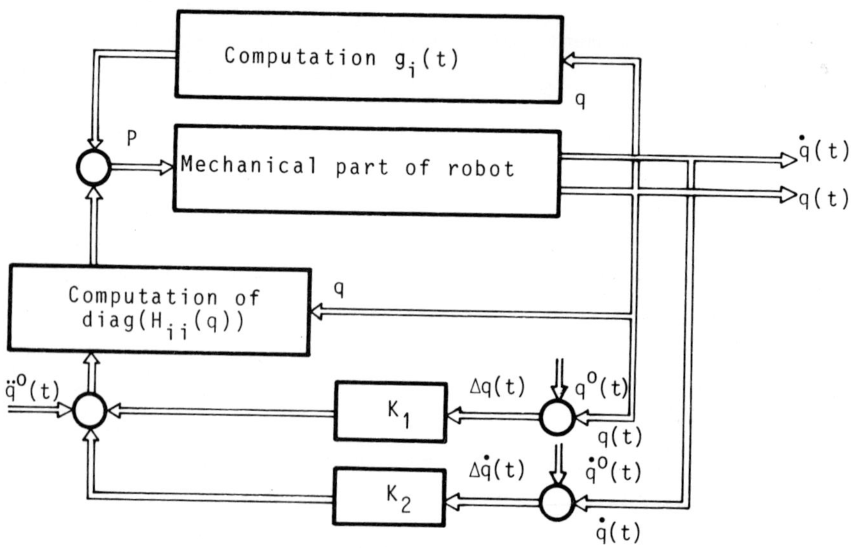

Fig. 2.7. Control scheme of "inverse problem" technique-simplified computation (after [19, 20])

the computation is considerably simplified, but it is still very cumbersome for some types of manipulators. On the other hand, it is questionable whether or not the efficiency of control is lost by these simplifications. Paul [19] has found that in his experiments with the Stanford manipulator, the contribution by the Coriolis and centrifugal terms is relatively insignificant. This is true when the manipulator approaches its goal position and orientation since the joint velocities are low during that time (so-called "fine" motions). However, no general analysis of the robot performance has been presented. Actually, in order to improve the quality of control for both "gross" and "fine" manipulator motions so that they can be executed with higher speed (but without loss in precision), we have to compensate for these velocity dependent terms.

A further reduction of computation time has been done by Bejczy. Bejczy [20] investigated a specific manipulator and found approximations to the diagonal terms in matrix H(q, d) and in gravity terms. These approximations lead to a tremendous reduction of computing time for the control scheme presented in Fig. 2.7. However, these approximations are valid for a specific manipulator only and for relatively slow motion conditions.

Paul [19] suggested that some terms in the dynamic model could be precalculated and stored, so as to reduce the time necessary for on-line calculation.

As we have already claimed, the problem of on-line computation of robot dynamic model is very difficult. Here, we shall not discuss various methods for on-line computation of dynamic model. Let us mention that in [22] one can find a study on the computational efficiency and comparison of various methods of computing the input torques which is based on a recent survey of numerical solution techniques made by Hollerbach [23]. On the other hand, Bejczy and Paul [24] have analyzed some approximations of robot dynamic equations using a symbolic state equation technique. However, the method for forming analytical equations of robot dynamics by computer which has been presented briefly in Chapter 1, leads to by far the fastest computation of robot dynamic models.

Raibert and Horn [25] have used a partial table look-up approach to simplify the computation automatically on a digital computer. Rather than compute the coefficients in (2.2.20) each time they are needed (every sampling period), their approach (called configuration space method) is to look them up in a pre-defined multi-dimensional memory organized by the positional variables q(t) (the so-called configuration space controller). Their control scheme is presented in Fig. 2.8. The disadvantage of this scheme is the requirement for a large memory.

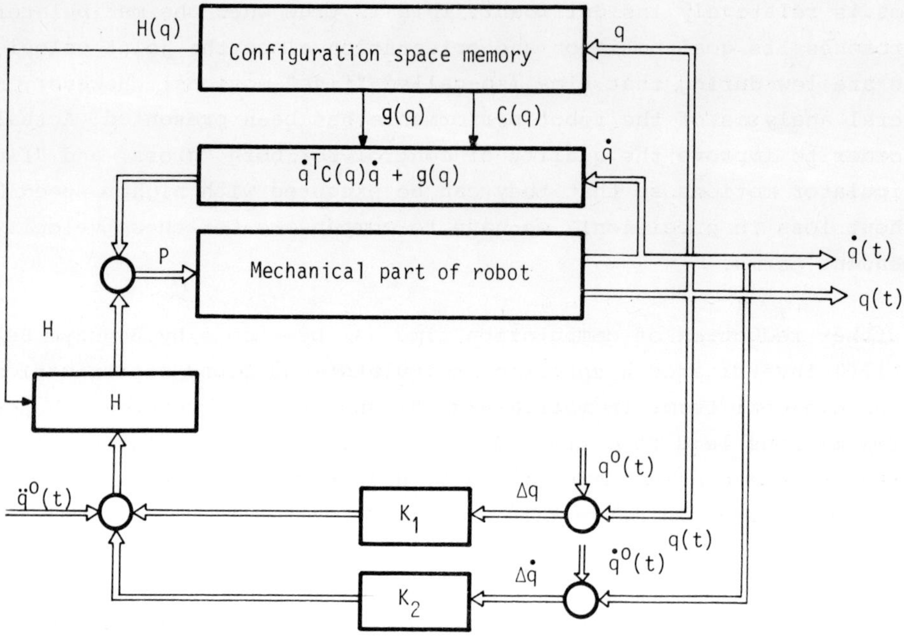

Fig. 2.8. Control scheme of the configuration space controller (after [25])

Thus, it is obvious that the main difficulty with the "inverse problem" technique is on-line computation of the dynamic model of the robot. Actually, the analysis of the complexity of the model that is needed for this control law in order to achieve sufficiently good tracking of nominal trajectory has not been given. However, there are several other problems with this approach. The implementation of control law (2.2.20) (or (2.2.21)) requires knowledge of the complete vector d of mechanism parameters; however, in def. (2.2.18) (or (2.2.19)) we have assumed that parameters d are not known in advance. So, it is questionable whether the control (2.2.20) (if synthesized for one chosen set of parameters d^o) is robust enough to withstand all parameter variations $d \in D$. Thus, the robustness of this control law has not been precisely analyzed.

Timofeyev has extended this approach to adaptive control in the case of unknown and variable parameters of the robot [26]; this approach will be described in Chapter 3. However, he has not analyzed whether non--adaptive control (2.2.20) can be applied to the manipulator with variable parameters.

The next difficulty with the "inverse problem" technique is the choice of gains in (2.2.20). In [21] it has been proved that the robot is stable if the gain matrices K_1 and K_2 are chosen as explained above. However, if we choose full matrices the control structure becomes complex and thus suffers from all disadvantages that were mentioned concerning the centralized linear regulator (Paragraph 2.3.2). Obviously, we can choose gain matrices in a diagonal form. However, the problem with such a choice lies in its suboptimality, i.e. such control may be very suboptimal from the standpoint of energy consumptions. Saridis and Lee [27] have proposed the method for suboptimal choice of the gains; the choice is done taking into account the complete nonlinear model of the robot. However, the method results in choice of full matrices K_1 and K_2.

Thus, there are several problems with the "inverse problem" technique, the main being its complex implementation and difficult maintenance. In addition, no systematic procedure for control synthesis of an arbitrary manipulator (choice of the model complexity, feedback structure and gains, analysis of robustness, etc.) has been elaborated up to now. In our opinion the problem with all of the above mentioned control schemes lies in their centralized approach to control problem; i.e. the robot is considered as a single plant which makes the implementation and maintenance of such control schemes very difficult and inpractical.

There were made few attempts to develop such control algorithms which include compensation for robot dynamics during motion but in which on--line computation of complex terms in the robot model is avoided. The force feedback offers such a possibility.

2.3.4. Force feedback

The forces (moments) that are acting in the robot joints can be directly measured. Thus, by introducing feedback by these forces we can compensate for dynamics of the robot and decouple the robotic manipulator into a set of separate subsystems. Force feedback allows easy compensation for the complete dynamics of the robot. This attractive idea was recognized and elaborated by the authors of this book [2, 7-9, 28, 29] and it will be presented and discussed in Paragraph 2.6.

The force feedback has recently been exploited by several authors. Hewit and Burdess [30] introduced force transducers in the joints of a manipulator, but they also introduced accelerometers which provide information about accelerations $\ddot{q}(t)$. They calculate on-line the matrix of inertia $H(q, d)$. Since by force transducers they measure the complete forces (moments) $P(t)$, from (2.2.1) they can calculate $h(q, \dot{q}, d)$ - gravity, friction, centrifugal and Coriolis forces. So their control scheme is relatively complex (although the computation is shorter than when the complete model has to be computed on-line). The control scheme is presented in Fig. 2.9. They have not analyzed the stability of the robot and they have not discussed whether the whole implemented equipment (transducers, accelerometers, computer for on-line calculation of the inverse of inertia matrix, etc.) is necessary to achieve good tracking of nominal trajectory. On the other hand, the problem of unknown parameters is not solved by this scheme since calculation of matrix $H(q, d)$ which depends on parameters, is also required.

There were made several attempts to include force feedback in servo control of joints of manipulators. Wu and Paul [31] made an experiment on a single-joint manipulator with a force sensor in the joint. Because the force feedback loop was closed around the joint and all the hardware was of analog type, the scheme completely avoided the computational difficulties, and provided a fast response. However, they have not analyzed stability.

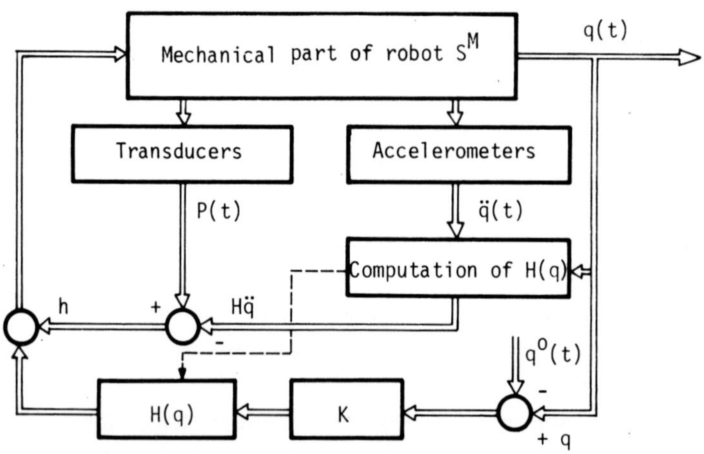

Fig. 2.9. Control scheme with force feedback and acceleration feedback (after [30])

Luh, Fisher and Paul [32] have made an experiment with Stanford manipulator: the first two joints have been redesigned and fabricated to include torque sensors. They analyzed each joint separately; by means of the method of describing functions they analyzed limit cycles of the nonlinear models of separate joints and they eliminated the limits by series compensation. However, the stability of the complete robot has not been analyzed in their paper. This analysis will be presented in Paragraphs 2.5 and 2.6 [2].

All above mentioned attempts concern the implementation of force sensors directly in the joints, i.e. direct force feedbacks around the joints of robot. However, there were many attempts to measure forces in the wrist of manipulator. A force sensing wrist has been suggested by Nevins and Whitney [33]. Paul and Shimano [34] developed a control scheme with sensing wrist so that load forces and moments with respect to the axis of the wrist coordinates could be measured on line. Force controllers employing the sensing wrist were implemented by Craig and Raibert [35]. Their control scheme contained two separate closed loops: one for positional control and the other for force control. However, since the signal from the sensing wrist is referenced to wrist coordinates it is necessary to perform the Jacobian transform to get joint loads every sampling period, i.e. some computational difficulties are present with this solution*). All these attempts mostly deal with application of the force feedback for assembling with manipulators. As we have already stated, we shall not treat assembling tasks in this book, so these and numerous other results from the force compliance will not be discussed here in more detail. We shall limit ourselves to the cases when force feedbacks are used to compensate for dynamics of robots (and not for sensing external forces acting upon the robot as is the case in the insertion process [2]).

The force feedback also offers a possibility to avoid the inverse of Jacobian matrix when the desired trajectory is imposed in external coordinates. Vukobratović and Stokić [2, 36] proposed a dynamic control scheme for the so-called "drink-test" (transfer of a vessel with liquid by a manipulator); we have implemented force transducers at the contact points between the hand and the object and by this information we can approximately solve the dynamic equilibrium of the object during motion,

---

*) Actually, the relationship beetwen the static wrist forces and static joint torques is given by transpose of Jacobian matrix. In the case of dynamic forces this relation is obviously much more complex (see the text to follow).

i.e. we can achieve the desired orientation of the object along the trajectory. Popov, Vereschagin and Zenkevich [37] also discussed implementation of force feedback for solving various robotic tasks and to avoid the inverse of Jacobian matrix*).

Wu and Paul [38] proposed position control of manipulator in external coordinates which they called resolved motion force control (RMFC). This control scheme adapts the force convergence concept to control the hand forces and (external) position instead of classical joint control. Actually, all methods described in the previous paragraphs assume that at the tactical control level the calculation of joint coordinates is performed and then at the executive level joint coordinates are controlled. Although this book concerns just executive control level, at this point we shall very briefly review some methods for obtaining joint coordinates $q^o(t)$ for imposed external coordinates $S^o(t)$ [38]:

a) Resolved motion position control [39]: In this method $q^o(t)$ is directly obtained from the inverse solution of (2.2.14). However, as we have already said, in the general case eq. (2.2.14) is very complex and may not be solved. If the manipulator is of simple structure then we can obtain $q^o(t)$ from given $S^o(t)$; in addition, joint velocities $\dot{q}^o(t)$ and accelerations $\ddot{q}^o(t)$ can be obtained by numerical differentiation.

b) Resolved motion rate control [40, 41]: In this method the actual manipulator external coordinates $S(t)$ are calculated from (2.2.14) and measured $q(t)$. By comparing actual $S(t)$ to desired nominal coordinates $S^o(t)$ we obtain corrective external rates $\dot{S}(t)$ in order to reduce any position error. Joint velocities are obtained by inverting (2.3.18):

$$\dot{q}(t) = J^{-1}(q(t))\dot{S}(t) \tag{2.3.22}$$

Then, joint accelerations $\ddot{q}(t)$ are obtained by numerical differentiation.

c) Resolved motion acceleration control [42]: This method calculates corrective external accelerations $\ddot{S}(t)$ (from the measured $q(t)$, $\dot{q}(t)$

*) All these attempts concern "mixed" tactical and executive levels, i.e. the same control loop solves both levels.

actual $\dot{S}(t)$, are calculated and compared to desired $\dot{S}^o(t)$). By differentiating (2.3.22) we obtain joint accelerations:

$$\ddot{q}(t) = J^{-1}(q(t)) \cdot \left[\ddot{S}(t) - \frac{\partial J}{\partial t}\dot{q}(t)\right] \tag{2.3.23}$$

where $\partial J/\partial t$ is the derivative of the Jacobian matrix with respect to time.

We have briefly reviewed these methods for obtaining joint coordinates $q^o(t)$, velocities $\dot{q}^o(t)$ and accelerations $\ddot{q}^o(t)$, in order to show complexity of the calculations that have to be performed at the tactical control level and to make a connection with the RMFC method. RMFC is based on the relationship between terminal link forces F and joint torques P, and consists of two parts. The first part is the position control of the terminal link (hand) and calculates the desired forces $F^o(t)$ to be applied to the hand in order to track the desired external trajectory $S^o(t)$. The second part of RMFC is the determination of the joint torques P, to be applied, such that hand has desired forces F^o applied as calculated in the first step. Desired hand forces $F^o(t)$ for correcting the external position and orrientation errors are obtained by closing a simple position-derivative (PD) controller, compensated with the desired external accelerations $\ddot{S}^o(t)$. Actually, if the desired external trajectory $S^o(t)$ is given, we can numerically obtain desired external velocities $\dot{S}^o(t)$ and accelerations $\ddot{S}^o(t)$. The actual external accelerations $\ddot{S}(t)$ are obtained by closing two servo feedback loops*):

$$\ddot{S}(t) = \ddot{S}^o(t) + K_1\Delta S(t) + K_2\Delta\dot{S}(t), \tag{2.3.24}$$

where $\Delta S(t) = S(t) - S^o(t)$ is the position error, $\Delta\dot{S}(t) = \dot{S}(t) - \dot{S}^o(t)$ is the velocity error, and $K_1 \in R^m$, $K_2 \in R^m$ are constant position and velocity feedback gains. Actual external coordinates $S(t)$ and velocities $\dot{S}(t)$ are obtained by measuring actual joint coordinates $q(t)$ and velocities and by applying (2.2.14) and (2.3.18). The gains K_1 and K_2 are calculated so that the characteristical roots of

$$\Delta\ddot{S}(t) = K_1\Delta S(t) + K_2\Delta\dot{S}(t)$$

---

*) Luh, Walker and Paul [42] used eq. (2.3.24) in their resolved acceleration control scheme to calculate actual external accelerations $\ddot{S}(t)$; then, they used (2.3.23) to calculate corresponding internal accelerations $\ddot{q}(t)$; using calculated $\ddot{q}(t)$ and measured $q(t)$, $\dot{q}(t)$, driving torques $P(t)$ are computed according to the control scheme of "inverse problem" type (Paragraph 2.3.3).

have negative real parts. The desired hand forces $F^o(t)$ to correct for any external position errors can be calculated by

$$F^o(t) = M\,\ddot{S}(t), \tag{2.3.25}$$

where $M = \mathrm{diag}(m^o, m^o, m^o, I_x^o, I_y^o, I_z^o)$, m^o is the total mass of the load and hand, I_x^o, I_y^o, I_z^o are the moments of inertia on the principal axes of the load and hand (we assume that the principal axes of the load are aligned with the hand's principal axes).

Now, in the second step we have to calculate joint torques $P(t)$ that will ensure realization of $F^o(t)$. If we assume that all members of the manipulator are massless except for the hand and load, the desired hand forces F^o can be resolved into applied joint torques by:

$$P(t) = J^T F^o(t) \tag{2.3.26}$$

This scheme involves only calculation of the transpose of Jacobian matrix $J(q(t))$. At this point, if the manipulator has no mass except for the hand and load, this scheme works perfectly*). For a manipulator in which the load clearly dominates, (the space shuttle manipulator with a 25000 kg load, for example), the scheme also works well. As the mass of the hand and load approaches the mass of the rest of the manipulator, the above simple control scheme no longer converges as the desired forces to be applied to the hand are reduced, as some portions of the applied joint torques obtained from (2.3.26) are used in accelerating the members of the manipulator.

In order to compensate for these effects, the second part of control scheme - force convergent control - has been developed. Force convergent control adopts the Robbins-Monro approximation method to determine an applied hand force F_a and corresponding applied joint torques such that the actual observed hand force F will converge to the desired hand force F^o obtained from (2.3.25). The actual applied force F at the hand is measured by a multi-element strain gauge force wrist sensor. RMFC is presented in Fig. 2.10. It contains two servos; the outer servo

---

*) It should be mentioned that Krutko and Lakota [43] used practically the same approach; they also neglected the masses of manipulator members and used eq. (2.3.24)-(2.3.26) to calculate driving torques P(t); however, they did not consider the case when the masses of the manipulator are not negligible with respect to load mass and did not use force feedback as in [38].

is the hand position control (2.2.24) and the inner servo is the force convergent control. Actually, the outer servo corresponds to the tactical control level, and the inner servo to the executive control level. Thus RMFC solves both control levels at the same time. Calculation of the inverse of Jacobian matrix and calculation of complicated dynamics of the manipulator are both avoided. However, the convergent rate of force convergent control depends on the structure of manipulator, so it might demand several iterations to achieve good accuracy of the method. Thus, calculation of dynamic model is replaced by an iterative stochastic method which also demands extensive computation. The problem of variable parameters has not been discussed in [38]. However, implementation of wrist force sensor offers a possibility of identify a payload (see Chapter 3), and thus we could easily extend this control scheme to include a variable payload. The robustness of this control scheme has not been analyzed.

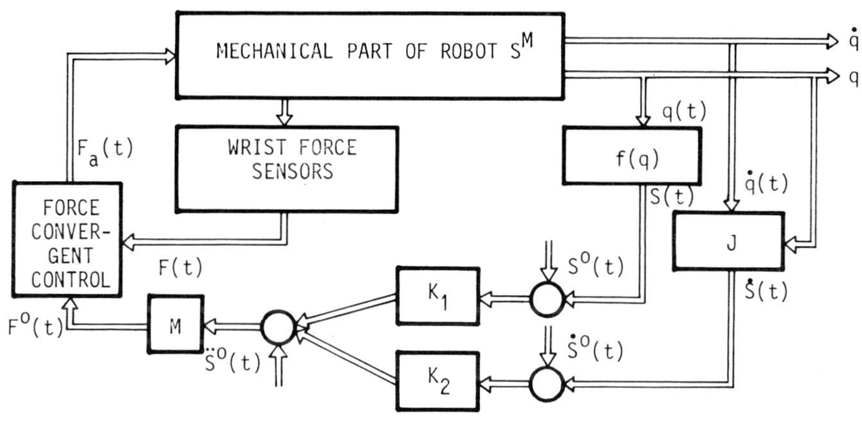

Fig. 2.10. Resolved motion force control scheme (after [38])

Thus, we have presented several control schemes which use force feedback. It is evident that the force feedback offers a good opportunity to simplify control law and to compensate for the robot dynamics. However, there are several drawbacks of implementation of force feedback. The price of the force transducers it not negligible and there are some technical problems regarding the noise appearing at the force sensors. Thus, a trade-off between the complexity and price of control computer and price of the sensor system has to be carefully studied. As we have already said we shall also use the force feedback in our non-adaptive and adaptive control schemes (see Paragraph 2.6 and Chapter 3).

2.3.5. Decoupled control

In all above presented approaches to control synthesis the main problem is to compensate for the influence of robot dynamics. In many control schemes the aim is to decouple the robotic system into the set of decoupled subsystems of lower order which can be controlled separately. Since in most cases the trajectory given in external coordinates is resolved into compatible joint coordinates trajectories, we ussually want to use these joint trajectories as inputs for servos which drive individual joints. Thus, we usually want to control each joint separately, i.e. to treat each joint servo as an independent subsystem. Obviously, this solution is the simplest and the most convenient, from the standpoint of control implementation. We shall discuss the synthesis of local servos in Paragraph 2.4.

A few attempts to decouple the robot dynamics were made. Yuan [44] tried to dynamically decouple a manipulation system by linear control. He observed a linearized model of the robot in which Coriolis, centrifugal and gravity forces were neglected. Actually, he considered the model of the mechanical part of the robot in the form:

$$H(q^o(0),\ d^o)\Delta\ddot{q} = \Delta P \tag{2.3.27}$$

i.e. he neglected the complete vector $h(q,\ \dot{q},\ d)$. In (2.3.27) $d^o \in D$ denotes some chosen nominal values of parameters. The linear models of actuators (2.2.2) were also taken into account. In the model (2.3.27) the coupling among the degrees of freedom of the robot are given by off-diagonal elements of the inertia matrix $H(q^o(0),\ d^o)$ i.e. by the cross-inertia terms. Thus, if we want to decouple the system (2.3.27) we just have to introduce linear control which will compensate for the cross-inertia terms in (2.3.27), i.e. a decoupling compensator has to be chosen in the form:

$$\Delta P = (I+R\cdot Q)\Delta\hat{P} \tag{2.3.28}$$

where $\Delta\hat{P} \in R^n$ is a vector of driving torques of the decoupled model of the robot, $I \in R^{n\times n}$ is the identity matrix and the matrices $R \in R^{n\times n}$, $Q \in R^{n\times n}$ are given by:

$$Q = \begin{bmatrix} 0 & H_{12} & \cdots & H_{1n} \\ H_{21} & 0 & \cdots & H_{2n} \\ \vdots & \vdots & \ddots & \vdots \\ H_{n1} & H_{n2} & \cdots & 0 \end{bmatrix}$$

$$R = \text{diag}\,(H_{11}, H_{22}, \ldots, H_{nn}),$$

where H_{ij} are the elements of matrix H. If we implement control (2.3.28) (Fig. 2.11) we decouple the model (2.3.27) into the set of subsystems:

$$H_{ii}\Delta\ddot{q}_i = \Delta\hat{P}_i, \quad \forall i \in I \tag{2.3.29}$$

each corresponding to one degree of freedom of the robot. Obviously, this is decoupling of the mechanical part of the robot. It is easy to extend (2.3.28) to include linear models of actuators (2.2.1) which are associated to each degree of freedom. Actually, we get the decoupled model of the robotic system (when actuators are taken into account) in the form:

$$\Delta\dot{x}^i = \hat{A}^i\Delta x^i + \hat{b}^i N(\hat{u}^i), \quad \forall i \in I$$

where $\hat{u}^i \in R^1$ is a new input of the decoupled model (corresponding to $\Delta\hat{P}$), and $\hat{A}^i$, $\hat{b}^i$ are the matrices of decoupled robot model given by (assumming $P_i = M_i$):

$$\hat{A}^i = (I_{ni} - f^i H_{ii} T^i)^{-1} A^i, \quad \hat{b}^i = (I_{ni} - f^i H_{ii} T^i)^{-1} b^i, \quad \forall i \in I$$

where by T^i we denote $T^i = \partial\hat{G}^i/\partial x^i$, so that, acc. to (2.2.13), $\ddot{q}_i = T^i\dot{x}^i$. Thus, by implementing the control scheme from Fig. 2.11 we get the model of robot in decoupled form (2.3.29), where $\hat{u}^i$ is input to each decoupled subsystem. Now, we can introduce a local controller for each decoupled subsystem (2.3.29), in the classical linear form

$$\hat{u}^i = k_i^T \Delta x^i, \quad \forall i \in I \tag{2.3.30}$$

where $k_i \in R^{n_i}$ is a local gain vector, which is chosen so as to ensure that the closed-loop local subsystem is stable. Namely, local controllers (2.3.30) can be chosen as local servos for each actuator of the robot, as they are usually implemented with robots in practice. Thus, by control (2.3.28) we compensate for coupling among the degrees of freedom of the manipulator, and we enable each actuator to be independently controlled.

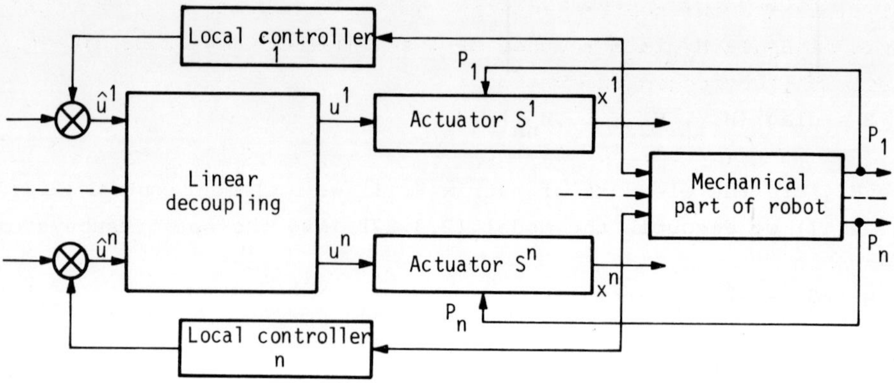

Fig. 2.11. Linear decoupling control scheme (after [44])

The decoupling scheme (2.3.28) is of open-loop type and very simple, and it is directly compatible with the independent joint control which is adopted in most manipulator control systems. However, this scheme suffers from many disadvantages. Since the model used (2.3.27) is very approximative, this control cannot compensate for many dynamical effects. Actually the simplicity and linear form of the control (2.3.28) result from the very rough approximation of the robot model. If the complete nonlinear model of the robot were taken into account the decoupled control would be much more complex. The problem of the effect of parameter variations on decoupling has not been treated in [44]. However, it is obvious that this scheme is very sensitive to parameter variation. This decoupling solution may happen to be structurally unstable.

Roessler [45] applied the nonlinear hierarchical control to decouple robot model. He "divided" the control vector into two terms which correspond to two control levels. At the level of local subsystems, local controllers are synthesized as usual to stabilize local decoupled subsystems. Subsystems in this case correspond to actuators' models*).

---

*) Actually, Roessler [45] considered the nonlinear models of subsystems, and he linearized them and compensated for nonlinearities by nonlinear local controllers. However, this has no consequence on his decoupled control so we omitted these nonlinear terms for simplicity.

$$\dot{x}^i = A^i x^i + b^i N(u^i) \qquad \forall i \in I \qquad (2.3.31)$$

where coupling among actuators (2.2.2), given by f^iM_i, has to be compensated for by the global control synthesized at the "coordination" level. This "second part" of control vector is intended to decouple the robotic manipulator into the set of local (subsystems (2.3.31)). This control u^G is proposed in the form:

$$u_i^G(t) = \begin{cases} (b^{iT}b^i)^{-1}b^{iT}f^iM_i(t_{\ell-1}) & \text{for} \quad \kappa_i(t_{\ell-1})>0 \\ (1-\varepsilon)(b^{iT}b^i)^{-1}b^{iT}f^iM_i(t_{\ell-1}) & \text{for} \quad \kappa_i(t_{\ell-1})<0 \end{cases}$$
$$\forall t \in (t_{\ell-1}, t_\ell) \qquad (2.3.32)$$

where by $\kappa_i(t_{\ell-1})$ we denote

$$\kappa_i(t_{\ell-1}) = d(f^iM_i(x(t), d))/dt\Big|_{t=t_{\ell-1}}$$

The control (2.3.32) is constant during the interval $(t_{\ell-1}, t_\ell)$. By ε we denote the number which satisfies:

$$||f^iM_i - b^iu_i^G|| \leqslant \varepsilon||f^iM_i|| \qquad (2.3.33)$$

The number ε determines the extent to which the coupling is reduced by the global control u_i^G. Obviously if $\varepsilon=1$ the system is not decoupled, if $\varepsilon=0$ the system is completely decoupled. The time intervals $(t_{\ell-1}, t_\ell)$ during which the global control is kept constant are determined by "coordination" algorithm: it checks in every sampling interval T_D whether the conditions (2.3.33) are satisfied with the global control (2.3.32) calculated by the moment t_ℓ:

$$||f^iM_i(t_{\ell-1}+rT_D)-b^iu_i^G(t_{\ell-1})|| \leqslant \varepsilon||f^iM_i(t_{\ell-1}+rT_D)|| \qquad (2.3.34)$$

where $r = 0,1,2,\ldots$. If the condition (2.3.34) is satisfied the control (2.3.32) is kept constant; if not, new global control is calculated for $t_\ell = t_{\ell-1} + rT_D$.

By this control scheme (Fig. 2.12) decoupling of the robot can be achieved. If we choose $\varepsilon=0$, we can totally decouple the robot into the set of subsystems (2.3.31) and then implement local controllers as is usually done in practice. Thus, the control (2.3.32) can compensate for the complete dynamic effects of the robot. However the control (2.3.32) requires the calculation of the complete dynamic model of the robotic system in each sampling interval T_D. This is a very severe condition since the model of robotic manipulator may be very complex. Actually, this control is similar to "the inverse problem" technique described in Paragraph 2.3.3, i.e. "the inverse problem" technique may also be regarded as decoupled control since this control also compensates for the effects of dynamics of the robot (coupling among degrees of freedom).

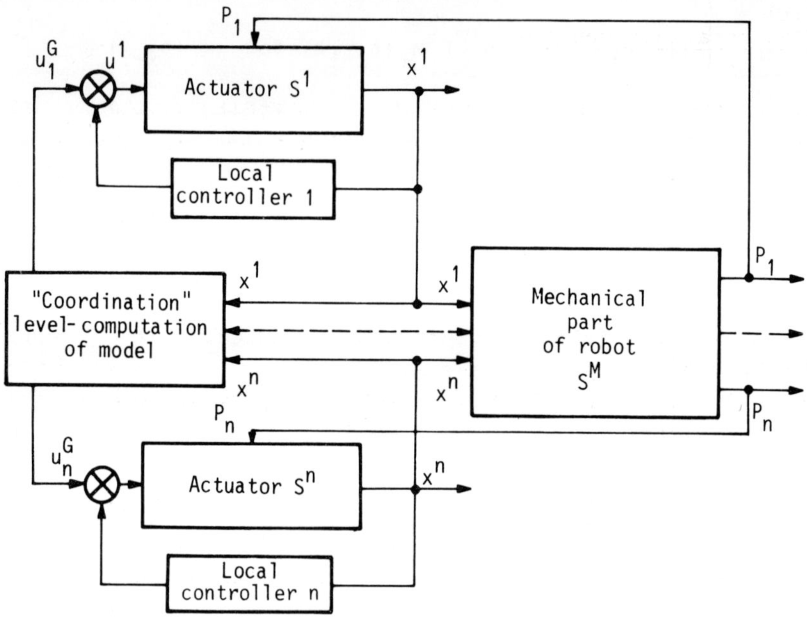

Fig. 2.12. Nonlinear hierarchical control scheme for decoupling of robot (after [45])

However, all problems concerning the implementation of "the inverse problem" technique are also encountered with the nonlinear hierarchical control of Roessler (cumbersome calculation of dynamic model, sensitivity to parameter variations etc.).

Nonlinear decoupling control of robot has also been proposed by Freund [46]. Starting from a very general control strategy for decoupling of nonlinear systems, he attempts to completely decouple the nonlinear model of robot dynamics in external coordinates. However, such control would be extremely complex and very hard to implement. Thus, when dealing with a particular robot the author simplifies the control by neglecting the accelerations in the coupling and he decouples the robot in joint coordinates. In essence, this approach also reduces to "inverse problem" technique. The crucial problem of the robot stability if such simplified controller is applied has not been considered in [46].

As we have already mentioned in previous paragraph force feedback can also be used to decouple dynamics of robotic system. Thus, force feedback can also be regarded as decoupled control of robots.

2.3.6. Decentralized control

Decoupled control presented in previous paragraph is intended to decouple the robotic system into the set of independent subsystems which can be stabilized by local servos. Thus, we arrive to the simplest and most commonly used control in practice-decentralized control.

The decentralized control structure is adopted as our approach to control synthesis in this book. We have already discussed this approach in our previous book [2] and we shall present it in detail in this chapter. Here, we shall briefly survey some results of other authors who have also considered decentralized control of robotic systems.

In this approach to control synthesis the robot is, at once, regarded as a set of decoupled subsystems, each corresponding to a separate joint and coupling among them is neglected. Starting from the model of robot, the robot is considered as a set of local decoupled actuators (2.3.31) in which coupling term $f^i M_i$ is neglected. For each subsystem (2.3.31) a local control is synthesized in order to stabilize free (decoupled) subsystem.

In many applications it has been assumed that local control can stabilize the whole robotic system. This most commonly used approach is effective for positioning of a manipulator and tracking of a relatively slow trajectory. However, if fast trajectory has to be precisely tracked this approach cannot be accepted. Medvedov, Leskov and Yuschenko [47] have analyzed overall system behaviour when only local controllers are implemented. They linearized the robot model around nominal trajectory and analyzed the stability of the linearized model with local controllers in frequency domain. The validity of this analysis is limited by the validity domain of the linearized model.

Paul [48] has presented several servosystem schemes in which compensation for some dynamic effects of the robot is introduced (but only at servosystem level). He presented compensation for the influence of gravity forces and Columb friction, compensation for the velocity and acceleration changes along the trajectory, etc. However, all these improvements have been made at the local level, and their effects might be very limited (this has not been analyzed in [48]).

Arimoto and Miyazaki [49] have proved that PID local controller can be effectively applied for the position control of manipulation robots. They proved the global asymptotic stability of the robot around a target point under the condition of a PID local feedback control scheme. However, the capability of the control to ensure tracking of nominal trajectory has not been considered in [49]. Arimoto and Miyazaki, following the results presented in [18], have modified their PID feedback scheme to suit the case when the target point is described in external coordinates (similarly to (2.3.18)). The asymptotic stability of such a PID sensory feedback scheme is also proved in [49].

In [50] Arimoto and co-authors attempted to extend the results of [49] to tracking of a desired path in space. Actually, they introduced such a local controller which enables the robot to move at a high speed and smoothly stop with high positioning accuracy. However, this scheme ensure the asymptotic stability of positioning, but tracking of some path in space is not reliable.

Thus, the conventional local servos ensure just positioning of the robot even at high speeds. The problem of tracking a desired trajectory by decentralized control, which is very complex, has been studied in [2]. In this book we shall for the first time consider decentralized control synthesis for manipulation robots with variable parameters, and we shall also add some global feedback loops if it is necessary to ensure stability of the robotic system and satisfy a stated control task.

2.3.7. Other approaches

We have briefly discussed some of the most characteristic results in dynamic control of manipulation robots. We have started with the most complex approach (optimal control) and arrived to the simplest (decentralized-joint servo control). The considered approaches are, in our opinion, the most characteristic and the most representative (but not always the most successful). Besides the considered approaches there were many other attempts in dynamic control of robots which should also be mentioned. Albus [51] attempted to control manipulators by control signals directly memorized in the computer memory. Young [52] attempted to synthesize the control for manipulators using the theory of variable structure systems, etc. There are also a few papers dealing with the problem of flexibility of the manipulator. Book, Maizza-Neto and Whitney [53] considered a simple two-beam, two-joint manipulation with distributed flexibility. A more general case was treated in a paper by Truckenbrodt [54] where a manipulator with rigid and flexible elements is properly modelled.

In [55] the preview tracking control algorithm is applied to the trajectory control of robots; the control algorithm consists of the integral control action of the afore predicted deviation. In [56] compensation for the robot dynamics is performed by nonlinear compensator and the control of external coordinates is established. Actually, the control scheme proposed has some similarities to "inverse problem" technique, except for the fact that control in external coordinates is proposed and that compensation for nominal acceleration along trajectory is not introduced.

In the text to follow, we shall present in detail our approach to synthesis of control of manipulation robots which satisfies the stated control task. As already mentioned, we shall first consider the decentralized control for robotic systems and, present stability analysis of the robots with variable parameters if local controllers are applied. Then, we shall also introduce additional global control to compensate for the robot dynamics in the case when the local controllers cannot withstand the influence of coupling between the robot joints. Actually, our aim is to find the simplest control law which can satisfy the given control task. Instead of treating the system as a single plant and introducing at once the centralized and complex control and, then, by various approximations trying to simplify the control algorithm, we shall consider the problem piece-by-piece. We first introduce local controllers and then add some global feedback loops until the desired performance of the system is achieved. In doing this, we shall consider the complete dynamic model of the manipulation robots with no simplifications and neglecting any nonlinear term.

2.4 Synthesis of Local Controllers

We shall start our control synthesis from the classical, simplest and most frequently encountered, in practice, approach via local controllers. In this approach, for each joint and its corresponding actuator, an independent local controller is synthesized which stabilizes the actuator and joint regardless of the rest of the robot. This approach is obviously the simplest possible from the point of view of control information structure; each local controller has the information about the coordinates of the corresponding joint only. This decentralized control structure is schematically presented in Fig. 2.13. The simplicity of such control structure is the reason why it is most commonly used in up-to-day robotic manipulators.

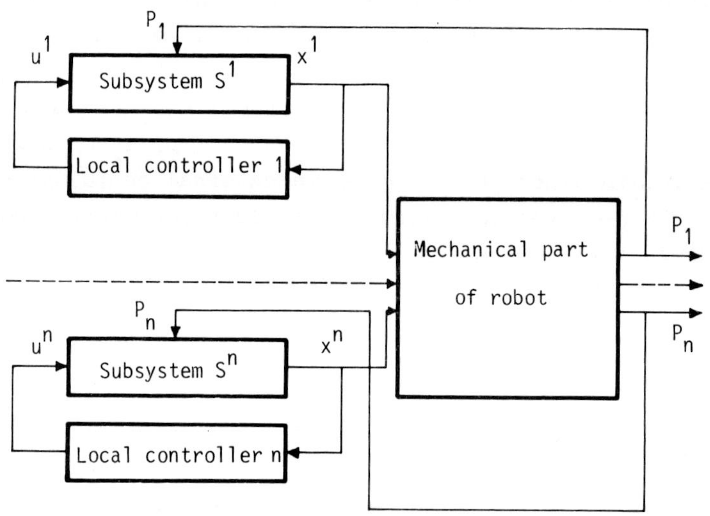

Fig. 2.13. Decentralized control structure

We shall not discuss various possibilities as to how to choose decentralized control structure. We adopt the control structure presented in Fig. 2.13 since it is physically the most appropriate. This means that we assume that each control input u^i has the information about the state coordinates of the corresponding actuator x^i, i.e. only:

$$u^i = \phi_i(x^i), \quad \forall i \in I \tag{2.4.1}$$

where ϕ_i: $R^{n_i} \rightarrow R^1$ is some function of the subsystem state vector x^i.

Here, we shall consider the control task at the executive level as stated in Paragraph 2.2.2. We assume that nominal trajectory $x^o(t) = (x^{o1T}(t), x^{o2T}(t), \ldots, x^{onT}(t))^T$ $\forall t \in T$ is synthesized at the tactical control level using some of the methods for calculating the joint coordinates when the nominal path is given in external coordinates [1]. Thus, we assume that we have $x^{oi}(t)$, $\forall t \in T$, $\forall i \in I$ which we have to track by the control law at the executive level.

There are several possibilities for synthesizing local controllers (2.4.1) for independent manipulator joints. We shall briefly discuss several approaches to the synthesis of local controllers.

Synthesis of decentralized control for large-scale systems has been widely investigated by many authors approaching this problem from the aspect of general systems theory. A very good survey of decentralized control methods for large-scale systems is given in [57].

The first question encountered when attaching this problem is whether the decentralized control has to minimize some criterion (2.3.1) or not. Namely, as we have explained in Paragraph 2.3, we might try to synthesize control in the form (2.4.1) which will minimize criterion (2.3.1) along with satisfying control task (2.2.18). However, numerous problems arise if optimal decentralized control is to be synthesized. In [2] we have analyzed these problems, so we shall not repeat this discussion.

The second question concerning the synthesis of decentralized control is: should we synthesize the decentralized control law (2.4.1) and its parameters using the complete centralized model of the robot (2.2.8), or should we first synthesize local controllers using the approximate decoupled model of the robot, and, then, analyze the stability of the overall system. The first approach has been considered by many authors but mainly for linear large-scale systems. They attempt to synthesize decentralized control (2.4.1) so that the overall linear model has certain properties (e.g. poleplacement). This means that we have at once to determine the decentralized control law and feedback gains which ensure stability of the complete system. Davison [58] considered robust decentralized servomechanism problem, i.e. he attempted to find a decentralized controller such that certain outputs of the system asymptotically track given reference inputs independent of any external disturbances which may affect the system, and independent of any variations in the plant parameters and gains of the system. He found necessary and sufficient conditions for a solution to the problem to exist, and found a complete characterization of all decentralized controllers which solves the problem. However, this approach is very difficult to apply to nonlinear systems. On the other hand, this "centralized" synthesis of decentralized control is not convenient for practical application: a designer can hardly use his own experience about the robot and tune the controller in accordance with his desires, since it is very difficult to think about dynamics of the complete robot as a single plant.

This is the reason why we have chosen the second approach: we first synthesize local controllers for independent (decoupled) subsystems and then check whether or not the overall robot is practically stable with these local controllers. This approach has certain advantages from the point of view of control designer since the synthesis is performed "piece-by-piece" [59]: local controllers are synthesized for low-order subsystems and many classical and well-elaborated methods for the synthesis of controller can be applied. Designer has to think about low order subsystems and synthesize local controllers for each manipulator joint regardless of the complete robot dynamics. Then, the stability of the complete robot is analyzed and if the system is not stable the local controllers have to be chosen again. If we cannot find local controllers which can stabilize overall robot, we have to introduce global feedback loops, i.e. the global control which should compensate for coupling among subsystems.

Thus, we shall first consider the synthesis of local controllers for local subsystems. However the choice of subsystems models is not unique, even if information control structure has already been chosen. As we have already explained above, the control structure (2.4.1) is the most adequate from the physical and practical points of view. However, the model associated to each independent joint of the manipulator has to be chosen. Each joint is driven by its actuator S^i the model of which is given by (2.2.2) where $f^i(\theta^i)M_i$ represents the influence of the manipulator dynamics upon the ith actuator. Thus, we can choose the models of decoupled (free) subsystems as:

$$\bar{S}^i\colon\ \dot{x}^i = A^i(\theta^i)x^i + b^i(\theta^i)N(u^i), \qquad \forall i \in I \tag{2.4.2}$$

where we have neglected coupling term $f^i(\theta^i)M_i$.

However, M_i representing the coupling among subsystems (2.4.2) is a function of all state coordinates of the manipulator. It is also a function of state coordinates x^i of the subsystem $\bar{S}^i$: Since the relation the mechanism joint coordinate q_i and the state vector x^i is given by (2.2.5) and since P_i is a function of q_i, $\dot{q}_i$, $\ddot{q}_i$ (2.2.1), it is clear that P_i is a function of x^i (relation between M_i and P_i is given by (2.2.7)). For example model (2.4.2) takes into account only the inertia of the actuator: inertia forces of the mechanical part around the ith joint are included in coupling term $f^i M_i(x)$. Similarly gravitational term around the ith joint which might depend upon x^i is not included in the model (2.4.2). Thus, we have to extract from $M_i(x)$ the part that depends on x^i and associate it to the subsystem model $\bar{S}^i$. If we look at the model of the mechanism dynamics (2.2.1) we can try to extract the terms that depend on x^i; for example, inertia term $H_{ii}(q, d)\ddot{q}_i$, gravity term $G_i(q, d)$ etc. However, in the general case it is not possible to extract the terms that are functions of x^i only, for example, H_{ii} is usually a function of all mechanism joint coordinates q and it is the same situation with gravity term G_i. On the other hand, these terms are, in the general case, highly nonlinear. The models of subsystems $\bar{S}^i$ are linear, except for nonlinearity of the amplitude saturation type upon the input $N(u^i)$. We would prefer to preserve linearity of the subsystems models in order to simplify the synthesis of local controllers*). Thus, we shall extend the subsystems models by constant inertia term $\bar{H}_{ii}\ddot{q}_i$ (similarly as in Paragraph 2.3.5) and constant gravity term $\bar{h}_i q_i$ so that we get the models of the subsystems in the form [2]:

$$\tilde{S}^i: \quad \dot{x}^i = \bar{A}^i(\theta^i)x^i + \bar{b}^i(\theta^i)N(u^i), \qquad \forall i \in I \tag{2.4.3}$$

where $\bar{A}^i \in R^{n_i \times n_i}$, $\bar{b}^i \in R^{n_i}$ are matrices of the subsystem $\tilde{S}^i$ given by:

$$\bar{A}^i(\theta^i) = (I_{n_i} - f^i(\theta^i) \cdot \bar{z}^i \bar{H}_{ii} T^i)^{-1} (A^i(\theta^i) + f^i(\theta^i)\bar{z}^i \bar{h}_i \hat{T}^i) \tag{2.4.4}$$

$$\bar{b}^i(\theta^i) = (I_{n_i} - f^i(\theta^i)\bar{z}^i \cdot \bar{H}_{ii} T^i)^{-1} b^i(\theta^i), \quad \forall i \in I$$

where by $\bar{z}^i$ we denote a constant number estimating the function $z^i(q_i)$ in (2.2.7), T^i is a constant number estimating $\partial G^i/\partial x^i$ in (2.2.13), $\hat{T}^i$ is a constant number estimating $G^i(x^i)$ in (2.2.5). The numbers $\bar{H}_{ii}$ estimate inertia terms $H_{ii}(q, d)$, and constant numbers $\bar{h}_i$ estimate gravity terms in $h_i(q, \dot{q}, d)$, i.e. estimate terms $g_i(q, d)$. In [2] we have already analyzed the introduction of these estimations into subsystems

---

*) In fact, we can choose nonlinear models of subsystems. However, as we want to synthesize linear controllers we should linearize subsystems models. Thus, we shall start at once with linear subsystems model.

models. We have analyzed the influence of choice of the numbers $\bar{H}_{ii}$ and $\bar{h}_i$ upon the synthesis of control parameters and the suboptimality of decentralized control. We have to choose numbers $\bar{H}_{ii}$ and $\bar{h}_i$ such that

$$\bar{H}_{ii} \geq \max(H_{ii}(q, d)), \qquad \bar{h}_i \leq \min|\partial g_i(q, d)/\partial q_i|$$

where the maximum and minimum are taken on the sets $x \in X^t$, $d \in D$. Otherwise, if we choose $\bar{H}_{ii} < \max(H_{ii}(q, d))$, then the overshoot might appear in the *i*th joint when the robot comes in position in which inertia around this joint is greater than $\bar{H}_{ii}$. Namely, if we choose $\bar{H}_{ii} < \max(H_{ii}(q, d))$, and synthesize local gains so that the subsystem is critically damped then the corresponding damping factor might be less than critical in the position which corresponds to $\max(H_{ii}(q, d))$; thus, when robot comes to a position in which inertia around the *i*th joint is $\max(H_{ii}(q, d))$, then the subsystem is underdamped and overshoot must appear. It is well known that overshoot must not occur in controlling robots under any circumstances. Thus, we shall choose $\bar{H}_{ii}$ and $\bar{h}_i$ as above. It should be noted that we have taken into account only linear dependent member in the Taylor's series of gravity term $g_i(q, d)$, while the rest members are taken into account through coupling term.

The model of subsystem $\tilde{S}^i$ (2.4.3) includes the model of actuator in the *i*th manipulator joint and the model of the independent *i*th joint dynamics*):

$$\tilde{P}_i = \bar{H}_{ii} T^i \dot{x}^i + \bar{h}^i \hat{T}^i x^i \tag{2.4.5}$$

where $\tilde{P}_i$ is an estimate of the driving torque in the *i*th joint. Thus, the models of subsystems (2.4.3) are better approximations of the robot dynamics than models (2.4.2) which take into account dynamics of actuators only. The approximative decoupled model (2.4.3) also takes into account dynamics of the mechanism but using linear decoupled model (2.4.5). Obviously, a more adequate decoupled model would take into account the nonlinear model of the mechanism (in a decoupled form), but, as explained above we want to deal with linear models of subsystems. The coupling among subsystems (2.4.3) is given by $f^i(\theta^i)z^i(q_i)\cdot\bar{P}_i(x, d)$ where $\bar{P}_i$ is given by

---

*) The physical explanation of the model (2.4.5) is obvious: it is assummed that all joints of the robot except for the *i*th are locked and only the *i*th joint can rotate with acceleration $\ddot{q}_i$; since $H_{ii}(q, d)$=const for q_j=const, for $j \neq i$, it can be taken $H_{ii}(q, d) = \bar{H}_{ii}$ and $h^i q_i$ is the approximation of the gravity force.

$$\bar{P}_i(x, d) = P_i(x, d) - \tilde{P}_i(q_i, \ddot{q}_i)\cdot\bar{z}^i/z^i(q_i) \quad (2.4.6)$$

If we compare the coupling among subsystems $\bar{S}^i$ (2.4.2) and the coupling among subsystems $\tilde{S}^i$ (2.4.3) we can see that the coupling in the latter case is considerably reduced for the value of estimation of driving torques $\tilde{P}_i(q_i, \ddot{q}_i)$. Since we shall synthesize local controllers for independent subsystems neglecting the coupling among them, it is obvious that subsystems $\tilde{S}^i$ with reduced coupling would lead to more adequate local controllers regarding the stability of the overall robotic manipulator. Actually, the approximative model (2.4.3) is closer to the real system model than (2.4.2), so the control synthesized using (2.4.3) has a better chance to accommodate the overall system than the control synthesized using (2.4.2).

The next problem is the variation of actuator parameters θ^i. As we have explained in Paragraph 2.2.2, we shall assume that the parameters of the actuators vary relatively slowly, so that there is no need to permanently take care about parameter variation. We shall synthesize local controllers assuming that parameters of the actuators are constant and known in advance. We shall consider the synthesis of local controllers assuming that the values of parameters of the actuators are given $\bar{\theta}^i \in R^{\ell_i}$ so that the models of subsystems $\tilde{S}^i$ (2.4.3) can be written in the form:

$$\tilde{S}^i:\quad \dot{x}^i = \tilde{A}^i x^i + \tilde{b}^i N(u^i) \quad (2.4.7)$$

where by $\tilde{A} \in R^{n_i \times n_i}$ and $\tilde{b}^i \in R^{n_i}$ we denote $\tilde{A}^i = \bar{A}^i(\bar{\theta}^i)$ and $\tilde{b}^i = \bar{b}^i(\bar{\theta}^i)$. So we shall synthesize local controllers for linear time-invariant subsystems (2.4.7) with constant parameters. From time to time the control system will identify the parameters of the actuators and if they change significantly the controllers gains must be calculated again (and changed). Thus, in the text to follow we shall deal with subsystems models in the form (2.4.7).

Now, we have to define the control task on the subsystems level in accordance with the control task for robotic manipulators (2.2.18), or (2.2.20). As we have explained in Paragraph 2.2.2, the regions of practical stability X^I and X^t are asummed to be in "decentralized" form. Thus, we can decouple control task (2.2.18) into subsystems subtasks. We shall synthesize local controllers for decoupled subsystems (2.4.7) so that the following control task has to be accomplished:

(2.4.8) Local control task (for the *i*th subsystem): The local control $u_L^i(t, x^i)$ has to be synthesized which will ensure that $\forall x^i(0) \in X^{I(i)}$ implies $x^i(t) \in X^{t(i)}$, $\forall t \in T$, where $X^{I(i)}$ and $X^{t(i)}$ are finite regions around the prescribed nominal trajectory $x^{oi}(t)$.

This definition of local control task corresponds to the control task for robotic manipulator (2.2.18). It is easy to extend this definition to correspond to definition (2.2.20). According to (2.2.19) the regions of practical stability for local subsystems are defined as $X^{I(i)} = \{x^i(0): ||\Delta x^i(0)|| \leq \bar{X}^{I(i)}\}$ and $X^{t(i)}(t) = \{x(t): ||\Delta x^i(t)|| \leq \bar{X}^{t(i)} \cdot \exp(-\alpha^{(i)} t)\}$, $\forall t \in T$, where $\bar{X}^{t(i)} > \bar{X}^{I(i)} > 0$, $\alpha_i > 0$, $\forall i \in I$. Our approach is to first synthesize the local controllers which will stabilize the local subsystems in accordance with local practical stability definition (2.4.8) and then analyze if such local controllers can stabilize the overall robotic manipulator according to the control task definition (2.2.18). If the result of the test is negative, we have to choose new local controllers regardless of the local task definition (2.4.8), or, in other words, we must synthesize local controllers with higher gains (or, we redefine local tasks by increasing $X^{I(i)}$, $X^{t(i)}$ and α_i and synthesize local controllers with higher feedback gains).

Before starting to consider various procedures for local control synthesis, we must once again underline the difference between the control strategy presented in our previous book [2] and the control scheme which we shall discuss here. In [2] we have considered the manipulators with constant and known parameters. Thus, we were able to synthesize nominal programmed control $u^o(t)$ using a centralized (complete) model of the manipulator, i.e. we synthesized programmed control $u^o(t)$ to satisfy (2.3.4) where d^o, θ^o are given (defined in advance) values of parameters. Then, we considered the model of deviation from nominal trajectory $x^o(t)$ and nominal control $u^o(t)$ (2.3.5). We considered the approximative decoupled model of the system at this level of deviation from nominal. The form of this decoupled model is analogous to (2.4.7) where Δx^i and Δu^i have to be substituted for x^i and u^i (since the deviations from nominal trajectory $x^{oi}(t)$, and control $u^{oi}(t)$ have been considered). The only difference is that in [2] the coupling among subsystems is given by $f^i(\theta^{io})[z^i(q)\cdot\bar{P}_i(q, d^o) - z^i(q^{oi})\bar{P}_i^o(q^o, d^o)]$ which is less that for subsystems (2.4.7). (The difference is in nominal driving torque $z^i(q^{oi})\bar{P}_i^o(q^o, d^o)$ which is compensated by nominal programmed control $u^o(t)$). Thus, the coupling among subsystems is considerably reduced if the nominal programmed control is applied. However, the application of nominal programmed control synthesized using the centralized model suffers from two main drawbacks:

(1) Synthesis of the programmed control on the basis of centralized model (2.3.4) is very complex, and it can hardly be computed in real time. This is the reason why in [2, 7, 10] we have proposed to synthesize the nominal programmed control off-line and store it in the computer memory. This approach is accepteble in many cases in robotic practice in which the control tasks and operating conditions are well defined in advance. However, in the general case, the task cannot be prepared in advance since all conditions cannot be anticipated. On the other hand, if a complex and rather long control task has to be performed, this approach demands large memory capacity. This approach has the same drawbacks as the approach by Raibert and Horn [25] or Albus [51], although some advantages of our strategy [2] concerning simplicity of the control scheme are evident.

(2) The synthesis of nominal programmed control $u^o(t)$ using the centralized model (2.3.4) assumes that all parameters of the robot are defined and precisely known in advance. If this assumption is not satisfied, application of the centralized programmed control is not efficient in general since the nominal control cannot compensate for robot dynamics when the parameters deviate from their nominal values d^o, θ^o (for which the nominal control has been synthesized).

Since in this book we consider the robotic manipulators with variable parameters the application of centralized nominal control will be omitted. Thus, we shall start at once with the approximate decoupled model of the robot (2.4.7) and directly synthesize local controllers.

We can attack the problem of local control synthesis (i.e. the synthesis of local controller for the subsystem $\tilde{S}^i$ (2.4.7) which will satisfy local control task (2.4.8)) by numerous approaches. Namely, we can apply various approaches to control synthesis which have been developed in the classical and modern control theory: optimal servosystem approach [60]; the tuning regulators [61]; the robust servomechanisms approach [62]; classical PID controllers, synthesized by various methods in frequency domain [63] or by root-locus methods [63] and so on; pole placement approach [64], etc. Certainly, we shall not survey various methods for the synthesis of local controllers for linear subsystems. We shall briefly present only two characteristic approaches: a) by minimization of the standard quadratic criterion, and b) by pole placement. In doing this, we shall consider the synthesis of the local controller: a) by complete state feedback b) by addition of integral feedback, c) by output feedback. However, numerous other methods can be applied; the stability analysis of the complete robotic system and introduction of global control are valid regardless of the method applied for local control synthesis.

Note that the local models of subsystems (2.4.7) are nonlinear with respect to inputs due to the presence of nonlinearities $N(u^i)$ of amplitude saturation type. However, in the synthesis of local controller we shall first consider the linear model of subsystem:

$$\tilde{S}^i:\quad \dot{x}^i = \tilde{A}^i x^i + \tilde{b}^i u^i \tag{2.4.9}$$

where input enters also linearly in the model, and afterwards we shall consider the influence of the input amplitude constraint upon the subsystem stability and overall system performance. Thus, we shall start with the local control synthesis for the linear models of subsystems (2.4.9).

2.4.1. Optimal servosystem

Let us suppose that we have to solve the local control task (2.4.8) by minimizing standard quadratic criterion in the form:

$$J_i(x^i(0), x^{oi}(t)) = \int_0^T [u^i r_i u^i + (x^i(t) - x^{oi}(t))^T Q_i (x^i(t) - x^{oi}(t))]dt \tag{2.4.10}$$

where $r_i>0$, is a positive number and $Q_i \in R^{n_i \times n_i}$ is a positive definite symmetric matrix. The theory of optimal linear servosystem is well known [60], so we shall not present it here. For the linear subsystem (2.4.9) and performance index (2.4.10) the solution of optimal servo-problem is given by

$$u^i(t) = -r_i^{-1} \tilde{b}^{iT} (\tilde{K}_i(t) x^i - g^i(t)), \quad \forall t \in T \tag{2.4.11}$$

where $\tilde{K}_i\colon T \to R^{n_i \times n_i}$ is the positive definite symmetric matrix which is the solution of the Riccati differential equation:

$$-\dot{\tilde{K}}_i(t) = (\tilde{A}^{iT}\tilde{K}_i(t) + \tilde{K}_i(t)\tilde{A}^i) + Q_i - \tilde{K}_i(t)\tilde{b}^i r_i^{-1}\tilde{b}^{iT}\tilde{K}_i(t)$$
$$\tilde{K}_i(\tau) = 0 \tag{2.4.12}$$

and $g^i: T\to R^{n_i}$ is the $n_i\times 1$ vector which is the solution of the equation:

$$-\dot{g}^i(t) = (\tilde{A}^i - \tilde{b}^i r_i^{-1}\tilde{b}^{iT}\tilde{K}_i(t))^T g^i(t) + Q_i x^{oi}(t), \quad g^i(\tau) = 0, \tag{2.4.13}$$

The control scheme of optimal linear servosystem (2.4.11) is presented in Fig. 2.14. Obviously, implementation of such a control law is too complex. Differential equations (2.4.12) and (2.4.13) have to be integrated backwards from the time instant τ, so these computations cannot be performed on-line. This means that we have to compute $\tilde{K}_i(t)$, $g(t)$ off-line and store it in the computer memory for each movement. Evidently such a solution is not acceptable in practice, since it demands too large computer memory, and the computational time even in the off-line mode is too long.

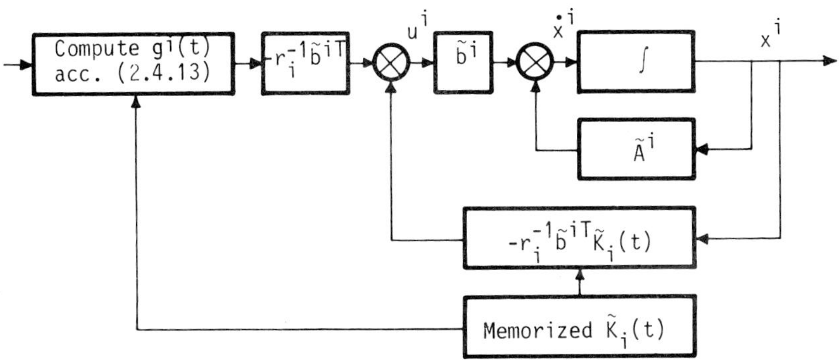

Fig. 2.14. Local, optimal servosystem for tracking desired trajectory

Therefore we shall consider a simplified solution of the optimal servo problem. Let us assume that there exists programmed control $\bar{u}^{oi}(t)$ satisfying:

$$\dot{x}^{oi}(t) = \tilde{A}^i x^{oi}(t) + \tilde{b}^i \bar{u}^{oi}(t), \quad \forall t\in T \tag{2.4.14}$$

This means that the desired trajectory x^{oi} is chosen so that such a $\bar{u}^{oi}(t)$ can be found which satisfies (2.4.14). Then, let us consider the model of the deviation of the subsystem state from the nominal trajectory $x^{oi}(t)$ and nominal programmed control $\bar{u}^{oi}(t)$. The model of state deviation is given by:

$$\Delta\dot{x}^i = \tilde{A}^i\Delta x^i + \tilde{b}^i\Delta u^i, \qquad \forall t\in T \tag{2.4.15}$$

where $\Delta x^i(t) = x^i(t) - x^{oi}(t)$ is the vector of dimension n_i which represents the deviation of the actual subsystem state from the nominal state $x^{oi}(t)$ at each moment $t\in T$, and $\Delta u^i(t) = u^i(t) - u^{oi}(t)$ is the scalar, $\Delta u^i \in R^1$, which represents the control which should be added to the nominal programmed control $u^{oi}(t)$ in order to stabilize subsystem (2.4.9). Now, we can consider stabilization of the model of deviation (2.4.15) around its origin $\Delta x^i(t) = 0$, i.e. around the nominal trajectory $x^i(t) = x^{oi}(t)$ and nominal control $\bar{u}^{oi}(t)$. To stabilize subsystem (2.4.15) we shall minimize the criterion:

$$\bar{J}_i(\Delta x^i(0)) = \int_0^\infty (\Delta u^i r_i \Delta u^i + \Delta x^{iT} Q_i \Delta x^i)\exp(2\beta_i t)dt \tag{2.4.16}$$

where $\beta_i > 0$ is the prescribed degree of the exponential stability of free subsystem. Here, we observe the criterion over an infinite time interval, although the practical stability during a finite time interval is desired. Actually, we have to choose β_i and weighting matrices Q_i and r_i, so as to ensure the practical stability of the local subsystem, i.e. to satisfy local control task (2.4.8). The choice of the weighting matrices Q_i and r_i and β_i will be explained in the text to follow. The solution of the optimal regulator problem (i.e. minimization of the standard quadratic criterion (2.4.16) under the constraint (2.4.15)) is well known [60]. The control law minimizing (2.4.16) is given by*)

$$\Delta u_i^L(t) = -r_i^{-1}\tilde{b}^{iT}K_i\Delta x^i(t), \qquad \forall t\in T \tag{2.4.17}$$

where $K_i \in R^{n_i \times n_i}$ is the positive definite matrix which represents the solution of the Riccati algebraic equation

$$K_i(\tilde{A}^i+\beta_i I) + (\tilde{A}^{iT}+\beta_i I)K_i + Q_i - K_i\tilde{b}^i r_i^{-1}\tilde{b}^{iT}K_i = 0 \tag{2.4.18}$$

Now, the control which is to stabilize free subsystems (2.4.9) around the nominal trajectory $x^{oi}(t)$ is given by

$$u^i(t) = \bar{u}^{oi}(t) + \Delta u_i^L(t) = \bar{u}^{oi}(t) - r_i^{-1}\tilde{b}^{iT}K_i\Delta x^i(t), \ \forall t\in T \tag{2.4.19}$$

---

*) Note that we have denoted the control which stabilizes the model of deviation by Δu_i^L in order to underline that it has local information only.

The control scheme of such solution of the local controller is presented in Fig. 2.15. In comparison to the control scheme of optimal servosystem (Fig. 2.13) this control scheme is much simpler and its implementation is considerably easier. However, the control (2.4.19) is evidently suboptimal with respect to the original criterion (2.4.10). Nevertheless, it is more convenient to adopt a suboptimal solution which can relatively easily be implemented. The computation of local nominal programmed control $\bar{u}^{oi}(t)$ can easily be performed on-line, since the orders of local subsystems (2.4.9) are relatively low (see Paragraph 4.2 for details about the computation of local programmed control which satisfies (2.4.14)). The computation of local regulator gains $D^i = -r_i^{-1}\tilde{b}^{iT}K_i$ is performed off-line and the implementation of control (2.4.17) reduces to simple multiplication of the system state deviation from nominal trajectory Δx^i by the gains D_i. However, this solution has some drawbacks regarding its robustness to parameter variations. Thus, we have to consider the synthesis of robust servo problem if we keep in mind the fact that this control should withstand variations of manipulator parameters.

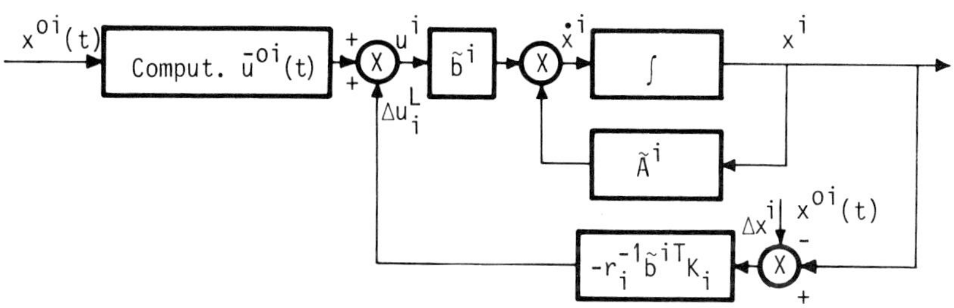

Fig. 2.15. Local optimal regulator and local nominal control

Actually the controller presented in Fig. 2.15 involves only static feedback (but it requires a feedback with respect to the complete subsystem state x^i). In order to achieve a robust controller capable of withstanding parameter variation we have to consider a dynamic feedback controller.

2.4.2. Robust servo-controller

Davison has considered the general robust servomechanism controller [62], i.e. he considered synthesis of a controller which can withstand arbitrarily large perturbations in the plant parameters. Davison has found necessary and sufficient conditions for there to exist a robust controller for a linear, time-invariant, multivariable system (plant) so that asymptotic tracking/regulation occurs independent of input disturbances and arbitrary perturbations in the plant parameters of the system. It has been shown that any robust controller must consist of two devices: 1) a servocompensator, and 2) a stabilizing compensator. The servocompensator is a feedback compensator with error input which depends on the disturbances and reference inputs to the system. The servocompensator is a generalization of the integral controller of clasical control theory. The purpose of the stabilizing compensator is to stabilize the resultant system obtained by applying the servocompensator to the plant.

However, such approach to the robust servomechanism problem is too general and complex. Thus, we shall consider a simplified form of the dynamic robust servo-controller.

First, we shall assume again that the local programmed control $\bar{u}^{oi}(t)$, satisfying (2.4.14), exists and that it is applied to the subsystem $\tilde{S}^i$. Now, the model of deviation (2.4.15) from the nominal trajectory and nominal programmed control is considered. We have to synthesize a robust controller which will stabilize subsystem (2.4.15) around the point $\Delta x^i = 0$ in the state space. Let us assume that we want to synthesize a dynamic controller in the form:

$$\begin{aligned} \dot{z}^i &= \hat{S}^i z^i + \hat{R}^i \Delta x^i \\ \Delta u^i &= \hat{Q}^i z^i + \hat{D}^{iT} \Delta x^i \end{aligned} \quad , \qquad \forall i \in I \tag{2.4.20}$$

where $z^i \in R^{m_i}$ is the state vector of the ith dynamic controller, $\hat{S}^i \in R^{m_i \times m_i}$ is the matrix of the ith controller, $\hat{R}^i \in R^{m_i \times n_i}$, $\hat{Q}^i \in R^{1 \times m_i}$ and $\hat{D}^i \in R^{n_i \times 1}$ are gain vectors, m_i is the order of the ith controller. The scheme of such a dynamic controller is presented in Fig. 2.16. We have to choose the elements of the matrices $\hat{S}^i$, $\hat{R}^i$, $\hat{Q}^i$ and $\hat{D}^i$ so as to stabilize subsystem (2.4.15) and ensure robustness of the subsystem to parameter variations. Actually, we have to stabilize the subsystem with the dynamic controller by a proper choice of the elements of the corresponding matrices. The augmented subsystem (plant $\tilde{S}^i$ + controller) is given by:

$$\Delta \dot{\hat{x}}^i = \hat{A}^i \Delta \hat{x}^i, \qquad \forall i \in I \tag{2.4.21}$$

where $\hat{A}^i \in R^{(n_i+m_i)\times(n_i+m_i)}$ is the matrix of the total closed-loop subsystem given by:

$$\hat{A}^i = \left[\begin{array}{c|c} \tilde{A}^i+\tilde{b}^i\hat{D}^{iT} & \tilde{b}^i\hat{Q}^i \\ \hline \hat{R}^i & \hat{S}^i \end{array}\right]$$

and $\Delta\hat{x}^i \in R^{n_i+m_i}$, $\Delta\hat{x}^i = (\Delta x^{iT}, z^{iT})^T$ is the state vector of the augmented subsystem. We have to choose the elements of $\hat{D}^i$, $\hat{Q}^i$, $\hat{S}^i$, $\hat{R}^i$ so as to achieve that the eigen-values of the closed-loop augmented subsystem lie to the left from the line $\lambda = -\beta_i$ in the complex plane, where β_i is the desired stability degree. Namely, we want to achieve that

$$\mathrm{Re}(\lambda_i(\hat{A}^i)) \leqslant -\beta_i, \; i=1,2,\dots,n_i+m_i \tag{2.4.22}$$

where λ_i (...) denotes the *i*th eigenvalue of the corresponding matrix. Even more, we want to achieve robustness of the plant (2.4.21) to parameter variations, i.e. we want to ensure regulation of the subsystem (2.4.15) when the parameters of the plant are perturbed. Actually, if the parameters of the plant are perturbed so that the subsystem matrices become $\tilde{A}^i \to \tilde{A}^i + \delta\tilde{A}^i$, $\tilde{b}^i \to \tilde{b}^i + \delta\tilde{b}^i$, the subsystem with the dynamic controller (2.4.20) should still be stable and, even more, condition (2.4.22) should still hold. Roughly speaking, we have to choose local controller (2.4.20) so that perturbations of the subsystems parameters cannot

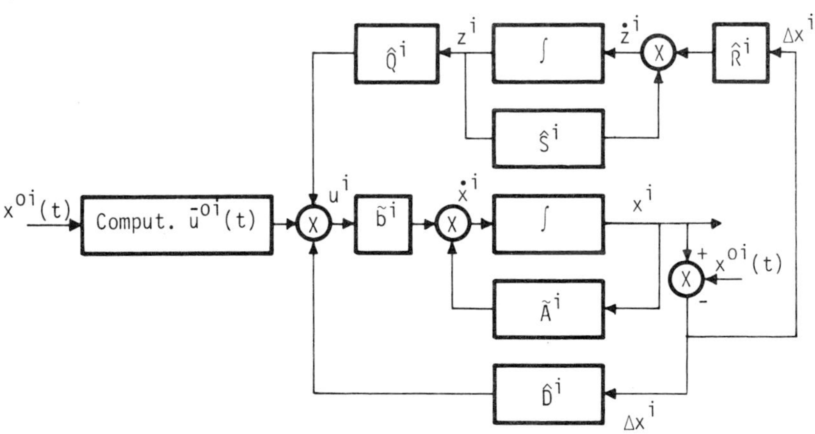

Fig. 2.16. Local dynamic controller

"destroy" subsystem stability. The robustness of the subsystem might require high feedback gains. However, by introduction of dynamic controller we get "more room" to stabilize the system with relatively low gains, by a proper choice of the dynamic part of the controller (by choosing matrices $\hat{S}^i$, $\hat{R}^i$, $\hat{Q}^i$). On the other hand, dynamic controller (2.4.20) might be too complex to be implemented if we choose too high order m_i of the controller. Thus, a trade-off between the controller complexity and its robustness should be analyzed.

Here, we shall consider the simplest dynamic controller, the order of which is $m_i=1$ and $\hat{S}^i=0$. Thus, the synthesis of dynamic controller reduces to choice of the gain matrices $\hat{Q}^i$ and $\hat{D}^i$, so as to stabilize the subsystem matrix $\hat{A}^i$. Thus, the local control is reduced to the local regulator with the addition of the integral feedback loop:

$$\Delta u_i^L = \hat{D}^{iT}\Delta x^i + \hat{Q}^i z^i = \hat{D}^{iT}\Delta x^i + \hat{Q}^i \int_0^t \hat{R}^i \Delta x^i dt, \qquad \forall i \in I \tag{2.4.23}$$

i.e. the local controller is reduced to a classical PID regulator (assuming that $z^i(0) = 0$). The scheme of such simple dynamic local controller is presented in Fig. 2.17.

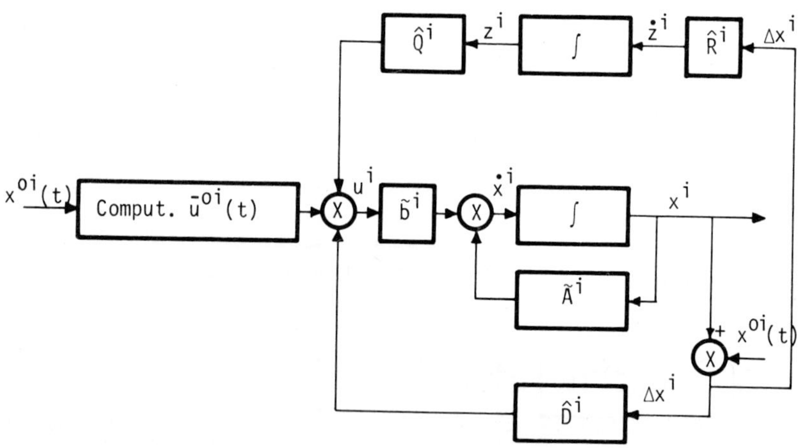

Fig. 2.17. Local controller with integral feedback

The total open-loop subsystem (plant $\tilde{S}^i$ + integrator) is given by

$$\Delta\dot{\hat{x}}^i = \hat{A}_I^i \Delta\hat{x}^i + \hat{b}_I^i \Delta u^i \tag{2.4.24}$$

where the matrix $\hat{A}_I^i$ and the vector $\hat{b}_I^i$ of the total subsystem are now given by:

$$\hat{A}_I^i = \left[\begin{array}{c|c} \tilde{A}^i & 0 \\ \hline \hat{R}^i & 0 \end{array}\right], \qquad \hat{b}_I^i = \begin{bmatrix} \tilde{b}^i \\ \\ 0 \end{bmatrix} \tag{2.4.25}$$

and $z^i(t) = \int_0^t \hat{R}^i \Delta x^i dt$, $\Delta\hat{x}^i = (\Delta x^{iT}, z^i)^T$.

The synthesis of the feedback gains $\hat{D}^i$ and $\hat{Q}^i$ has to ensure satisfaction of the nonequality (2.4.22). This can be done either by minimization of the standard quadratic criterion, or by pole placement (or, by some other classical method for stabilization of linear time-invariant system, but, as already mentioned, we shall consider just these two methods). The linear optimal controller has already been presented in the previous Paragraph. Here, we shall extend the results from the previous Paragraph to include the dynamic controller in minimization of standard quadratic criterion.

Instead of the standard quadratic criterion in the form (2.4.16), for the augmented subsystem (2.4.24), (2.4.25) we shall consider the criterion:

$$\hat{J}_i(\Delta\hat{x}(0)) = \int_0^\infty (\Delta u^i r_i \Delta u^i + \Delta\hat{x}^{iT}\hat{Q}_i\Delta\hat{x}^i)\exp(2\beta_i t)dt \tag{2.4.26}$$

where $\hat{Q}_i \in R^{(n_i+1)\times(n_i+1)}$ is a positive definite symmetric matrix, which can be chosen in the form

$$\hat{Q}_i = \left[\begin{array}{c|c} Q_i & 0 \\ \hline 0 & \hat{q}^i \end{array}\right], \qquad \hat{q}_i \in R^1$$

where Q_i is defined in (2.4.10). Thus, we have to synthesize control Δu^i which should minimize criterion (2.4.26) under the constraint given by the total subsystem (2.4.24), (2.4.25). The solution of this problem is analogous to the solution of the problem considered in the previous Paragraph. The control minimizing (2.4.26) is given by:

$$\Delta u_i^L = -r_i^{-1}\hat{b}_I^i\hat{K}_i\Delta\hat{x}^i = -\hat{k}^i\Delta\hat{x}^i \tag{2.4.27}$$

where $\hat{K}_i \in R^{(n_i+1)\times(n_i\times 1)}$ is the positive-definite solution of the corresponding Riccati algebraic equation:

$$\hat{K}_i(\hat{A}_I^i+\beta_i I)+(\hat{A}_I^{iT}+\beta_i I)\hat{K}_i+\hat{Q}_i-\hat{K}_i\hat{b}_I^i r_i^{-1}\hat{b}_I^{iT}\hat{K}_i = 0 \tag{2.4.28}$$

In (2.4.27) by $\hat{k}^i \in R^{n_i+1}$ we have denoted the feedback gains of the augmented subsystem $\hat{k}^i=(\hat{D}^{iT}, \hat{Q}^i)^T$. Since $\hat{b}_I^i$ is given by (2.4.25), and $\Delta\hat{x}^i= =(\Delta x^{iT}, z^i)^T$, the control (2.4.27) can be written in the form (2.4.23), where the feedback gains $\hat{D}^i$ and $\hat{Q}^i$ are given by:

$$\hat{D}^i = -r_i^{-1}\tilde{b}^{iT}\tilde{K}_i, \qquad \hat{Q}^i = -r_i^{-1}\tilde{b}^{iT}\tilde{k}_i \tag{2.4.29}$$

where $\tilde{K}_i$, $\tilde{k}_i$ are defined by:

$$\hat{K}_i = \left[\begin{array}{c|c} \tilde{K}_i & \tilde{k}_i \\ \hline \cdots & \cdot \end{array}\right]$$

$\tilde{K}_i \in R^{n_i \times n_i}$, $\tilde{k}_i \in R^{n_i}$. Thus, we again get the standard linear regulator (2.4.17) with the addition of integral feedback loop (2.4.23). Johnson [65] has shown that such a controller can withstand constant perturbations acting upon the subsystem, but it is also more robust to parameter variations than the static controller presented in Fig. 2.15*). Up to now, we have assumed that the information on the complete state vector x^i of the subsystem is available, i.e. that we can measure all state coordinates x^i and implement feedback loops by them. Thus, both static (Fig. 2.15) and dynamic controllers (Fig. 2.17) assume feedback loops by all state coordinates of the subsystems. However, it is not always convenient to measure all state coordinates of the robot. For example, technogenerators are not always available to meassure speeds of the joints, feedbacks by robot currents (if D.C. electro-motors are implemented), or feedback by pressure (if hydraulic actuators are used) is not convenient to be introduced. In these cases, if we still want to apply optimal local regulators we can solve these problems in two ways: by considering output regulators (i.e. trying to minimize standard quadratic local criteria by output feedback) [60], or by introducing observers to reconstruct the subsystem state from the output. Namely, we assume that outputs of the subsystems y^i are measurable:

$$y^i = C^i x^i \tag{2.4.30}$$

---

*) It should be noted that we might consider subsystem with constant estimation of the gravity term in coupling, and synthesize PID regulator, but the real values of gravity term will be taken into account in stability analysis.

where $y^i \in R^{k_i}$ is the output of the ith subsystem, k_i is the order of the ith output vector, $k_i < n_i$, $C^i \in R^{k_i \times n_i}$ is the output matrix. Thus, we can minimize criteria (2.4.16) or (2.4.26) by output-feedback, but this usually leads to complex problems regarding the synthesis of feedback gains.

The second approach, by introduction of observer, has already been considered in [2, 66]. Thus, we shall not repeat the synthesis of decentralized observer for robotic systems. Let us note, that we can introduce a local observer for each subsystem $\tilde{S}^i$ in the form of Luanberger's minimal observer [67]. Using these observers we get an estimate $\Delta\tilde{x}^i$ of the subsystem of state vector which we apply to compute local regulator (2.4.17), or (2.4.23) instead of real state Δx^i. Under the assumption that the subsystems are observable by output y^i (which is always satisfied if we assume that joint position is measurable), it has been shown that the subsystem is stabilized by such a controller in which estimation of the subsystem state $\Delta\tilde{x}^i$ is applied instead of real state Δx^i. The control scheme with local observers is presented in Fig. 2.18. Obviously, such local controllers, with local observers, are suboptimal with respect to local criteria (2.4.16). In [2, 66] the stability of the global system is analyzed if local regulators and local observers are applied.

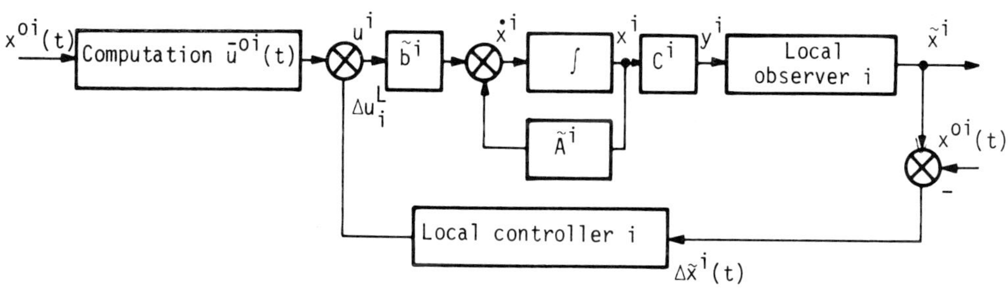

Fig. 2.18. Local controller with local observer

So far we have considered synthesis of the local controller as static or dynamic optimal linear regulators, i.e. by minimization of local standard quadratic criteria. This analytic approach to local control synthesis certainly has some drawbacks: it is relatively difficult to establish a connection between the weighting matrices in the criteria and requirements of the local control task (2.4.8). Thus, the choice of the criteria is to some degree heuristic. However, once the weighting matrices are chosen, the synthesis of local controller is straightforward (in both cases of static or dynamic controller). The problem of choice of the weighting matrices will be considered in Paragraph 2.4.4.

2.4.3. Synthesis of local controller by pole-placement method

The second method that might be applied for synthesis of either static or dynamic local controllers is by pole-placement. This is a well-known procedure for synthesis of control for linear systems. Let us consider the synthesis of dynamic output controller, i.e. let us assume that the output y^i (given by (2.4.30)) of the subsystem is measurable and that we want to introduce dynamic controller of the first order. This means that we shall consider the augmented model of the subsystem (2.4.24), (2.4.25) in which an integrator is added, but we assume that feedback by the output y^i can be implemented, only (together with integral feedback).

We shall assume again that local nominal control $\bar{u}^{oi}(t)$ satisfying (2.4.14) is introduced, and we shall consider the model of deviation from the nominal trajectory $x^{oi}(t)$ and control $\bar{u}^{oi}(t)$ but in its augmented form (2.4.24), (2.4.25).

The idea of the pole-placement method is simple. We want to apply linear feedback control by the subsystem output and by the integral feedback in the form:

$$\Delta u^i = k_p^{iT} \Delta y^i + \hat{Q}^i z^i = k_p^i C^i \Delta x^i + \hat{Q}^i \int_0^t \tilde{R}^i C^i \Delta x^i dt, \tag{2.4.31}$$

where $\Delta y^i = y^i - y^{oi} = C^i \Delta x^i$, $k_p^i \in R^{k_i}$ is the vector of output feedback gains $k_p^{iT} C^i = \hat{D}^i$, and $\tilde{R}_i C^i = \hat{R}_i$, $\tilde{R}_i \in R^{1 \times k_i}$, i.e. we assume that the dynamic part of the controller represents integral of the subsystem output. The problem is to choose feedback gains k_p^i and $\hat{Q}_i$. We observe closed-loop subsystem in the form:

$$\Delta \dot{\hat{x}}^i = (\hat{A}_I^i - \hat{b}_I^i k_I^{iT} \hat{C}^i) \Delta \hat{x}^i, \tag{2.4.32}$$

where $k_I^i \in R^{(k_i+1)}$ is the vector of feedback gains of the dynamic controller given by: $k_I^i = (k_p^{iT}, \hat{Q}^i)^T$ and $\hat{C}^i \in R^{(k_i+1) \times (n_i+1)}$ is the output matrix of the augmented subsystem defined by:

$$\hat{C}^i = \begin{bmatrix} C^i & \vdots & 0 \\ \cdots & + & \cdots \\ 0 & \vdots & 1 \end{bmatrix}$$

The state vector of the augmented subsystem has been defined above.

The gains k_I^i should be chosen so as to achieve that the poles of the closed-loop subsystem (2.4.32) are placed in desired positions, i.e. we have to choose k_I^i so that the eigenvalues of the closed-loop matrix $(\hat{A}_I^i-\hat{b}_I^i k_I^{iT}\hat{C}^i)$ are in certain prescribed positions in the complex plane. Since the closed-loop matrix of the subsystem (2.4.32) is of dimensions $(n_i\times 1)\times(n_i\times 1)$, it has n_i+1 eigenvalues. However, the number of eigenvalues that can be placed in arbitrary positions depends on the matrix structure and feedback loops introduced, i.e. it depends on the measurable output of the subsystem. Since the structure of the subsystem matrices $\hat{A}_I^i$, $\hat{b}_I^i$ in (2.4.32) is known (see Appendix 2.A), if all state coordinates of the subsystem are measurable (i.e. if $y^i = x^i$, $C^i = I_{ni}$), it is clear that we can place all n_i+1 poles in desired positions. However, if the number k_i of measurable state coordinates (outputs y^i) is less than the subsystem order n_i, then we can place just k_i+1 poles of the subsystem in desired positions in the complex plane. This means, that we can choose k_I^i so that:

$$\lambda_j(\hat{A}_I^i-\hat{b}_I^i k_I^{iT}\hat{C}^i) = -\sigma_1^D, -\sigma_2^D, \ldots, -\sigma_{k_i+1}^D \tag{2.4.33}$$

where $-\sigma_j^D$, $j=1,2,\ldots,k_i+1$ are the desired positions of subsystems poles. Obviously these poles might be real or complex (i.e. various combinations of pole types can be imposed). For example if $k_i=2$ one real pole and a pair of conjugate complex poles or three real poles might be set, or two real poles and the real part of the conjugate-complex pair might be prescribed etc. However, the remaining poles of the subsystem (i.e. n_i-k_i poles) cannot be specified in advance and their positions are free. The position of the poles that are not pre-specified, depends on the specified poles σ_j^D. Thus, the problem is how to specify k_i+1 poles of the subsystem so as to achieve that the remaining (nonspecified) poles of the subsystem also lie in some allowable regions in the complex plane.

Actually, the problem of the connection between the requirements in the local control task (2.4.8) and positions of the subsystem poles arises here. This problem will be discussed in the next paragraph. Let us note that one of the obvious constraints upon the placement of the subsystem poles is that they must lie to the left from the line $\lambda = -\beta_i$ where β_i is desired exponential stability degree, i.e.

$$|\mathrm{Re}(\sigma_j)| > \beta_i, \qquad j=1,2,\dots,n_i+1 \tag{2.4.34}$$

This means that we have to place all poles of the subsystem far enough in the left part of the complex plane in order to ensure stabilization of the local subsystem (and global system, too). However, the condition (2.4.34) is not always sufficient to satisfy all requirements of local control task (2.4.8) (but this will be discussed in the next paragraph).

Due to the structure of the subsystems matrices $\hat{A}_I^i$, $\hat{b}^i$ and under the assumption that feedback with respect to position of the joint must always be introduced, we may conclude that we can stabilize the local subsystems even if $k_i=1$. However, the stability degree β_i cannot be arbitrarily high when $k_i<n_i$. Here, we shall not discuss various possibilities for pole placement depending on the order of the subsystem output k_i.

Once the output of the subsystem is determined and poles σ_j^D, $j=1,2,\dots$ $\dots,k_i+1$ are specified, the problem is to compute feedback gains k_I^i that satisfy (2.4.33), i.e. to compute k_I^i so that the poles of the subsystem are placed in desired positions. There are many various methods for computing feedback gains, if poles of the system are specified [64]. However, these methods solve this problem in the general case for arbitrary structures of the subsystems matrices of arbitrary order, and for various output matrices. In our case, as we have explained in Paragraph 2.2, the order of the subsystem is at most $n_i=3$, and the structure of the subsystems matrices is fixed (see Appendix 2.A). Thus, the computation of feedback gains if poles of the subsystem are given is very simple and there is no need to apply some general procedure for pole placement. Actually, the feedback gains k_I^i are calculated so as to satisfy the following equation

$$\det(\hat{A}_I^i-\hat{b}_I^i k_I^{iT}\hat{C}_i^i-\sigma I_{(n_i+1)}) = k_a \sum_{j=1}^{n_i+1} (-\sigma_j^D-\sigma) \tag{2.4.35}$$

where $I_{(n_i+1)}$ denotes $(n_i+1)\times(n_i+1)$ unit matrix, k_a is a proportionality coefficient, $-\sigma_j^D$ are desired poles (real or complex). Obviously, eq. (2.4.35) corresponds to the case when $k_i=n_i$, i.e. when all n_i+1 poles of the subsystem can be specified and, thus, n_i+1 gains k_I^i can be determined from (2.4.35). We can write similar equations in the cases if

$k_i<n_i$, and if a smaller number of poles are specified. Thus, the calculation of k_I^i in all these cases is trivial.

In this way we get the control scheme for the ith local subsystem as presented in Fig. 2.19, which is quite similar to the control scheme

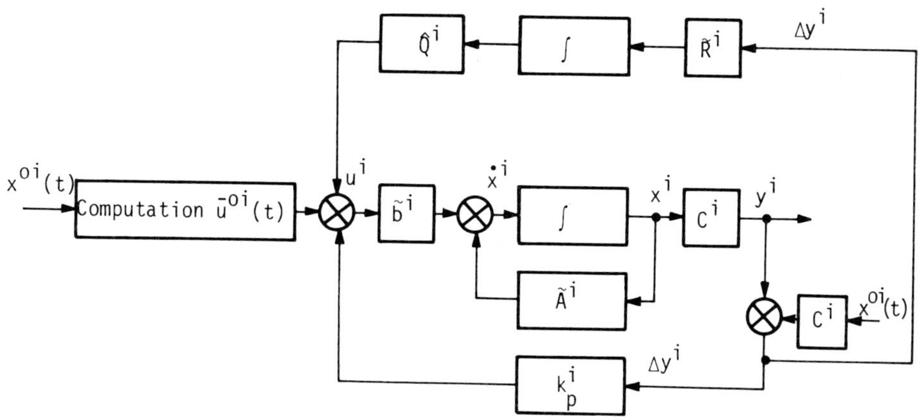

Fig. 2.19. Local dynamic controller with output feedback

in Fig. 2.17 (the optimal regulator with integral feedback loop), except that in the former scheme output feedback is implemented, instead of feedback by the complete state vector. Actually, the only difference between these two control procedures is in the way in which feedback gains are calculated, but the control laws are equal (if $k_i=n_i$).

Two remarks have to be added. First, it is obvious that, if we want to apply the static controller, we may also use pole placement method to compute the feedback gains, but in (2.4.35) we should put matrices $\tilde{A}^i$, $\tilde{b}^i$, C^i instead of $\hat{A}^i$, $\hat{b}^i$, $\hat{C}^i$ and the maximum number of poles that might be specified is just k_i, instead of k_i+1 as in the case of controller with the integral feedback loop. All considerations on pole-placement method stated above can be directly applied to the static controller.

The second remark concerns the application of observer if the order of output of the subsystem k_i is less than the order of subsystem n_i. Namely, instead of applying output feedback (2.4.31) we can apply the local observer to reconstruct the subsystem state from its output, as explained in the previous Paragraph. However, the synthesis of feedback gains can be performed by pole-placement instead of by minimizing the standard quadratic criterion. Thus, the control scheme in Fig. 2.18 can be synthesized using pole-placement method. Obviously, the output feedback in Fig. 2.19 is much simpler to implement than the local observer and complete state feedback (the number of additions and multiplications that have to be performed in each sampling period is by far smaller with the controller in Fig. 2.19 than with the controller in Fig. 2.18), but the performance of the latter is better.

2.4.4. Local control task

Up to now, we have presented two well-known procedures for synthesis of local control: minimizing the standard quadratic criterion and pole placement. Both methods lead to equal linear control laws (2.4.19), or (2.4.20), or (2.4.31). As we have mentioned in Paragraph 2.4, there are many other methods for synthesis of local controllers for linear subsystems. Most of them lead to equal or similar linear control laws, but nonlinear control laws might also be applied. Here, we shall not survey all these methods, since they are well-known and can easily be found in literature. We have briefly presented two above methods in order to demonstrate local control synthesis. However, stability analysis and synthesis of global control which will be presented in the text to follow, can be applied regardless of the method used for local control synthesis and regerdless of the form of the local control law (i.e. they can be applied even with nonlinear local control laws).

Regardless of the method used for local control synthesis, there arises the problem of connection between the imposed local control task (2.4.8) and the parameters that have to be specified in each particular method for control synthesis. Namely, the synthesis of optimal linear regulator requires weighting matrices Q_i, r_i and stability degree β_i in criterion (2.4.16) to be prescribed; pole-placement method requires poles of the subsystem σ_i^D in (2.4.33) to be specified. Similarly, other methods that can be used for local control synthesis also require some data to be specified. The problem is to establish a connection between the stability regions $X^{I(i)}$, $X^{t(i)}$ that are specified in local control task (2.4.8) and data that are required by the chosen method for local control synthesis.

In the general case it is very difficult, to establish an analytic relation between these two sets of parameters. For example, it is difficult to find a relation between the weighting matrices Q_i, r_i in the local criterion and finite stability regions $X^{I(i)}$, $X^{t(i)}$. There are some connections that are evident. For example, it is well known that if the local control is synthesized by minimizing standard quadratic criterion (2.4.16) with prescribed stability degree β_i, then the solution of the closed-loop subsystem (i.e. the subsystem with control (2.4.19)) must satisfy the following inequality:

$$||\Delta x^i(t)|| < \Pi_i ||\Delta x^i(0)|| \exp(-\beta_i t), \quad \forall t \in (0, \infty), \; \Pi_i > 0 \qquad (2.4.36)$$

In local control task (2.4.8) it is required that the state of the subsystem satisfies:

$$||\Delta x^i(t)|| \leqslant \bar{X}^{t(i)} \exp(-\alpha^{(i)} t), \qquad \forall t \in T \qquad (2.4.37)$$

where $\bar{X}^{t(i)} > \bar{X}^{I(i)} \geqslant ||\Delta x^i(0)||$. It is obvious that the sufficient conditions to be satisfied are:

$$\beta_i \geqslant \alpha^{(i)}, \quad \bar{X}^{t(i)} \geqslant \Pi_i ||\Delta x^i(0)|| \qquad (2.4.38)$$

If (2.4.38) are satisfied, (2.4.36) implies (2.4.37). Similarly, if we want to apply pole-placement method, it is obvious that we have to place all poles of the subsystem so that the following condition is satisfied:

$$|\mathrm{Re}(\sigma_j^i)| \geqslant \alpha^{(i)}, \quad \bar{X}^{t(i)} \geqslant \Pi_i ||\Delta x^i(0)|| \qquad (2.4.39)$$

If (2.4.39) holds, condition (2.4.37) is satisfied.

However, the conditions (2.4.38), (2.4.39) are not necessary for (2.4.37) to be satisfied, since the condition (2.4.37) concerns a finite time interval T, and $\Delta x^i(t)$ is constrained by $\bar{X}^{t(i)} \exp(-\alpha^{(i)} t)$ where $\bar{X}^{t(i)}$ is larger than $||x^i(0)||$. Thus, conditions (2.4.38), or (2.4.39) might be too severe. Actually, the local control can satisfy condition (2.4.37), even if conditions (2.4.38), or (2.4.39) are not met. On the other hand, if the finite regions $X^{t(i)}$ are not specified in the form (2.4.37) the relation between them and the stability degree is not simple. Even if we use conditions (2.4.38), or (2.4.39) the specification of the weighting matrices Q_i, r_i in (2.4.16), or of the poles in (2.4.33) is not unique. Finally, if some other method is used for local control synthesis (e.g. Bode's method in frequency domain, or nonlinear bang-bang control etc.) the relation between the finite regions $X^{I(i)}$, $X^{t(i)}$ and the parameters that have to be specified in the method applied, might be difficult to establish.

It should be noted that in the case of linear subsystems we may solve the above problems in a direct way by some classical analytical approaches. However, we shall establish a general algorithm for solving the local control task which can be used as part of the procedure for control synthesis of robotic systems.

We shall present the general numerical solution of the stated problem. Actually, we shall use the method for practical stability analysis to synthesize the local controller which ensures satisfaction of local control task (2.4.8). First, we shall establish a method for analyzing whether the requirements of the local control task are met by the applied control. Then, if these requirements are not met, we shall iteratively change the control parameters until we reach the specified goal.

Now, the problem is to check whether or not the subsystem $\tilde{S}^i$ with applied control satisfies the conditions of practical stability set in (2.4.8). To analyze the practical stability of the linear subsystem $\tilde{S}^i$ (2.4.9) we shall use the method of Michel [68]. This method can be described as follows.

The subsystem $\tilde{S}^i$ is practically stable with respect to $(X^{I(i)}, X^{t(i)}, \tau)$, as defined by (2.4.8), if there exists a real valued continuously differentiable function $v_i(t, x^i)$ and a real-valued function of time $\psi_i(t)$ which is integrable over time interval T, such that

$$\dot{v}_i(t, x^i) < \psi_i(t), \quad \forall x^i \in \tilde{X}^{t(i)}(t), \; \forall t \in T \tag{2.4.40}$$

$$\int_0^t \psi_i(t')dt' \leq v_{im}^{\partial X^{t(i)}(t)}(t) - v_{iM}^{\partial X^{I(i)}}(0), \qquad \forall t \in T \tag{2.4.41}$$

where:

$v_{im}^{\partial X^{t(i)}(t)}(t)$ denotes $\min\limits_{x^i \in \partial X^{t(i)}(t)} v_i(t, x^i)$

$v_{iM}^{\partial X^{I(i)}}(0)$ denotes $\max\limits_{x^i \in \partial X^{I(i)}} v_i(0, x^i)$

$\partial X(t)$ denotes the boundary of the corresponding region and $\tilde{X}^{t(i)}(t)$ denotes $\tilde{X}^{t(i)}(t) = X^{t(i)}(t) - \bar{X}^{I(i)} \exp(-\alpha^{(i)} t)$.

In (2.4.40) $\dot{v}_i$ denotes the time derivative of function $v_i(t, x^i)$ along the solution of the subsystem $\tilde{S}^i$ (2.4.9) with the chosen local control. The conditions (2.4.40), (2.4.41) are sufficient conditions for the practical stability of subsystem $\tilde{S}^i$ (but not necessary), i.e. if the conditions (2.4.40), (2.4.41) are satisfied then the subsystem $\tilde{S}^i$ is practically stable with respect to $(X^I, X^{t(i)}, \tau)$, but vice versa is not always satisfied (if subsystem $\tilde{S}^i$ is practically stable then conditions (2.4.40), (2.4.41) might not always be satisfied).

The proof of the above stated method for testing the practical stability is given in Appendix 2.B [69]. Here, we shall just explain the application of this method. We shall present how this method might be used for synthesizing the local controllers which ensure local practical stability (2.4.8).

The main problem with the application of the above stated method for practical stability testing lies in the choice of the function $v_i(t, x^i)$ since this choice is not unique and no general recommendations can be given. On the other hand, efficiency of the stability test largely depends on the choice of this function. The method stated gives only sufficient conditions for practical stability. Thus, the test might be conservative: if we make an improper choice of function $v_i(t, x^i)$ the result of the test (2.4.40), (2.4.41) may be negative although the subsystem is practically stable. Thus, the choice of the function $v_i(t, x^i)$ has to be done carrefully in order to ensure validity of the stability test. Otherwise, we might get very restrictive results, i.e. we might choose a local controller with feedback gains too high.

Although there are no general solutions for the choice of functions $v_i(t, x^i)$, we shall suggest one possible choice of these functions. This approach follows the recommendations given by Šiljak [59]. He has shown*) that "the best choice" of these functions (from the standpoint of the test of the overall system) is the one by which the stability degree of the subsystem $\tilde{S}^i$ is precisely estimated; this means that we have to choose function $v_i(t, x^i)$ such that the following condition is satisfied:

$$\dot{v}_i(t, x^i) \leqslant -\gamma_i v_i(t, x^i), \quad \forall x^i \in X^{t(i)}(t), \quad \forall t \in T \qquad (2.4.42)$$

---

*) Here, we shall not prove this statement, since its proof is not essential for our application. Those who wish to deeply understand this problem are recommended to read [59].

where γ_i denotes $\min|\mathrm{Re}(\sigma_j^i)|$, σ_j^i, $j=1,2,\dots,n_i$ are the poles of the closed-loop subsystem $\tilde{S}^i$ (i.e. of the subsystem $\tilde{S}^i$ when the local controller is applied). $\dot{v}_i$ again stands for the derivative of function v_i along the solution of the closed-loop subsystem $\tilde{S}^i$. The inequlity (2.4.42) means that by the function $v_i(t, x^i)$ we get the right estimate of the subsystem stability degree γ_i (i.e. the slowest mode of the subsystem $\tilde{S}^i$ must be faster than $||x^i(0)||\exp(-\gamma_i t)$). According to Šiljak [59], if we find such a function $v_i(t, x^i)$, that its derivative along the solution of closed-loop subsystem $\tilde{S}^i$, satisfies the inequality (2.4.42), then this function v_i is the best choice for the stability test (i.e. this choice has the best chance to prove the stability of global system). Thus, we have to find out the function $v_i(t, x^i)$ which, for the given closed-loop subsystem, satisfies the inequality (2.4.42). Obviously, this choice, in the general case, is not trivial. However, in the case of linear subsystems (2.4.9) $\tilde{S}^i$ with the linear local controller, this problem can easily be solved in the following way. Let us choose the function $v_i(t, x^i)$ in the form:

$$v_i(t, x^i) = (\Delta x^{iT} H_i \Delta x^i)^{1/2} \tag{2.4.43}$$

where $H_i \in R^{n_i \times n_i}$ is a positive definite matrix. The derivative of the function v_i (2.4.43) can be written as:

$$\dot{v}_i(t, x^i) = \frac{\partial v_i}{\partial t} + (\mathrm{grad} v_i)^T \Delta\dot{x}^i =$$

$$= \frac{1}{2}\frac{1}{v_i}(\Delta x^{iT} H_i \Delta\dot{x}^i + \Delta\dot{x}^{iT} H_i \Delta x^i) \tag{2.4.44}$$

First, we consider the case when the static controller is applied. Let us assume that the local control is given by (2.4.19), i.e. by local nominal control $\bar{u}^{oi}(t)$ and standard quadratic regulator $\Delta u_i^L(t)$. Then, the closed-loop subsystem is given by:

$$\dot{x}^i = \tilde{A}^i x^i + \tilde{b}^i(\bar{u}^{oi}(t) - r_i^{-1}\tilde{b}^{iT} K_i \Delta x^i)$$

Since nominal local control $\bar{u}^{oi}(t)$ satisfies (2.4.14), it can be written:

$$\Delta\dot{x}^i = (\tilde{A}^i - \tilde{b}^i r_i^{-1}\tilde{b}^{iT} K_i)\Delta x^i \tag{2.4.45}$$

Substituting (2.4.45) in (2.4.44) we get:

$$\dot{v}_i(t, x^i) = \frac{1}{2v_i}\{\Delta x^{iT}[H_i(\tilde{A}^i - \tilde{b}^i r_i^{-1}\tilde{b}^i K_i) + (\tilde{A}^i - \tilde{b}^i r_i^{-1}\tilde{b}^i K_i)^T H_i]\Delta x^i\} \tag{2.4.46}$$

Since we want to find function $v_i(t, x^i)$ which satisfies inequality (2.4.42), we have to find out the matrix H_i which satisfies the matrix equation [59]:

$$H_i^T(\tilde{A}^i-\tilde{b}^i r_i^{-1}\tilde{b}^{iT}K_i)+(\tilde{A}^i-\tilde{b}^i r_i^{-1}\tilde{b}^{iT}K_i)^T H_i = 2\mathrm{diag}(\sigma_1^i,\sigma_2^i,\ldots,\sigma_{ni}^i)H_i$$

$$\leqslant -2\min_j|\sigma_j^i|H_i \qquad (2.4.47)$$

where σ_j^i $(j=1,2,\ldots,n_i)$ are the eigenvalues of the closed-loop matrix of the subsystem $(\tilde{A}^i-\tilde{b}^i r_i^{-1}\tilde{b}^{iT}K_i)$. Thus, we have to solve the equation (2.4.47) by the matrix H_i. Since the order of the subsystem n_i is low, it is not too difficult to determine the matrix H_i which satisfies (2.4.47). Obviously, we must first determine the eigenvalues of the closed-loop matrix σ_j^i; neither this is too complex a problem due to the low order of the subsystem.

If we have found the matrix H_i which satisfies (2.4.47), equation (2.4.46) becomes:

$$\dot{v}_i(t, x^i) = \frac{1}{2v_i}(\Delta x^{iT} 2\ \mathrm{diag}(\sigma_1^i, \sigma_2^i,\ldots,\sigma_{ni}^i)\cdot H_i\Delta x^i) \leqslant$$

$$\leqslant -\frac{1}{v_i}\min|\sigma_j^i|\ v_i^2 = -\min|\sigma_j^i|v_i$$

assuming that $\sigma_j^i<0$ for $j=1,2,\ldots,n_i$.

Evidently, from the last equation it follows:

$$\dot{v}_i(t, x^i) \leqslant -\gamma_i v_i$$

where γ_i is $\min_j|\sigma_j^i|$ $(j=1,2,\ldots,n_i)$. Thus, if we find the matrix H_i which satisfies (2.4.47), then the function $v_i(t, x^i)$ (2.4.43) satisfies the inequality (2.4.42), i.e. it gives "the exact estimate" of the subsystem stability degree. Thus, we have shown how we can determine the function $v_i(t, x^i)$ which has "the best chance" to prove the practical stability of subsystem.

Having chosen the function $v_i(t, x^i)$, let us consider how the conditions (2.4.40) and (2.4.41) can be applied with this particular form of function v_i (2.4.43). From the last inequality, it follows:

$$\dot{v}_i(t, x^i) \leqslant -\gamma_i\lambda_m^{1/2}(H_i)\,||\Delta x^i|| \qquad (2.4.48)$$

where we have used the well-known inequality:

$$\lambda_m^{1/2}(H_i)\,||\Delta x^i|| \leqslant (\Delta x^{iT} H_i \Delta x^i)^{1/2} = v_i$$

Since the condition (2.4.40) should be satisfied for the region $\tilde{X}^{t(i)}$ defined by $\{x^i\colon x^i \in X^{t(i)}(t).\text{AND}.||\Delta x^i|| \geqslant \bar{X}^{I(i)}\exp(-\alpha^{(i)}t),\ \forall t \in T\}$, we can write (2.4.48) in the form:

$$\dot{v}_i(t,\, x^i) \leqslant -\gamma_i \lambda_m^{1/2}(H_i)\bar{X}^{I(i)}\exp(-\alpha^{(i)}t) = \psi_i(t) \tag{2.4.49}$$

i.e. we can determine the function $\psi_i(t)$ as given by (2.4.49). From the elementary theory of matrices it is well-known that:

$$v_{im}^{\partial X^{t(i)}(t)}(t) \geqslant \lambda_m^{1/2}(H_i)\bar{X}^{t(i)}\exp(-\alpha^{(i)}t)$$

$$v_{im}^{\partial X^{I(i)}} \leqslant \lambda_M^{1/2}(H_i)\bar{X}^{I(i)}$$

By substituting these inequalities in (2.4.41) and taking into account the form of function $\psi_i(t)$ given by (2.4.49), the condition (2.4.41) can be written in the form:

$$\begin{aligned}
\int_0^t \psi_i(t')dt' &= -\int_0^t \gamma_i \lambda_m^{1/2}(H_i)\bar{X}^{I(i)}\exp(-\alpha^{(i)}t')dt' = \\
&= -\gamma_i \lambda_m^{1/2}(H_i)\bar{X}^{I(i)}(1-\exp(-\alpha^{(i)}t))/\alpha^{(i)} \leqslant \\
&\leqslant \lambda_m^{1/2}(H_i)\bar{X}^{t(i)}\exp(-\alpha^{(i)}t) - \lambda_M^{1/2}(H_i)\bar{X}^{I(i)},\ \forall t \in T
\end{aligned} \tag{2.4.50}$$

Thus, to test the practical stability of the closed-loop subsystem $\tilde{S}^i$, we have to check whether or not the condition (2.4.50) is satisfied. In (2.4.50) numbers $\bar{X}^{t(i)}$, $\bar{X}^{I(i)}$ and $\alpha^{(i)}$ are given by the requirements of the local control task (2.4.8), and numbers $\lambda_m(H_i)$, $\lambda_M(H_i)$ and γ_i are determined by the synthesized control. After synthesizing the local control, we have to determine the matrix H_i which satisfies (2.4.47). Then we can determine γ_i which is $\min_j |\mathrm{Re}(\sigma_j^i)|$, $j=1,2,\ldots,n_i$, and $\lambda_m(H_i)$, $\lambda_M(H_i)$. Then, it is easy to check whether the test (2.4.50) is positive or not, and get the answer as to whether the subsystem is practically stable. Once again we have to underline that, if the test (2.4.50) is not successfull, we cannot conclude whether the subsystem is stable or not; the subsystem may be stable even when the test (2.4.50) is not positive.

In this way, we have made a basis for establishing an algorithm for the synthesis of local control which satisfies local control task (2.4.8). However, before we establish this algorithm, we have to make several remarks and extensions of the tests presented above.

(I) The first remark concerns the solution of the matrix equation (2.4.47). In the special case when the standard quadratic regulator (2.4.19) is applied, we can avoid the problem of solving the equation (2.4.47). Then, we can choose the function v_i as:

$$v_i(t, x^i) = (\Delta x^{iT} K_i \Delta x^i)^{1/2} \tag{2.4.51}$$

where K_i is the solution of the corresponding Riccati equation (2.4.18). Namely, we choose matrix H_i in (2.4.43) as $H_i = K_i$. Since the matrix K_i satisfies equation (2.4.18) it can be written:

$$K_i(\tilde{A}^i - \tilde{b}^i r_i^{-1} \tilde{b}^{iT} K_i) + (\tilde{A}^i - \tilde{b}^i r_i^{-1} \tilde{b}^{iT} K_i)^T K_i =$$

$$= -Q_i - 2\beta_i K_i - K_i \tilde{b}^i r_i^{-1} \tilde{b}^{iT} K_i = -W_i - 2\beta_i K_i \tag{2.4.52}$$

Then, the derivative of the function v_i (2.4.51) can be written as:

$$\dot{v}_i(t, x^i) = -\Delta x^{iT}(W_i + 2\beta_i K_i)\Delta x^i / 2v_i \leqslant -\left(\frac{1}{2} \frac{\lambda_m(W_i)}{\lambda_M(K_i)} + \beta\right) v_i$$

$$\dot{v}_i(t, x^i) \leqslant -\left(\frac{1}{2} \frac{\lambda_m(W_i)}{\lambda_M^{1/2}(K_i)} + \beta_i \lambda_m^{1/2}(K_i)\right) ||\Delta x^i|| \tag{2.4.53}$$

Thus, we get the estimate of the subsystem stability degree as $\gamma_i = (\lambda_m(W_i)/2\lambda_M(K_i) + \beta_i)$, i.e. this form of function $v_i(t, x^i)$ satisfies inequality (2.4.42). Namely, equation (2.4.52) corresponds to equation (2.4.47). Instead of solving equation (2.4.47) to find matrix H_i, we may take the matrix K_i for which we know that it satisfies (2.4.52), and that the function v_i (2.4.51) gives an estimate of the subsystem exponential stability degree. However, it should be noted that this is not "the best estimate" of the subsystem stability degree since in the general case $\lambda_m(W_i)/2\lambda_M(K_i) + \beta_i \leqslant |\mathrm{Re}(\sigma_j^i)|$ for $j=1,2,\ldots,n_i$. Namely the optimal linear regulator places the subsystem poles to the left from the line $\mathrm{Re}(\sigma) = -2\beta_i$. Thus, the estimate of stability degree (2.4.53) might be poor. This is the reason why we suggest to choose function $v_i(t, x^i)$ in the form (2.4.43) where matrix H_i satisfies (2.4.47) even in the case when the local optimal regulator is applied. However, we have to solve equation (2.4.47).

Let us state that, if we choose function $v_i(t, x^i)$ as (2.4.51), then, according to (2.4.53), we get function $\psi_i(t)$ in the form:

$$\psi_i(t) = -\left(\frac{1}{2}\frac{\lambda_m(W_i)}{\lambda_M^{1/2}(K_i)} + \beta_i\lambda_m^{1/2}(K_i)\right)\bar{X}^{I(i)}\exp(-\alpha^{(i)}t)$$

In this case, the condition (2.4.41) can be rewritten as:

$$-\left(\frac{\lambda_m(W_i)}{2\lambda_M^{1/2}(K_i)} + \beta_i\lambda_m^{1/2}(K_i)\right)\bar{X}^{I(i)}\frac{1}{\alpha^{(i)}}[1-\exp(-\alpha^{(i)}t)] \leqslant \tag{2.4.54}$$

$$\leqslant \lambda_m^{1/2}(K_i)\bar{X}^{t(i)}\exp(-\alpha^{(i)}t) - \lambda_M^{1/2}(K_i)\bar{X}^{I(i)}, \quad \forall t\in T$$

(II) The second remark concerns some additional condition which obviously follows from (2.4.50) (or from (2.4.54)). Namely, for t=0 it is clear that the following relation must be satisfied:

$$\lambda_m^{1/2}(H_i)/\lambda_M^{1/2}(H_i) \geqslant \bar{X}^{I(i)}/\bar{X}^{t(i)} \tag{2.4.55}$$

(III) Now, we have to extend the above stability test to include *dynamic* controllers. Up to now we have considered stability of the local subsystem if only *static* controllers are applied. To extend the test, we must note that dynamic controllers increase the order of the closed-loop subsystem. Let us consider the local dynamic controller in the form (2.4.31), i.e. let us assume that the local control is given by:

$$u^i(t) = \bar{u}^{oi}(t) + \Delta u_i^L(t) = \bar{u}^{oi}(t) - k_I^{iT}\hat{C}^i\Delta\hat{x}^i \tag{2.4.56}$$

where $\Delta\hat{x}^i$ is the augmented state vector of the subsystem $\tilde{S}^i$ as explained in connection to (2.4.24), k_I^i is the vector of feedback gains of the dynamic controller, $\hat{C}^i$ is the output matrix of the augmented subsystem (k_I^i, $\hat{C}^i$ are explained in connection with (2.4.32)). We shall again restrict our attention to considering the first order dynamic controller, i.e. the controller which includes just the integral feedback loop in addition to state (or output) feedback loops. In Paragraphs 2.4.2 and 2.4.3 we have also considered this local controller with the integral feedback loop (either a dynamic optimal regulator, or an output dynamic controller synthesized by pole-placement method). This controller increases the order of the subsystem by one only. The closed loop subsystem (i.e. the subsystem with controller (2.4.56)) is given by (2.4.32). Obviously, if we consider the optimal controller with full-

-state feedback, the matrix $\hat{C}^i$ is equal to a unit matrix ($\hat{C}^i = I_{(n_i+1)}$). We have to consider stability of the closed-loop subsystem (2.4.32).

Now, instead of function $v_i(t, x^i)$, which is a function of time and state vector x^i, we have to consider function $v_i(t, \hat{x}^i)$ which is a function of time and state vector of the augmented subsystem $\hat{x}^i$. Let us consider this function in form:

$$v_i(t, \hat{x}^i) = (\Delta\hat{x}^{iT}\hat{H}_i\Delta\hat{x}^i)^{1/2} \tag{2.4.57}$$

where $\hat{H}_i \in R^{(n_i+1)\times(n_i+1)}$ is a positive definite matrix. Using this function we can test the practical stability of the subsystem with the dynamic controller analogously as we have done with the static controller. Let us assume that the matrix $\hat{H}_i$ satisfies the following equation (which is analogous to (2.4.47)):

$$\hat{H}_i^T(\hat{A}_I^i - \hat{b}_I^i k_I^{iT}\hat{C}^i) + (\hat{A}_I^i - \hat{b}_I^i k_I^{iT}\hat{C}^i)^T\hat{H}_i = 2\,\mathrm{diag}(\sigma_1^i, \sigma_2^i, \ldots, \sigma_{n_i+1}^i)\cdot\hat{H}_i \tag{2.4.58}$$

where σ_j^i $(j=1,2,\ldots,n_i+1)$ are the eigenvalues of the closed-loop matrix of the augmented subsystem $(\hat{A}_I^i - \hat{b}_I^i k_I^{iT}\hat{C}^i)$. Then the derivative of the function v_i (2.4.57) along the solution of subsystem (2.4.32) is given by:

$$\dot{v}_i(t, \hat{x}^i) = (\mathrm{grad}\,v_i)^T\Delta\dot{\hat{x}}^i = \frac{1}{2v_i}\Delta\hat{x}^{iT}[\hat{H}_i^T(\hat{A}_I^i - \hat{b}_I^i k_I^{iT}\hat{C}) + (\hat{A}_I^i - \hat{b}_I^i k_I^{iT}\hat{C}^i)^T\hat{H}_i]\Delta\hat{x}^i = \frac{1}{2v_i}\Delta\hat{x}^{iT}2\mathrm{diag}(\sigma_1^i, \sigma_2^i, \ldots, \sigma_{n_i+1}^i)\hat{H}_i\Delta\hat{x}_i$$

$$\dot{v}_i(t, \hat{x}^i) \leqslant -\gamma_i v_i \tag{2.4.59}$$

where $\gamma_i = \min_j |\sigma_j^i|$, $(j=1,2,\ldots,n_i+1)$.

Now, we have to solve the problem of stability regions for the augmented subsystem (subsystem with the dynamic controller). Namely, the augmented subsystem has one state coordinate more than the original subsystem. As explained in the previous text this state coordinate is defined by:

$$\Delta x_{n_i+1}^i = \int_0^t \tilde{R}^i C^i \Delta x^i dt$$

We shall assume that the initial value of this state coordinate is always equal to zero, i.e. $\Delta x_{n_i+1}^i(0) = 0$, which is a realistic assumption since this is an additional coordinate determined by the control system. Thus, we can write that the region of the initial conditions for the augmented subsystem $\hat{X}^{I(i)}$ is defined by:

$$\hat{X}^{I(i)} = \{\hat{x}^i(0): ||\Delta x^i(0)|| \leq \bar{X}^{I(i)}; \quad \Delta x^i_{n_i+1}(0) = 0.\} \tag{2.4.60}$$

We can estimate the additional state coordinate in the region $X^{t(i)}$ in the following way:

$$|\Delta x^i_{n_i+1}| \leq (\tilde{R}^i C^i)_M \int_0^t ||\Delta x^i|| dt \leq (\tilde{R}^i C^i)_M \int_0^t \bar{X}^{t(i)} \exp(-\alpha^{(i)} t) dt$$

$$|\Delta x^i_{n_i+1}| \leq (\tilde{R}^i C^i)_M \bar{X}^{t(i)} (1-\exp(-\alpha^{(i)} t))/\alpha^{(i)} \tag{2.4.61}$$

where $(\tilde{R}^i C^i)_M$ denotes $\max(\tilde{R}^i C^i)_k$, $(k=1,2,\ldots,n_i)$, $(\tilde{R}^i C^i)_k$ is the kth member of the vector $\tilde{R}^i C^i \in R^{1\times n_i}$. Now, the region of allowed states $\hat{X}^{t(i)}$ for the extended subsystem can be estimated as:

$$\hat{X}^{t(i)} = \{\hat{x}^i(t): ||\Delta x^i(t)|| \leq \bar{X}^{t(i)} \exp(-\alpha^{(i)} t), \ |\Delta x^i_{n_i+1}| \leq$$
$$\leq (\tilde{R}^i C^i)_M \bar{X}^{t(i)} (1-\exp(-\alpha^{(i)} t))/\alpha^{(i)}\}, \quad \forall t \in T \tag{2.4.62}$$

or, we may estimate it by:

$$\hat{X}^{t(i)}(t) = \{\hat{x}^i(t): ||\Delta \hat{x}^i(t)|| \leq \hat{\bar{X}}^{t(i)} \exp(-\alpha^{(i)} t) +$$
$$+ (\tilde{R}^i C^i)_M \bar{X}^{t(i)}/\alpha^{(i)}\}, \quad \forall t \in T \tag{2.4.63}$$

where $\hat{\bar{X}}^{t(i)}$ is given by

$$\hat{\bar{X}}^{(i)} = \bar{X}^{t(i)} - (\tilde{R}^i C^i)_M \bar{X}^{t(i)}/\alpha^{(i)}$$

Obviously, estimation (2.4.62) is better than (2.4.63) but the latter is more compact and simpler for stability testing.

Starting from (2.4.59) and (2.4.63) we can determine function $\psi_i(t)$, analogously to (2.4.49), which satisfies:

$$\dot{v}_i(t, \hat{x}^i) \leq -\gamma_i v_i \leq -\gamma_i \lambda_m^{1/2}(\hat{H}_i)[\hat{\bar{X}}^{I(i)} \exp(-\alpha^{(i)} t) +$$
$$+ (\tilde{R}^i C^i)_M \bar{X}^{I(i)}/\alpha^{(i)}] = \psi_i(t) \tag{2.4.64}$$

where $\hat{\bar{X}}^{I(i)} = \bar{X}^{I(i)} - (\tilde{R}^i C^i)_M \bar{X}^{I(i)}/\alpha^{(i)}$.

The test of practical stability (2.4.41) for the augmented subsystem gets the following form:

$$-\frac{\gamma_i}{\alpha^{(i)}}\lambda_m^{1/2}(\hat{H}_i)[\hat{\bar{X}}^{I(i)}(1-\exp(-\alpha^{(i)}t))+(\tilde{R}^i C^i)_M \bar{X}^{I(i)}t] \leq \tag{2.4.65}$$

$$\leq \lambda_m^{1/2}(\hat{H}^i)[\hat{\bar{X}}^{t(i)}\exp(-\alpha^{(i)}t)+(\tilde{R}^i C^i)_M \bar{X}^{t(i)}/\alpha^{(i)}]-\lambda_M^{1/2}(\hat{H}^i)\bar{X}^{I(i)}$$

Thus, we obtain a test which is analogous to (2.4.50) for the static controller. Using test (2.4.65) we can check the practical stability of local subsystem $\tilde{S}^i$ with the dynamic controller. If the dynamic controller is synthesized by minimization of standard quadratic criterion (Paragraph 2.4.2), the test of practical stability can be written in the form analogous to (2.4.54), using the above explained estimations of the stability region.

(IV) The last remark concerns the amplitude constraints upon the inputs of actuators. The adopted models of subsystems (2.2.2) include the amplitude constraint upon the actuator input of saturation type (2.2.3). However, as we have explained in Paragraph 2.4.1, during the synthesis of local controller we have neglected these input constraints, and have considered the models of subsystems in the form (2.4.9). Thus up to now we have analyzed the practical stability of the local subsystems with unconstrained inputs. Now, we shall take into account the input constraint of amplitude saturation type. This problem has been partially considered in our previous book [2], but we shall here, for the first time present how this problem can be solved concerning the practical stability of the system.

Let us consider the case when the dynamic controller is applied and control is defined by (2.4.56). It is simple to apply the following procedure to the static controller. The amplitude constraint upon the subsystem input (2.2.3) defines an infinite region $X_i(t) \subset R^N$ in the following way:

$$X_i = \{\hat{x}^i : |\bar{u}^{oi}(t) - k_i^{iT}\hat{C}^i \Delta\hat{x}^i| \leq u_m^i\}, \forall t \in T \tag{2.4.66}$$

If the state of the subsystems $\hat{x}^i$ belongs to the region X_i, i.e. if $\hat{x}^i \in X_i$, then the input amplitude constraint is not reached and the subsystem behaves as presented above. However, if $\hat{x}^i \notin X_i$, the input amplitude constraint is violated by the desired control and the above analysis of the subsystem stability is not valid. Namely, if the regions $\hat{X}^{I(i)}$ and $\hat{X}^{t(i)}(t)$, $\forall t \in T$ of the practical stability are within the region X_i, i.e. if:

$$\hat{X}^{I(i)} \subset X_i \text{ .AND. } \hat{X}^{t(i)}(t) \subset X_i, \qquad \forall t \in T \tag{2.4.67}$$

then the above analysis is valid, and if, the test is positive, we can guarantee the practical stability of the subsystems for the given regions $\hat{X}^{I(i)}$ and $\hat{X}^{t(i)}(t)$, $\forall t \in T$.

However, if the conditions (2.4.67) are not satisfied we have to consider the stability of the subsystem when the input amplitude is saturated, and the desired control cannot be realized for the states which are outside the region X_i. Namely, the poles of the closed-loop subsystem $\sigma_1^i, \sigma_2^i, \ldots, \sigma_{n_i+1}^i$ are the eigenvalues of the matrix $(\hat{A}_I^i - \hat{b}_I^i k_I^{iT} \hat{C}^i)$, only for the states that belong to region X_i. For the states which are not within the region X_i the poles of the closed-loop subsystem move from their positions. However, the subsystem is still exponentially stable but with a stability degree which is less than γ_i $(\min|\mathrm{Re}(\sigma_j^i)|$, $j=1,2,\ldots,n_{i+1})$. In order to determine the stability degree of the subsystem for states which are not in the region X_i, let us apply the procedure explained in [2]. Let us introduce the number λ_i defined as:

$$\lambda_i = \begin{cases} 1 & \text{for} \quad \forall \hat{x}^i \in X_i \\ (u_m^i \pm \bar{u}^{oi}(t)) / |k_I^{iT} \hat{C}^i \Delta \hat{x}^i| & \text{for} \quad \hat{x}^i \notin X_i, \ \forall t \in T \end{cases} \tag{2.4.68}$$

where we take plus or minus depending on the sign of $k_I^{iT}\hat{C}^i\Delta\hat{x}^i$. It is evident that this consideration is valid only if $|\bar{u}^{oi}(t)| < \bar{u}_m^i$, $\forall t \in T$, otherwise the region X_i is empty and we cannot ensure the stability of the subsystem. This means that the nominal trajectory is such that we cannot determine even local nominal control $\bar{u}^{oi}(t)$ which drives a decoupled joint along trajectory, the actuator is too weak for the desired trajectory, or, the trajectory is too fast for chosen actuator (see Chapter 4).

Now, if we introduce λ_i as defined by (2.4.68), we can write the closed-loop subsystem (2.4.32) in the form:

$$\Delta\dot{\hat{x}}^i = (\hat{A}^i - \lambda \hat{b}_I^i k_I^{iT} \hat{C}^i) \Delta \hat{x}^i \tag{2.4.69}$$

This form of the subsystem model includes both cases: if the state $\hat{x}^i$ is within X_i and if it is not (as can be seen from definition of λ_i

(2.4.68)). The poles of the closed-loop subsystem (2.4.69) are given by the eigenvalues of the matrix $(\hat{A}^i-\lambda_i\hat{b}^i_I k^{iT}_I\hat{C}^i)$. Since the number λ_i depends on the state $\hat{x}^i$, it is obvious that the poles of the subsystem depend on the state $\hat{x}^i$ which we observe. In order to determine the exponential stability degree of the subsystem (if the conditions (2.4.67) are not satisfied), we have to determine the number λ_{ig} for which the following relation holds:

$$\lambda_{ig} \leqslant |u^i_m \pm \bar{u}^{oi}|/|k^{iT}_I\hat{C}^i\Delta\hat{x}^i|, \quad \text{for} \quad \forall\hat{x}^i\in\hat{X}^{I(i)}, \qquad \forall\hat{x}^i\in\hat{X}^{t(i)}(t) \tag{2.4.70}$$

When we have determined the number λ_{ig} satisfying the relation (2.4.70), we can compute the eigenvalues of the closed-loop subsystem for the regions $\hat{X}^{I(i)}$ and $\hat{X}^{t(i)}(t)$, $\forall t\in T$, i.e. we can compute:

$$\sigma^{iG}_1, \sigma^{iG}_2, \ldots, \sigma^{iG}_{n_i+1} = \lambda(\hat{A}^i-\lambda_{ig}\hat{b}^i_I k^{iT}_I\hat{C}^i) \tag{2.4.71}$$

Then, we can determine the exponential stability degree γ_{ig} of the sybsystem in the regions $\hat{X}^{I(i)}$, $\hat{X}^{t(i)}(t)$, $\forall t\in T$, as:

$$\gamma_{ig} = \min|\mathrm{Re}(\sigma^{iG}_j)|, \qquad j=1,2,\ldots,n_i+1 \tag{2.4.72}$$

Obviously γ_{ig} is less than γ_i (which is valid for linear control mode, i.e. in region X_i) since $\lambda_{ig}<1$.

Now, it is very simple to show that the stability test can be written in the form analogous to (2.4.65) where γ_{ig} is substituted for γ_i. If we choose function $v_i(t, \hat{x}^i)$ in the form (2.4.57) we can easily show that the following relation holds:

$$\dot{v}_i(t, \hat{x}^i) \leqslant -\gamma_{ig}v_i$$

if the matrix $\hat{H}^i$ is chosen to satisfy:

$$\hat{H}^T_i(\hat{A}^i_I-\lambda_{ig}\hat{b}^i_I k^{iT}_I\hat{C}^i)+(\hat{A}^i_I-\lambda_{ig}\hat{b}^i_I k^{iT}_I\hat{C}^i)^T\hat{H}_i = 2\mathrm{diag}(\sigma^{iG}_1,\sigma^{iG}_2,\ldots,\sigma^{iG}_{n_i+1})\hat{H}_i \tag{2.4.73}$$

Then, analogously to (2.4.65) we get the following test for the practical stability of local subsystem which takes into account nonlinear constraint upon the input:

$$-\frac{\gamma_{ig}}{\alpha(i)}\lambda^{1/2}_m(\hat{H}_i)[\hat{\bar{X}}^{I(i)}(1-\exp(-\alpha^{(i)}t))+(\tilde{R}^i C^i)_M\bar{X}^{I(i)}t] \leqslant$$

$$< \lambda_m^{1/2}(\hat{H}^i)[\hat{\bar{X}}^{t(i)}\exp(-\alpha^{(i)}t)+(\tilde{R}^i C^i)_M \bar{X}^{t(i)}/\alpha^{(i)}]-\lambda_M^{1/2}(\hat{H}_i)\bar{X}^{I(i)} \qquad (2.4.74)$$

Thus, in order to test the stability of the local subsystem if the condition (2.4.67) is not satisfied, we have to determine λ_{ig} in (2.4.70) and then γ_{ig} by (2.4.72) and to check test (2.4.74). The only problem is to determine λ_{ig} which satisfies (2.4.70) since we have to examine the whole regions $\hat{X}^{I(i)}$ and $\hat{X}^{t(i)}(t)$, $\forall t \in T$. However, due to definitions of these regions and to the fact that $\lambda_1 < \lambda_2$ if $||\Delta\hat{x}^i||_1 > ||\Delta x^i||_2$ (where λ_1 corresponds to $||\Delta x^i||_1$ and λ_2 to $||\Delta\hat{x}^i||_2$) instead of by (2.4.70) we can determine λ_{ig} by:

$$\lambda_{ig} < |u_m^i \pm \bar{u}^{io}|/|k_I^{iT}\hat{C}^i\Delta\hat{x}^i|, \quad \forall\hat{x}^i \in \partial\hat{X}^{I(i)}, \quad \forall\hat{x}^i \in \partial\hat{X}^{t(i)}(t), \ \forall t \in T \qquad (2.4.75)$$

This means that to determine the number λ_{ig} we have to examine the states along the boundaries of the regions $\partial\hat{X}^{I(i)}$ and $\partial\hat{X}^{t(i)}(t)$, $\forall t \in T$. Obviously, this is by far easier to perform than to examine complete regions $\hat{X}^{I(i)}$ and $\hat{X}^{t(i)}(t)$, $\forall t \in T$.

Now, we can establish the algorithm for iterative synthesis of the local controllers which ensure satisfaction of local control task (2.4.8), for local (decoupled) subsystems. The global flow-chart of the algorithm is presented in Fig. 2.20. By this procedure we can choose local controllers (i.e. compute feedback gains) which ensure the practical stability of the local subsystem in the sense of local control task definition (2.4.8).

The algorithm can be performed according to the following steps: (for the *i*th subsystem)

(1) The local control task (2.4.8) is set, i.e. the regions $X^{I(i)}$ and $X^{t(i)}(t)$, $\forall t \in T$ are set, by defining the numbers $\bar{X}^{I(i)}$, $\bar{X}^{t(i)}$, $\alpha^{(i)}$ and τ. We assume that the nominal trajectory $x^{oi}(t)$, $\forall t \in T$ has already been set, i.e. that the trajectories of all state coordinates for the *i*th subsystem are defined.

(2) Local nominal control $\bar{u}^{oi}(t)$, $\forall t \in T$ is synthesized in order to satisfy (2.4.14); namely, we have to compute $\bar{u}^{oi}(t)$, $\forall t \in T$ so that it drives the decoupled local subsystem $\tilde{S}^i$ along nominal trajectory $x^{oi}(t)$, $\forall t \in T$ if no perturbation is acting upon the subsystem and if its model is perfect; we have to check whether or not the con-

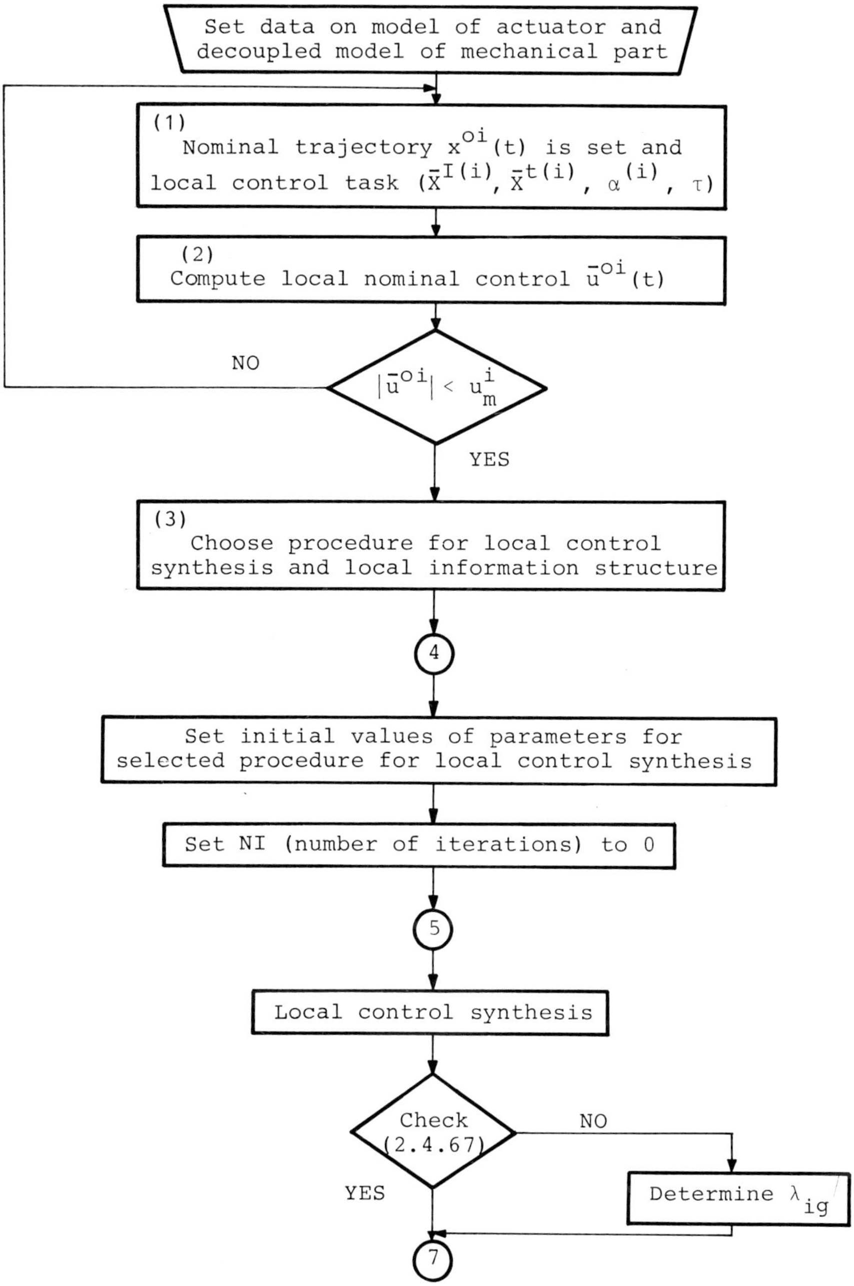

Fig. 2.20. Flow-chart of the algorithm for synthesis of local controller for the ith local subsystem

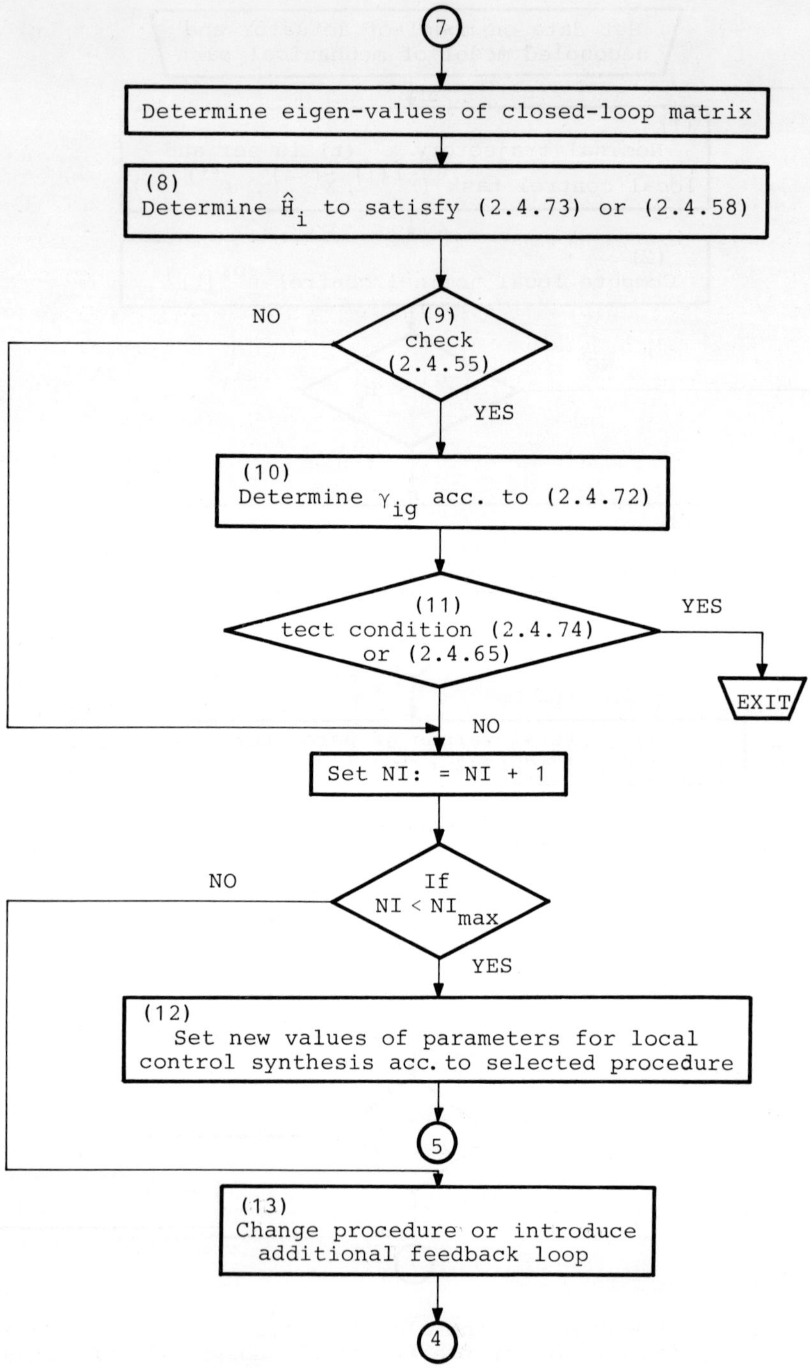
7
Determine eigen-values of closed-loop matrix
(8)
Determine $\hat{H}_i$ to satisfy (2.4.73) or (2.4.58)
NO
(9)
check
(2.4.55)
YES
(10)
Determine γ_{ig} acc. to (2.4.72)
(11)
tect condition (2.4.74)
or (2.4.65)
YES
EXIT
NO
Set NI: = NI + 1
NO
If
$NI < NI_{max}$
YES
(12)
Set new values of parameters for local
control synthesis acc. to selected procedure
5
(13)
Change procedure or introduce
additional feedback loop
4

Fig. 2.20. (Cont.)

dition $|\bar{u}^{oi}(t)| \leq u_m^i$, $\forall t \in T$ is satisfied; if not, we cannot realize the desired trajectory by the chosen actuator and we must select another trajectory, or another actuator (i.e. choose a slower trajectory, or change the actuator)

(3) The procedure for the local control synthesis has to be selected. We must define whether we shall synthesize the local controller by minimization of standard quadratic criterion (i.e. to synthesize the optimal quadratic controller), or by pole-placement method, or by some other method. We must also decide whether we shall introduce feedback loops by complete state vector x^i or by output of the subsystem y^i only. Finally, we have to decide whether a static or dynamic controller will be synthesized, i.e. whether or not integral feedback loop will be introduced. If we decide to synthesize the dynamic controller we must determine region $\hat{X}^{t(i)}(t)$ according to (2.4.63).

(4) Set initial values for the parameters of the local controller which have to be set in the selected procedure. For example, if we have selected to synthesize the local controller by minimization of standard quadratic criterion we have to set weighting matrices Q_i, r_i and prescribed stability degree β_i; in this, condition (2.4.38) may be taken into account. Similarly, if we have selected the pole--placement method for local control synthesis, we must define the desired poles of the closed-loop subsystem (obviously, condition (2.4.39) must be satisfied).

(5) Synthesize the local controller according to the selected procedure. If we have desided to synthesize the local controller by minimization of standard quadratic criterion, we synthesize the controller according to the procedure described in Paragraph 2.4.1 (or, in Paragraph 2.4.2 if we have introduced the integral feedback loop). If pole-placement has been selected we have to compute feedback gains according to the procedure described in Paragraph 2.4.2; obviously, many other procedures may be used for local control synthesis.

(6) Check whether conditions (2.4.67) are satisfied; if not, we have to determine λ_{ig} which satisfies (2.4.75).

(7) Determine the eigenvalues σ_j^i, $j=1,2,\ldots,n_i+1$ of the closed-loop

matrix of the subsystem $(\hat{A}_I^i-\hat{b}_I^i k_I^{iT}\hat{C}^i)$, or σ_j^{iG}, j=1,2,...,n_i+1 of the matrix $(\hat{A}_I^i-\lambda_{ig}\hat{b}_I^i k_I^{iT}\hat{C}^i)$; obviously, if we have chosen to apply the static controller we have to determine eigenvalues σ_j^i, j=1,2,...,n_i of the closed-loop matrix $(\tilde{A}^i-\tilde{b}^i r_i^{-1}\tilde{b}^{iT}K_i)$.

(8) Determine the matrix $\hat{H}_i$ which satisfies equation (2.4.73), or (2.4.58) (if conditions (2.4.67) are met); if the static controller has been selected we have to compute matrix H_i which satisfies equation (2.4.47) (as we have already noted, if the standard quadratic regulator has been selected we may avoid this step by choosing the function $v_i(t, x^i)$ as (2.4.51), but some drawbacks of this approach have been recognized).

(9) Check whether condition (2.4.55) is satisfied; if not, we cannot prove the practical stability of the subsystem $\tilde{S}^i$, so we must go to step (12).

(10) Determine the number γ_{ig} which satisfies (2.4.72), or γ_i which is $\min|\mathrm{Re}(\sigma_j^i)|$, j=1,2,...,$n_i$+1.

(11) Perform the test of practical stability (2.4.74), or (2.4.65); when the static controller is selected we have to perform the test (2.4.50); if the test is satisfied, we can accept the synthesized local controller and be sure that the local control provides the practical stability in accordance with local control task (2.4.8); by this the local controller is synthesized.

(12) If the test of practical stability is not satisfied, we cannot prove that the local subsystem is practically stable. We have to select again the local controller, i.e. to synthesize again feedback gains of the local controller. This means that we have to set new values of parameters which have to be specified in the selected procedure. For example, if we synthesize local optimal regulator we have to select again the weighting matrices Q_i, r_i and prescribed stability degree β_i (obviously, we should try to increase β_i and diagonal elements in the matrix Q_i, or to descrease r_i). Similarly, if pole-placement has been chosen we have to select again the desired poles of the closed-loop subsystem; then we have to go to step (5).

(13) If too many trials to select local controller have been made and we

have not achieved satisfaction of the test of practical stability, then probably we cannot ensure the practical stability of the subsystem by the chosen local controller structure. Then, we must either introduce an additional feedback loop (if we have not already introduced all state feedback loops and integral feedback loop), or select again the procedure for the synthesis of local controller*); thus, we go back to step (4).

Thus, using this procedure we can synthesize local control which ensures the satisfaction of local control task (2.4.8). Obviously this algorithm can easily be implemented by a digital computer to obtain computer-aided synthesis of local controller. This algorithm is a part of general algorithm for computer aided synthesis of control for the manipulation robots which will be presented in Chapter 4.

The critical point in the above presented algorithm is the repeated selection of the parameters that have to be specified in the particular procedure for control synthesis. If the procedure is to be performed completely automatically it is obvious that the specified parameters have to be iteratively changed in such a way as to ensure an increase of subsystem exponential stability degree γ_i. In the case of optimal regulator synthesis, this can be achieved by increasing the prescribed stability degree β_i, or by increasing the diagonal elements of matrix Q_i, or by decreasing r_i. If pole-placement method is used, we have to move the desired poles of the closed-loop subsystem to the left of the complex plane. However, it is questionable how fast this procedure leads to the desired result. If we have chosen output feedback, it may happen that this procedure cannot converge to the desired local controller. Then, as we have explained in connection with step (13) of the algorithm, we must select again the information structure of the local controller (i.e. introduce additional feedback loops). In automatic mode, the user has to specify the number of iterations that is allowed to the algorithm to search for the local controller; if this number of iterations is exceeded, the algorithm should go to re-selection of local information pattern, or the procedure for local control synthesis may be changed. Obviously, various improvements of this algorithm are possible. The algorithm might compare the exponential stability

---

*) In principle, we can synthesize equal local controllers with various procedures since local subsystems are linear. Thus, there is no need to select again the procedure for control synthesis. However, sometimes one procedure approaches the desired result faster than the order, and thus, this reselection might be effective.

degrees of the subsystem which are achieved in two successive iterations. We may thus conclude now fast the algorithm approaches the desired solution, and decide whether this direction is effective or it should be changed.

The best improvement of the algorithm can be achieved if we include a robot designer in this procedure, i.e. if we establish an interactive algorithm. In this case the designer can by himself choose again the control parameters, according to his own experience and, thus, speed-up the iterative procedure. This will be discussed in detail in Chapter 4.

2.5 Stability Analysis of Manipulation Robots

In the preceding paragraph we have presented synthesis of local controllers for decoupled subsystems of the robotic system. Namely, we have considered the robotic system as a set of decoupled subsystems (each subsystem is associated to a single joint of the robot and its corresponding actuator). We have synthesized a local controller for each subsystem independently from the rest of the system. Our aim is to synthesize a decentralized controller which should stabilize the complete robotic system. The structure of the chosen controller is presented in Fig. 2.13. The following question is posed: can such a decentralized controller, synthesized using the decoupled model of the robot, stabilize the complete robotic system when interactions between the joints are taken into account. In other words, we want to synthesize a decentralized controller which will ensure the practical stability of the robot in accordance with the definition of control task (2.2.18). Up to now, we have synthesized local controllers which ensure satisfaction of local control tasks (2.4.8) for decoupled subsystems (i.e. when interconnections between the joints are neglected). It is obvious that, if the interconnections between joints exist, satisfaction of local control task (2.4.8) is not sufficient to ensure satisfaction of global control task (2.2.18) for the complete robotic system. In this Paragraph we shall present a method for testing the practical stability of the complete robotic system. Using this method we shall establish a procedure for the synthesis of decentralized controller which satisfies the set control task (2.2.18) for the complete robotic system. We shall use the test of practical stability of the complete robotic system, to establish a procedure for iterative improvement of the local controllers.

The model of the robotic system S is given by (2.2.8). During the synthesis of local controllers, in Paragraph 2.4, we have considered the approximate decoupled model of the system in form (2.4.7). For each decoupled subsystem we have synthesized a local controller in form (2.4.19) - a static controller, or in form (2.4.56) - a dynamic controller, where in both cases local nominal control is also introduced. In Paragraph 2.4.4 we have analyzed the local practical stability of the decoupled subsystem if either static or dynamic controller has been implemented. Now, we shall analyze whether the complete model of the robot is stable if local controllers are applied. We have to analyze the stability of the robotic system with the decentralized controller. In case of static local controllers, the model of the closed-loop system is given by:

$$S^i: \quad \dot{x}^i = A^i(\theta^i)x^i - \lambda_{ig} b^i(\theta^i) r_i^{-1} \tilde{b}^{iT} K_i \Delta x^i + b^i(\theta^i)\bar{u}^{oi}(t) + f^i(\theta^i)\cdot M_i, \quad \forall i \in I \qquad (2.5.1)$$

This model is obtained by combining the model of the robot (2.2.2) and static controller (2.4.19), and λ_{ig} is introduced to take into account the amplitude constraint upon the actuators input (λ_{ig} is defined by (2.4.68), but for the static controller).

In Paragraph 2.4.4 the synthesis of local controller which ensures the practical stability of the local decoupled subsystem has been shown, i.e. the local control is synthesized in such a way that the closed-loop subsystems:

$$\tilde{S}^i: \quad \dot{x}^i = \tilde{A}^i x^i - \lambda_{ig} \tilde{b}^i r_i^{-1} \tilde{b}^{iT} K_i \Delta x^i + \tilde{b}^i \bar{u}^{oi}(t), \quad \forall i \in I \qquad (2.5.2)$$

are practically stable in the sense of the local control task (2.4.8).

If we compare decoupled subsystems (2.5.2) with the real system we can recognize that they differ in coupling terms (which are neglected in the approximate decoupled model of the system). If we recall the definition of the decoupled subsystem (2.4.7), i.e. the definitions of matrix $\tilde{A}^i$ and vector $\tilde{b}^i$, we can find out that the difference between the models (2.5.1) and (2.5.2) is given by:

$$\tilde{f}^i(\theta^i) z^i(q_i) \bar{P}_i(q, \dot{q}, \ddot{q}, d) = \tilde{f}^i(\theta^i) z^i(q_i) \cdot [P_i(q, \dot{q}, \ddot{q}, d) - \tilde{P}_i \bar{z}^i / z^i(q_i)] \qquad (2.5.3)$$

where $\bar{P}_i$ is defined by (2.4.6), and $\tilde{P}_i$ is given by (2.4.5), while P_i is the driving torque given by the model of the mechanical part of the robot (2.2.1). Here we have used the relation (2.2.7) between M_i and P_i. By $\tilde{f}^i(\theta^i)$ we denote:

$$\tilde{f}^i(\theta^i) = (I_{n_i} - f^i(\theta^i)\bar{z}^i\bar{H}_{ii}T^i)^{-1}f^i(\theta^i)$$

$\tilde{P}_i$ defined by (2.4.5), represents the decoupled model of the robot mechanism which is also included in the decoupled model of the system (2.5.2). Thus, the difference between the actual model of the robot (2.5.1) and the decoupled model (2.5.2) is practically given by the difference between the real model of mechanism (2.2.1) and decoupled model (2.4.5) (this difference is multiplied by factor $\tilde{f}^i(\theta^i)z^i(q_i)$).

The local controllers ensure the practical stability of the decoupled model of the robot (2.5.2). In order to analyze the stability of the actual model of the robot (2.5.1), we have to investigate the influence of the coupling terms (2.5.3) upon robot stability. However, the model of the coupling term (2.5.3) is given by the model of the mechanical part of the robot (2.2.1), which is, in general case, highly nonlinear as explained in Chapter 1. Thus, the analysis of the practical stability of the robot is related to numerous problems.

Several methods for stability analysis of nonlinear systems have been developed but we shall not survey them here. A very common procedure, as already mentioned in Paragraph 2.3, is to linearize the nonlinear model and then examine the stability of the linearized model. However, the stability of the linearized model of a nonlinear system is only a necessary condition for the stability of nonlinear system. This means that if the linearized model of the system is not stable, the nonlinear system is also unstable. However, vice versa does not hold: if the linearized model is stable, we cannot guarantee the stability of the nonlinear system. Nevertheless, there is a good reason to first examine the stability of the linearized model of the robot. Since the stability analysis of the nonlinear system is often complex and time consuming, and the analysis of the linearized model is much simpler, it is reasonable to first examine the stability of the linearized model. If it is not stable, there is no need to examine the stability of the nonlinear model of the system. In this way we can avoid unnecessary and complex analysis of the nonlinear model. Obviously, if the linearized model of the robot is stable, we must analyze the stability of the nonlinear model, too, in order to guarantee the stability of the robot.

Thus, we shall first briefly present stability analysis of the linearized model of the robot. We have already explained in Paragraph 2.3.2 how we can obtain the linearized model of the system. However, in Paragraph 2.3.2 we have considered the linearized model of the robot if nominal programmed control $u^o(t)$, synthesized using the complete centralized model of the robot (2.3.4), is applied. Let us briefly describe how we obtain the linearized model if decentralized controller (2.4.19) is applied (i.e. if nominal programmed control $\bar{u}^{oi}(t)$ is synthesized at subsystems level).

Let us consider the model of deviation of the state around the nominal trajectory $x^o(t)$, $\forall t \in T$. Starting from the model (2.5.1), we obtain the model of deviation of the state as:

$$\Delta\dot{x}^i = (\tilde{A}^i - \lambda_{ig}\tilde{b}^i r_i^{-1}\tilde{b}^{iT}K_i)\Delta x^i + \tilde{f}^i(\theta^i)\cdot z^i(q_i)\bar{P}_i(q, \dot{q}, \ddot{q}, d) \quad (2.5.4)$$

To obtain (2.5.4) we have employed the fact that local nominal control $\bar{u}^{oi}(t)$, $\forall t \in T$ satisfies (2.4.14). The relation (2.2.7) for connection between M_i and P_i is also applied. Actually, the model (2.5.4) represents the difference between the real model of the robot (2.5.1) with local controllers (2.4.19) and the local "nominal model" (2.4.14).

Now, let us linearize the model of deviation (2.5.4) around the point $\Delta x^i = 0$, $\forall i \in I$, (i.e. around the nominal trajectory in the state space $x^o(t)$, $\forall t \in T$). The complete linearized model is obtained in the form:

$$\Delta\dot{x} = A_L^o(\theta, d, t)\Delta x + 0(\theta, d, t) \quad (2.5.5)$$

where the matrix $A_L^o(\theta, d, t)$: $R^L \times D \times T \to R^{N\times N}$ is defined as:

$$A^o = \mathrm{diag}[\tilde{A}^i - \lambda_{ig}\tilde{b}^i r_i^{-1}\tilde{b}^{iT}K_i + \tilde{f}^i(\theta^i)\left(\frac{\partial z^i}{\partial q_i}\right)_{q_i = q_i^{oi}}$$
$$\cdot \left.\frac{\partial G^i}{\partial x^i}\right|_{x^i = x^{oi}} \cdot \bar{P}(q^o, \dot{q}^o, \ddot{q}^o, d) - \bar{z}^i\bar{H}_{ii}\frac{\partial \ddot{q}_i}{\partial x^i} - \bar{z}^i\bar{h}_i\frac{\partial q_i}{\partial x^i})] +$$
$$+ F(\theta)z'(q^o)[(\frac{\partial H(q^o, d)}{\partial q}\ddot{q}^o + \frac{\partial h(q^o, \dot{q}^o, d)}{\partial q})\frac{\partial G(x^o)}{\partial x} +$$
$$+ \frac{\partial h(q^o, \dot{q}^o, d)}{\partial \dot{q}}\frac{\partial \hat{G}(x^o)}{\partial x} + H(q^o, d)\frac{\partial \ddot{q}}{\partial x}]$$

where $\frac{\partial H(q^o, d)}{\partial q}$ denotes a three dimensional matrix n×n×n (a derivative of inertia matrix H(q, d) with respect to q), $\frac{\partial h(q^o, \dot{q}^o, d)}{\partial q}$ is the n×n matrix (derivatives of vector $h(q, \dot{q}, d)$ with respect to q), and $\frac{\partial h(q^o, \dot{q}^o, d)}{\partial \dot{q}}$ denotes n×n matrix (derivatives of vector $h(q, \dot{q}, d)$ with respect to $\dot{q}$). Matrices F(θ), z'(q) have been defined in relation to the robot model (2.2.8) and vectors G(x), $\hat{G}(x)$ are defined by (2.2.11) and (2.2.12), while $\frac{\partial G(x^o)}{\partial x}$ denotes an n×N matrix (derivatives of vector G(x) with respect to x) and the similar holds for $\frac{\partial \hat{G}(x^o)}{\partial x}$. As explained in Paragraph 2.3.2, the main problem in obtaining linearized model (2.5.5) lies in the computation of matrices $\frac{\partial H}{\partial q}$, $\frac{\partial h}{\partial q}$ and $\frac{\partial h}{\partial \dot{q}}$, i.e. in the linearization of the mechanical part of robot model. As we have mentioned, these matrices may be obtained by some numerical procedure [2], or analytically [16] or by means of various identification methods. In (2.5.5) O(θ, d, t) denotes the term which is not compensated by the local nominal programmed control, and which is given by:

$$O(\theta, d, t) = F(\theta)\cdot z'(q^o)\cdot \bar{P}(q^o, \dot{q}^o, \ddot{q}^o, d) \tag{2.5.6}$$

The linearized model (2.5.5) differs from the linearized model (2.3.7), just in this term (2.5.6), since in (2.3.7) this term is compensated by nominal centralized control $u^{oi}(t)$ ∀t∈T. Also, in (2.5.5) we have already introduced local static controller (2.4.19).

Now, we have to examine the stability of the linearized model of the robot (2.5.5). If we want to examine the practical stability of the linearized model we can apply the method presented in the previous Paragraph. Namely, in Paragraph 2.4.4 we have examine the practical stability of decoupled linear subsystem. It is obvious that we can apply the same method to analyze the complete linearized model (2.5.5) of the robot. However, since the stability of the linearized model of the robot is only a necessary condition for the stability of the actual model of the robot, there is no need for a precise examination of the practical stability of the linearized model. It is sufficient to test the exponential stability of the linearized model, since afterwards we shall test the practical stability of the nonlinear model of the robot. To test the exponential stability of the linearized model we have to determine the eigenvalues of matrix $A_L^o(\theta, d, t)$, ∀t∈T. Let us denote the eigenvalues of the matrix A_L^o by $\sigma_j(\theta, d, t)$, j=1,2,...,N. They depend on actuator parameters θ, mechanism parameters d, and time t (i.e.

$\sigma_j(\theta^*, d^*, t^*)$, $j=1,2,\ldots,N$ correspond to the eigenvalues of matrix $A_L^O(\theta^*, d^*, t^*)$, for particular values $\theta^* \in \Theta$, $d^* \in D$ and $t^* \in T$). Since the region $X^t(t)$, $\forall t \in T$ of practical stability is defined as exponential shrinkage of subregions $X^{t(i)}(t)$ with exponential degrees $\alpha^{(i)}$, $i \in I$, (see Paragraph 2.2.2), we have to test whether the following condition holds*):

$$\min_j |Re(\sigma_j(\theta, d, t))| \geq \max_{i \in I} \alpha^{(i)}, \quad \forall \theta \in \Theta, \forall d \in D, \forall t \in T \qquad (2.5.7)$$

where the minimum on the left side is taken over all eigenvalues σ_j, $j=1,2,\ldots,N$ of the matrix A_L^O and for all allowable values of parameters θ, d and over time interval T. Condition (2.5.7) means that all eigenvalues of the linearized model of the robot (in complex plane) must be on the left from the maximum exponential degree of the finite regions shrinkage $\alpha^{(i)}$. This must hold for all allowable values of parameters and during the desired time period. Thus, we have to test the condition (2.5.7). If it is not satisfied, there is no need to test practical stability of the nonlinear model of the robot, since it is probable that the control (2.4.19) cannot ensure fulfilment of the desired control task (2.2.18), i.e. we may assume that the robot is not practically stable with decentralized controller (2.4.19).

However, if the condition (2.5.7) is satisfied, we may proceed with testing the practical stability of the nonlinear model of the robot, since the satisfaction of condition (2.5.7) is not sufficient to prove the practical stability of the robot. Now, we shall present the method for analysing the practical stability of the nonlinear model of the robot. This method is directly related to the analysis of the practical stability of the local decoupled subsystems, as presented in Paragraph 2.4.4. Actually, the method we are going to present originates from the aggregation-decomposition methods for stability analysis of large-scale systems, established by several authors. As we have already explained in Paragraph 2.4.4, we have used the method for practical stability analysis of large-scale systems developed by Michel [68], but we have modified it to suit the form of the model of the robotic systems [69].

We have to analyze the practical stability of the nonlinear model of the robot (2.5.1) if static local controllers (2.4.19) are applied. We have to test whether these local controllers can accommodate the stated control task (2.2.18).

---

*) Obviously, all eigenvalues $\sigma_j(\theta, d, t)$ of the matrix A_L^O must be on the left side of complex plane.

In Paragraph 2.4.4. we have presented how it is possible to test the practical stability of local decoupled subsystems (2.5.2) if local controllers (2.4.19) are applied. As explained above, the difference between the actual model (2.5.1) and decoupled model (2.5.2) is given by terms (2.5.3). We have to analyze the influence of these nonlinear terms upon the robot stability.

Let us again consider functions $v_i(t, x^i)$ associated to each subsystem $\tilde{S}^i$. Let us consider the time derivative of these functions along the solution of the actual nonlinear model of robot (2.5.1). The derivatives of these functions may be written as:

$$\dot{v}_i(t, x^i)_{(\text{along solution } (2.5.1))} = \frac{\partial v_i}{\partial t} + (\text{grad}v_i)^T \dot{x}^i = \frac{\partial v_i}{\partial t} +$$

$$+ (\text{grad}v_i)^T \dot{x}^i_{(\text{for decoupled subsystem } (2.5.2))} +$$

$$+ (\text{grad}v_i)^T \cdot (\text{coupling term } (2.5.3)) \tag{2.5.8}$$

The equation (2.5.8) means that the time derivative of the functions v_i may be divided into two parts: the first part concerns the derivative of the state vector for decoupled subsystems (2.5.2) and the second part is related to the difference between the actual model (2.5.1) and decoupled model (2.5.2) (which is given by terms (2.5.2)). The first part has been considered in Paragraph 2.4.4. Now, we have to consider the second part.

Let us write again the terms (2.5.3) in more convenient forms. We may write them in the form:

$$z^i(q^i)\bar{P}_i(q, \dot{q}, \ddot{q}, d, t) = \Delta z^i(t, \Delta q^i)\Delta\bar{P}_i(\Delta q, \Delta\dot{q}, \Delta\ddot{q}, d, t) +$$

$$+ z^i(q^{oi})\bar{P}_i^o(q^o, \dot{q}^o, \ddot{q}^o, d, t) +$$

$$+ \Delta z^i \bar{P}_i^o + z^i(q^{oi})\Delta\bar{P}_i \tag{2.5.9}$$

where by $\bar{P}_i^o$ we denote the value of $\bar{P}_i$ for $q = q^o$, $\dot{q} = \dot{q}^o$, $\ddot{q} = \ddot{q}^o$, (the so-called nominal "driving torque"*) for arbitrary parameters $d \in D$), and by $\Delta\bar{P}_i = \bar{P}_i - \bar{P}_i^o$ the deviation of "driving torque" $\bar{P}_i$ from its

---

*) We put driving torque into quotation marks, since P_i differs from the actual driving torque P_i by the value $\tilde{P}_i$ which represents the driving torque of the decoupled model of the mechanism.

nominal value. Similarly, we have written $z^i(q_i)$ as $z^i(q^{oi})+\Delta z^i(t, \Delta q_i)$, where Δz^i denotes the deviation of $z^i(q_i)$ from its nominal value $z^i(q^{oi})$. Further, we may rewrite (2.5.9) in a more compact form:

$$z^i(q_i)\cdot\bar{P}_i = (\Delta z^i(q_i)\cdot\Delta\bar{P}_i+\Delta z^i\cdot\bar{P}_i^o+ z^i(q^{oi})\Delta\bar{P}_i) + z^i(q^{oi})\cdot\bar{P}_i^o \quad (2.5.10)$$

Now, we may rewrite the second term in (2.5.8) in the following manner:

$$(\text{grad}v^i)^T\tilde{f}^i(\theta^i)z^i(q_i)\bar{P}_i = (\text{grad}v^i)^T\tilde{f}^i(\theta^i)(z^i(q_i)\Delta\bar{P}_i + \\ + \Delta z^i\bar{P}_i^o + z^i(q^{oi})\Delta\bar{P}_i) + \\ + \text{grad}(v^i)^T\tilde{f}^i(\theta^i)z^i(q^{oi})\cdot\bar{P}_i^o \quad (2.5.11)$$

The aim of this trivial rewriting is to separate the part of coupling which depends on the deviations of state coordinates from the nominal trajectory from the nominal part of coupling which depends only on the nominal trajectory (i.e. on time). Since $\bar{P}_i$ converges to $\bar{P}_i^o$ as q, $\dot{q}$, $\ddot{q}$ converge to nominal values q^o, $\dot{q}^o$, $\ddot{q}^o$, and since Δq, $\Delta\dot{q}$, $\Delta\ddot{q}$ converge to zero as Δx^j converge to zero $\forall j\in I$, we can write:

$$\lim_{\Delta x^j\to 0,\ \forall j\in I} \tilde{f}^i(\theta^i)(\Delta z^i(q_i)\Delta\bar{P}_i+\Delta z^i\bar{P}_i^o+z^i(q^{oi})\Delta\bar{P}_i)\to 0,$$

$$\forall i\in I,\ \forall t\in T,\ \forall d\in D,\ \forall\theta\in\Theta \quad (2.5.12)$$

In (2.5.12) we have also used the fact that Δz^i converges to zero as Δx^i converges to zero. The relation (2.5.12) means that this part of coupling, which depends on the deviation of state vector from the nominal trajectory, converges to zero as the state vector converges to the nominal trajectory (i.e. the deviation of "driving torques" from nominal should vanish if the state vector approaches nominal trajectory $x^o(t)$, $\forall t\in T$). Obviously, (2.5.12) holds for each set of values of parameters $d\in D$.

Due to (2.5.12), we can find the numbers ξ_{ij} which satisfy the following inequalities:

$$(\text{grad}v_i)^T\tilde{f}^i(\theta^i)(\Delta z^i(q_i)\Delta\bar{P}_i+\Delta z^i\bar{P}_i^o+z^i(q^{oi})\Delta\bar{P}_i) \leq \sum_{j=1}^{n}\xi_{ij}||\Delta x^j||$$

$$\forall(t, x)\in T\times X^t(t),\ \forall i\in I,\ \forall d\in D, \quad (2.5.13)$$

Actually, we can determine the numbers ξ_{ij} so that the sums on the right sides of inequalities (2.5.13) estimate the terms on the left sides. This represents linearization of the nonlinear coupling term, but this linearization holds in the whole region $X^t(t)$, $\forall t\in T$ to which the state of the robot must belong during tracking of nominal trajectory.

Now, we can write the time derivative of function $v_i(t, x^i)$ starting from (2.5.8), (2.5.11) and (2.5.13), in the following manner:

$$\dot{v}_{i(\text{along } (2.5.1))} \leq \dot{v}_{i(\text{along } (2.5.2))} + \sum_{j=1}^{n} \xi_{ij} ||\Delta x^j|| + (\text{grad}v_i)^T \tilde{f}^i(\theta^i) z^i(q^{oi}) \cdot \bar{P}_i^o,$$

$$\forall i\in I,\ \forall (t, x)\in T\times X^t(t),\ \forall d\in D, \tag{2.5.14}$$

The derivative of function $v_i(t, x^i)$ for local subsystem has to satisfy (2.4.40). Along the boundary of the region $X^t(t)$, $\forall t\in T$ (2.5.14) can be rewritten:

$$\dot{v}_{i(\text{along } (2.5.1))} \leq \psi_i(t) + \sum_{j=1}^{n} \xi_{ij} \bar{X}^{t(j)} \exp(-\alpha^{(j)} t) + (\text{grad}v_i)^T \tilde{f}^i(\theta^i) z^i(q^{oi}) \cdot \bar{P}_i^o(t, d) =$$

$$= \psi_i'(t, d) \quad \forall x\in \partial X^t(t),\ \forall t\in T,\ \forall d\in D \tag{2.5.15}$$

In (2.5.15) we have taken into account the fact that for boundary of the region $X^t(t)$, $\forall t\in T$, it holds $||\Delta x^j(t)|| = \bar{X}^{t(j)} \exp(-\alpha^{(j)} t)$, $\forall j\in I$, $\forall t\in T$.

The conditions for the practical stability of the ith subsystem are given by (2.4.40) and (2.4.41). Similarly, we may establish conditions for the practical stability of the complete robotic system. The derivatives of the functions $v_i(t, x^i)$ along the solution of the nonlinear model of the robot (2.5.1) are bounded by functions $\psi_i'(t)$ given by (2.5.15). Analogously to condition (2.4.41), the conditions for the practical stability of the robotic system (2.5.1) can be written as:

$$\int_0^t \psi_i'(t', d)\,dt' \leq v_{im}^{\partial X^{t(i)}(t)}(t) - v_{iM}^{\partial X^{I(i)}}(0),\ \forall i\in I,\ \forall t\in T,\ \forall d\in D, \tag{2.5.16}$$

The proof of this statement is given in Appendix 2.B, so we shall here just rewrite these conditions to show how we can use them to test the practical stability of the nonlinear model of the robot. Introducing relation (2.5.15) for $\psi_i^!(t)$ in (2.5.16), we get sufficient conditions for the practical stability of the complete robotic system, if local controllers (2.4.19) are applied:

$$\int_0^t \psi_i(t')dt' + \sum_{j=1}^{n} \xi_{ij} \frac{\bar{X}^{t(j)}}{\alpha^{(j)}}[1-\exp(-\alpha^{(j)}t)] +$$

$$+ \int_0^t (\text{grad}v_i)^T \tilde{f}^i(\theta^i) z^i(q^{oi}) \bar{P}_i^o(t, d)dt \leqslant$$

$$\leqslant v_{im}^{\partial X^{t(i)}(t)}(t) - v_{iM}^{\partial X^{I(i)}}(0), \quad \forall i\in I, \forall t\in T, \forall d\in D \qquad (2.5.17)$$

Let us suppose that functions $v_i(t, x^i)$ are chosen as (2.4.43), where matrices H_i satisfy (2.4.47)*). We have shown in Paragraph 2.4.4 that for these functions we may choose functions $\psi_i(t)$ as (2.4.49). Let us introduce the real-valued functions $\rho_i(t, d)>0$ which satisfy:

$$\int_0^t z^i(q^{oi})\bar{P}_i^o(q^o, \dot{q}^o, \ddot{q}^o, d, t')dt' \leqslant \rho_i(t, d), \forall t\in T, \forall i\in I \quad (2.5.18)$$

By introducing $\psi_i(t)$ defined by (2.4.49), and v_i defined by (2.4.43), and $\rho_i(t, d)$ defined by (2.5.18), into (2.5.17), we get these tests in a more convenient form:

$$-\gamma_{ig}\lambda_m^{1/2}(H_i)\bar{X}^{I(i)}(1-\exp(-\alpha^{(i)}t))/\alpha^{(i)} + \sum_{j=1}^{n} \xi_{ij}\frac{\bar{X}^{t(j)}}{\alpha^{(j)}}[1-\exp(-\alpha^{(j)}t)] +$$

$$+ \frac{\lambda_M(H_i)}{\lambda_m^{1/2}(H_i)}||\tilde{f}^i(\theta^i)||\rho_i(t, d) \leqslant \lambda_m^{1/2}(H_i)\bar{X}^{t(i)}\exp(-\alpha^{(i)}t) -$$

$$- \lambda_M^{1/2}(H_i)\bar{X}^{I(i)}, \quad \forall t\in T, \forall i\in I, \forall d\in D \qquad (2.5.19)$$

Thus, to test the practical stability of the nonlinear model of robotic system we have to test the practical stability of decoupled (local) subsystems by examining conditions (2.4.50), and then determine the numbers ξ_{ij} in (2.5.13) and functions $\rho_i(t, d)$ in (2.5.18). Then it is simple to test conditions (2.5.19). Since the conditions (2.5.19) are sufficient, but not necessary for practical stability, if they are not satisfied, we cannot say whether the robot is practically stable or not. Nevertheless, we can use these tests to improve local controllers so as to ensure the practical stability of the robot. If the local

---

*) Actually, H_i satisfies the equation analogous to (2.4.73), but for the static controller. We have already taken into account the amplitude constraint on actuators inputs through λ_{ig} in (2.5.2).

controllers stabilize the decoupled subsystems $\tilde{S}^i$ (which is proved by testing conditions (2.4.50)), but do not satisfy conditions (2.5.19), we may try to compute again local feedback gains and test again the stability of the robot by examining conditions (2.5.19). Thus, we may establish an iterative algorithm for reselection of the local controllers until the conditions (2.5.19) are satisfied. In this way we can develop an algorithm for the synthesis of decentralized controller which will fulfil the required control task (2.2.18) for the actual nonlinear model of the robotic system.

Before we present this algorithm, we have to make a few remarks concerning the test of the practical stability of the nonlinear model of robot.

(I) We have developed the conditions (2.5.19) to test the practical stability of robot if static local controllers (2.4.19) are applied. In Paragraph 2.4.4 we have shown how we can test the stability of decoupled subsystems $\tilde{S}^i$ if dynamic local controllers (2.4.56) are applied. It is very simple to extend the conditions (2.5.19) to include the case when dynamic local controllers are applied. Here, we shall not present extension of the stability test in detail (since it is quite trivial if we take into account the results obtained in Paragraph 2.2.4). We shall just write tests for the practical stability of the robot if dynamic local controllers are applied. Analogously to local stability test (2.4.74), and to test (2.5.19), we can write the conditions for the practical stability of the robot in the form:

$$
\begin{aligned}
&- \frac{\gamma_{ig}}{\alpha^{(i)}}[\hat{\bar{X}}^{I(i)}(1-\exp(-\alpha^{(i)}t) + (\tilde{R}^i C^i)_M \bar{X}^{I(i)} \cdot t] + \\
&+ \sum_{j=1}^{n} \xi_{ij}\{\frac{\hat{\bar{X}}^{t(j)}}{\alpha^{(j)}}[1-\exp(-\alpha^{(j)}t)] + (\tilde{R}^j C^j)_M \frac{\bar{X}^{t(j)}}{\alpha^{(j)}} t\} + \\
&+ \frac{\lambda_M(\hat{H}_i)}{\lambda_m^{1/2}(\hat{H}_i)} ||\tilde{f}^i(\theta^i)||\rho_i(t, d) < \lambda_m^{1/2}(\hat{H}_i)[\hat{\bar{X}}^{t(i)} \exp(-\alpha^{(i)}t) + \\
&+ (\tilde{R}^i C^i)_M \bar{X}^{t(i)}/\alpha^{(i)}] - \lambda_M^{1/2}(\hat{H}_i)\bar{X}^{I(i)}
\end{aligned}
\qquad (2.5.20)
$$

where we have chosen functions $v_i(t, \hat{x}^i)$ in the form (2.4.57), such that matrices $\hat{H}_i$ satisfy (2.4.73), γ_{ig} is given by (2.4.72), and the stability regions are defined by (2.4.60) and (2.4.63). Obviously, we have to consider extended subsystems with extended state vectors $\hat{x}^i$, due to the addition of integral feedback loop.

If we apply local optimal regulators (2.4.19), we may, as explained in Paragraph 2.4.4, choose functions $v_i(t, x^i)$ in the form (2.4.51), where K_i is the solution of the corresponding Riccati equation. The tests of practical stability (2.5.19) can be rewritten in this case analogously to (2.4.54).

(II) We have already introduced the numbers λ_{ig} (through numbers γ_{ig}) into conditions (2.5.19) in order to take into account the amplitude constraints upon actuators inputs. If conditions (2.4.67) are fulfilled for one of the subsystems, then, for this subsystem we can take $\lambda_{ig}=1$ and $\gamma_{ig} = \gamma_i$. Thus, by (2.5.19) (or (2.5.20)) we can test the stability in both cases: if the actuators inputs are within the given constraints, and if they reach the amplitude limits.

(III) In book No. 2 of this series [2] we have analyzed the stability of the robotic system around the nominal trajectory when nominal programmed control $u^o(t)$, $\forall t \in T$ is applied, where this nominal control has been synthesized using the centralized nonlinear model of robot (2.3.4). If such nominal control is applied, and assuming that all robot parameters are precisely defined in advance, the system is in equilibrium along the nominal trajectory $x^o(t)$, $\forall t \in T$ (i.e. for $\Delta x(t)=0$ and $\Delta u(t)=0$ it follows that $\Delta\dot{x}(t)=0$, $\forall t \in T$). Thus, we were able to test the asymptotic stability of the robot around nominal trajectory, instead of directly examining its practical stability. The test of the asymptotic stability is simpler than the test of practical stability.

However, in this book we have not introduced nominal centralized control $u^o(t)$, $\forall t \in T$ but only nominal local control $\bar{u}^{oi}(t)$, $\forall t \in T$, which satisfies (2.4.14). The reasons for this have been explained in Paragraph 2.4. We are not able to examine practical stability by testing the asymptotic stability of the system as we did in [2]. Here, for the first time, we directly examine the practical stability of the robot.

Even more, we are in a position to examine whether the control synthesized can withstand parameter variations. Namely, the tests (2.5.19), or (2.5.20) take into account variable robot parameters (θ parameters of actuators and d parameters of the mechanical part of the robot). We have to test conditions (2.5.19) for each allowable value of parameters θ or d, and if the tests are satisfied for all parameter values we can guarantee that the chosen local regulators are robust enough to withstand all parameter variations. It should be noted that up to now we have not been able to test stability of the robot if its parameters are variable. Here, for the first time, we can test robustness of the control applied to robotic systems.

(IV) We have mentioned several times that the above-presented conditions for the practical stability of the robot, are only sufficient and not necessary. We want to underline once again that this is the reason why this method may be conservative. It may happen that the tests (2.5.19), or (2.5.20), are not satisfied, but, actually, the robot may be stable with the control applied. This may lead to synthesis of too strong local controllers and unacceptable conclusions. Moreover, it may happen that, due to conservatism of the method, we cannot find the decentralized controller which can stabilize the robot and accommodate all allowable parameter variations, although it actually exists. However, the main source of conservatism of the method lies in determination of the numbers ξ_{ij} which satisfy (2.5.13). If we ensure that the numbers ξ_{ij} are good estimations of the interactions between the subsystems, we may expect that the control synthesized using this method is quite acceptable and that all conclusions that follow from these analyses are valid.

Now, we shall establish the algorithm for synthesis of the decentralized control which stabilizes the robotic system around the given nominal trajectory $x^o(t)$, $\forall t \in T$, for all allowable parameters of the mechanism and actuators. The algorithm uses the above presented tests of the practical stability of the nonlinear model of the robot. The aim of this algorithm is to synthesize the decentralized controller which satisfies control task (2.2.18). A brief flow-chart of this algorithm is presented in Fig. 2.21. The algorithm includes the algorithm, shown in Fig. 2.20 for the synthesis of local controllers which stabilize local (decoupled) subsystems.

The algorithm can be performed according to the following steps:

(1) The global control task (2.2.18) is set, i.e. the regions X^I and $X^t(t)$, $\forall t \in T$ are set, by defining the numbers $\bar{X}^{I(i)}$, $\bar{X}^{t(i)}$, $\alpha^{(i)}$ for all n degrees of freedom (actuators) and time interval τ; also, the sets of allowable parameters Θ and D have to be defined; nominal trajectory $x^o(t)$, $\forall t \in T$ is given.

(2) Set initial values of the actuator parameters $\theta = \theta_o \in \Theta$.

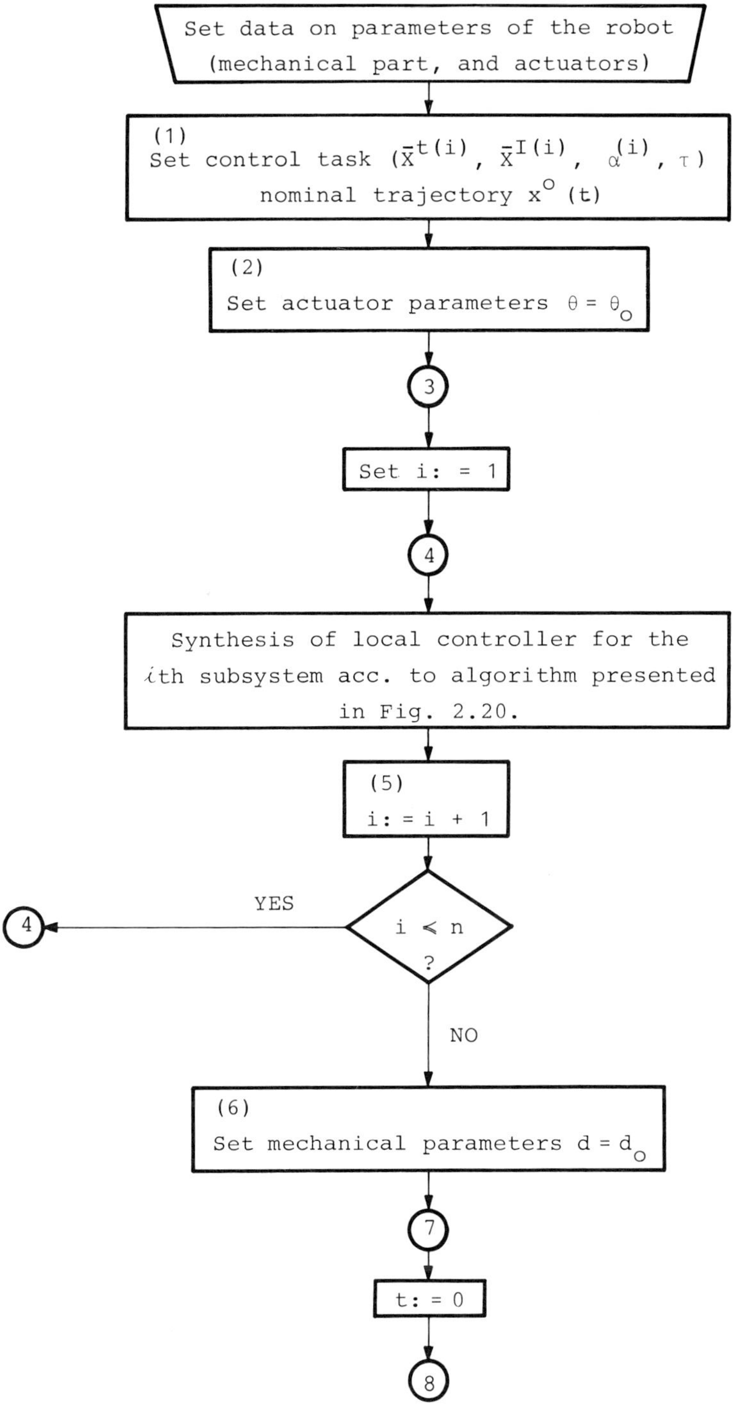

Fig. 2.21. Flow-chart of the algorithm for synthesis of decentralized controller on the basis of practical stability analysis

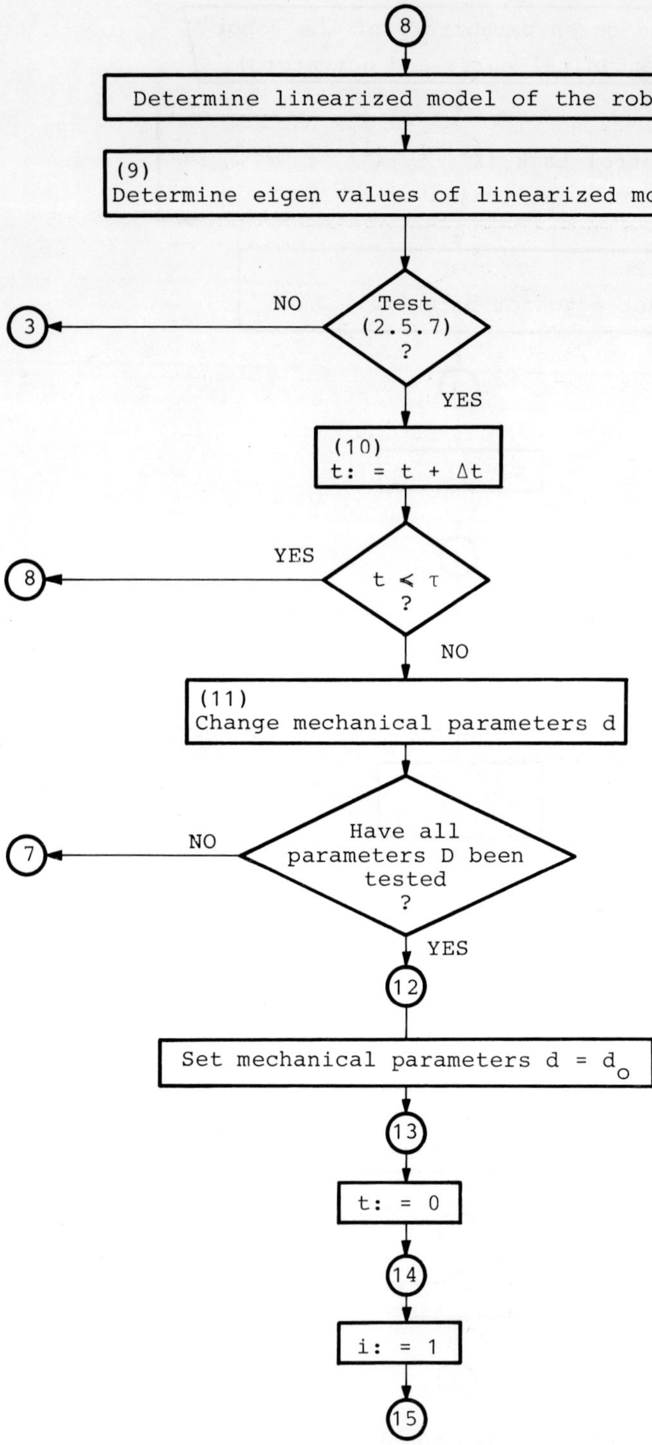
8
Determine linearized model of the robot
(9)
Determine eigen values of linearized model
Test
(2.5.7)
?
NO
3
YES
(10)
t: = t + Δt
t ≤ τ
?
YES
8
NO
(11)
Change mechanical parameters d
Have all
parameters D been
tested
?
NO
7
YES
12
Set mechanical parameters d = d_o
13
t: = 0
14
i: = 1
15

Fig. 2.21. (Cont.)

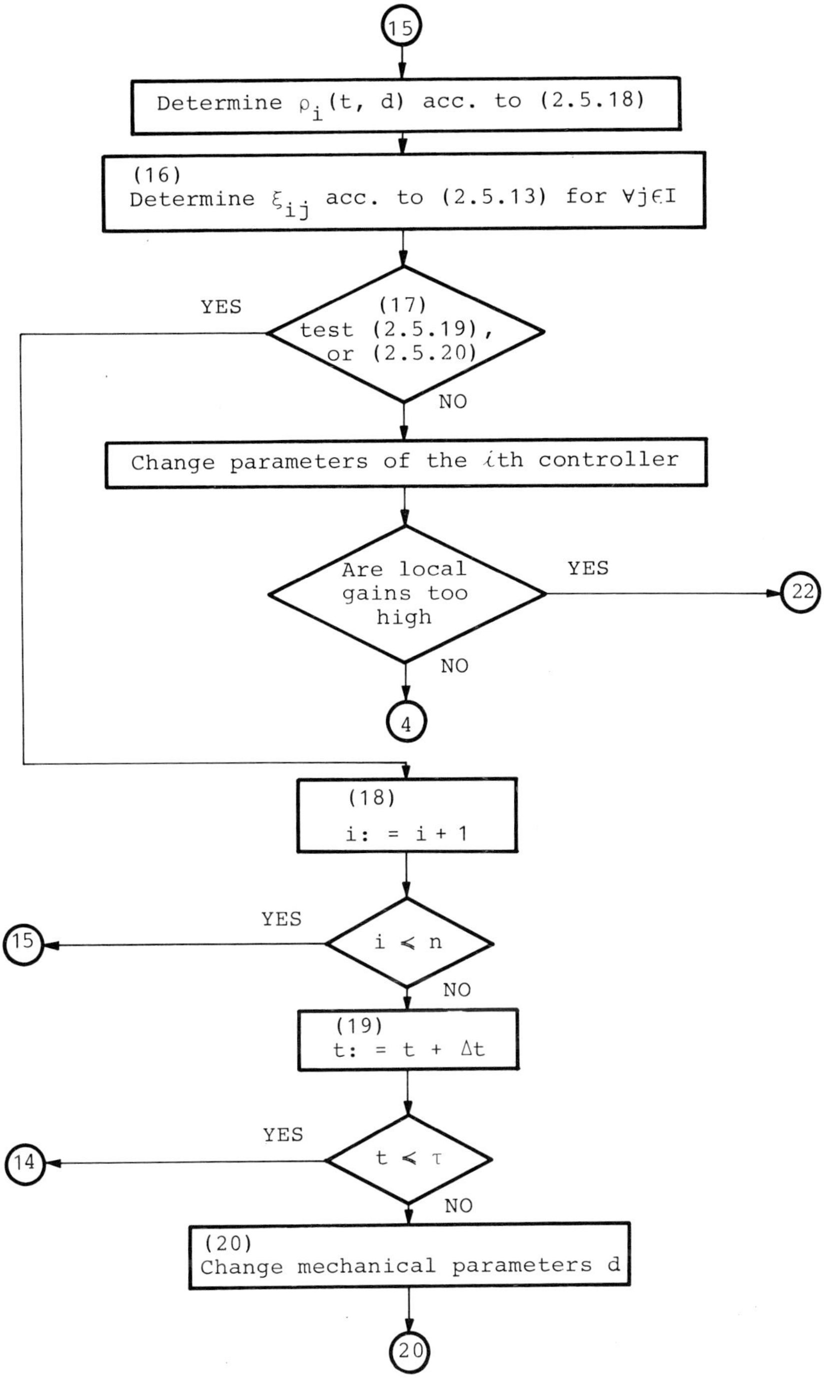

15
Determine $\rho_i(t, d)$ acc. to (2.5.18)
(16)
Determine ξ_{ij} acc. to (2.5.13) for $\forall j \in I$
YES
(17)
test (2.5.19),
or (2.5.20)
NO
Change parameters of the ith controller
Are local
gains too
high
YES
22
NO
4
(18)
i: = i + 1
YES
15
i ≤ n
NO
(19)
t: = t + Δt
YES
14
t ≤ τ
NO
(20)
Change mechanical parameters d
20

Fig. 2.21. (Cont.)

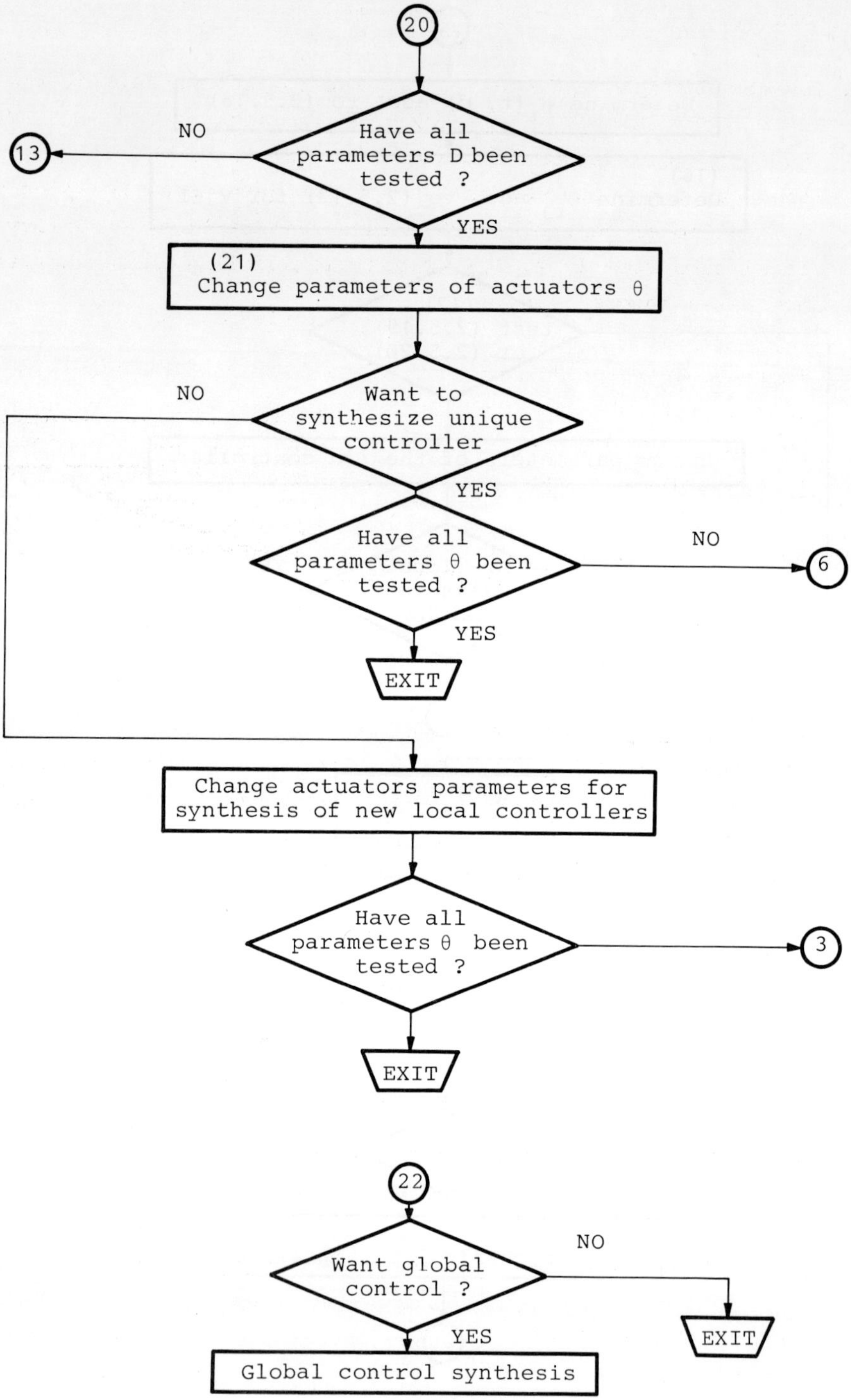
20
Have all parameters D been tested ?
NO
13
YES
(21) Change parameters of actuators θ
Want to synthesize unique controller
NO
YES
Have all parameters θ been tested ?
NO
6
YES
EXIT
Change actuators parameters for synthesis of new local controllers
Have all parameters θ been tested ?
3
EXIT
22
Want global control ?
NO
EXIT
YES
Global control synthesis

Fig. 2.21. (Cont.)

(3) Set index i to 1 (i=1) - (subsystem index).

(4) Synthesize the *i*th local controller so as to stabilize the local decoupled *i*th subsystem using the algorithm presented in Fig. 2.20; in this step the complete algorithm for the local controller is used; as the output of this step we obtain local feedback gains and nominal local control which ensure the practical stability of the local subsystem (2.4.7) - where the coupling among subsystems is not taken into account.

(5) Increase index i by one; if i is smaller than or equal to n - go back to step (4) and synthesize local controller for the next subsystem; if i is larger than n go to the next step; at the end of this step local controllers are synthesized for all subsystems.

(6) Set initial values of the parameters of the mechanical part of the robot $d = d_o \in D$.

(7) Set initial time instant (t=0.).

(8) Determine the linearized model of the complete model of the robot according to (2.5.5); for this purpose we have to determine matrices $\partial H/\partial q$, $\partial h/\partial q$ and $\partial h/\partial \dot{q}$ using the algorithm for automatic linearizaiton of the mechanical part of the robotic system; we determine these matrices for the nominal values of coordinates q^o, $\dot{q}^o$ for the time instant t and parameters d.

(9) Determine the eigenvalues of the matrix A_L^o of the linearized model of robot; test the condition (2.5.7); if this condition is not satisfied, go back to step (3) and choose again the local controllers (increase the stability degree, or move the desired poles of closed-loop subsystems to the left in the complex plane); if the condition (2.5.7) is met, go to the next step.

(10) Increase time instant by Δt, $t: = t+\Delta t$, where Δt is the desired time interval in which we want to examine system stability; check whether $t>\tau$, if yes, go to the next step (this means that we have analyzed the stability of the linearized model over the whole desired time interval T); if $t\leqslant\tau$, go back to step (8).

(11) Change parameters of the mechanical part of the robot $d \in D$; check

whether we have analyzed the stability of the linearized model for allowable values of mechanism parameters D; if yes, go to the next step; otherwise, go to step (7).

(12) From this step we start to analyze the stability of the nonlinear model of the robot; set initial values of the parameters of the mechanical part of the robot $d = d_o \in D$.

(13) Set initial time instant (t=0.).

(14) Set index i to 1 (i=1).

(15) Determine the number $\rho_i(t, d)$ such that (2.5.18) is satisfied for the *i*th joint of the robot.

(16) Determine numbers ξ_{ij} that satisfy inequality (2.5.13) for all $j \in I$. This is the most critical step in the algorithm. We have to carefully determine these numbers in order to make the analysis less conservative. A numerical procedure should be developed which searches along the complete finite region $\tilde{X}^t(t)$, $\forall t \in T$, to find maximum values of the coupling between subsystems. In this way we ensure a good (linear) estimation of the robotic system.

(17) Test condition (2.5.19), or (2.5.20) for the *i*th subsystem depending on whether we have chosen static local controller (2.4.19) or dynamic local controller (2.4.56); if the condition is not satisfied we have to change the *i*th local controller; thus, we should change parameters of the *i*th controller so as to increase the stability degree of the *i*th local subsystem; then, we should go back to step (4) and synthesize again the *i*th controller; however, we should constrain the stability degree of subsystems; in other words, we have to avoid too high local feedback gains; thus, we have to test whether the local gains (local stability degree) reach the imposed limits; if yes, we cannot find the acceptable local controller which can satisfy test of practical stability (2.5.19), or (2.5.20) and we must stop the algorithm; if local gains are within the given limits, go back to step (4) and try with a new local controller for the *i*th subsystem; if the test (2.5.19), or (2.5.20) is satisfied (for the *i*th joint) go to the next step of the algorithm.

(18) Increase index i by one ($i: = i+1$); test whether $i \leq n$; if yes, go back to step (15); if $i > n$ go to the next step.

(19) Increase time instant for Δt ($t: = t+\Delta t$); test whether $t \leq \tau$; if yes go back to step (14); if $t > \tau$, go to next step (this means that we have tested the practical stability of the robot over the complete time interval T).

(20) Change parameters of the mechanical part of the robot $d \in D$; check whether we have analyzed the practical stability of the nonlinear model of the robot for all allowable values of parameters $d \in D$; if yes go to the next step; otherwise, go back to step (13).

(21) Change parameters of the actuators $\theta \in \Theta$; in principle, we may try to synthesize unique local controllers which stabilize the robotic system for all allowable parameters of the actuators $\theta \in \Theta$; in that case we should preserve the already synthesized local controllers and go back to step (6) and test the practical stability of the robot for new values of parameters θ, but with old parameters of controllers; we have to check whether such controllers can accommodate all parameter variations; however, in general case, it is difficult to find unique local controllers which can stabilize the robotic system for various parameters of mechanism d and for various parameters of the actuators θ; on the other hand, as we have explained in Paragraph 2.4 the actuator parameters usually vary slowly and there is no need for synthesizing unique control for all allowable actuators parameters; it is sufficient to synthesize several local controllers corresponding to various values of actuator parameters and, from time to time, identify these parameters and if necessary, change, local feedback gains; in that case we should go back to step (2) and synthesize new local controllers for new values of actuators parameters θ; in this step we have to check whether all allowable actuators parameters have been already tested; if yes, we may terminate the algorithm.

This algorithm can be used to synthesize the decentralized controller (static or dynamic, with complete state feedback or with output feedback), which should satisfy control task (2.2.18): ensure the practical stability of the robot around given nominal trajectory $x^0(t)$, $\forall t \in T$. We obtain *unique* local controllers which accommodate all nonlinearities of the model and all expected parameter variations. This algorithm can be easily implemented by a digital computer to obtain computer-aided synthesis of decentralized controller for robotic system (see Chapter 4).

There are few evident numerical problems with the above algorithm, which have already been mentioned: search over finite region $X^t(t)$, $\forall t \in T$, to compute numbers ξ_{ij} satisfying (2.5.15), repeated synthesis of local controllers (how to define new parameters required in local control synthesis) etc. As already mentioned in Paragraph 2.4.4, the interaction with robotic system designer throughout the execution of the algorithm is of great importance. In our opinion, the completely automatic synthesis of control for robotic system without any interaction with designer is not recommendable. We suggest the development of an interactive algorithm, which should utilize designer's experience as much as possible (see Chapter 4). The designer should decide whether the algorithm should continue to iteratively search for the decentralized controller, which can accommodate the imposed control task (2.2.18), or the iteration should be stopped, and try to solve the problem by some different control. Also the designer might be "included" in the loop to directly change the parameters of the local controllers in an appropriate way to achieve the practical stability of the robot and, thus, speed up the algorithm.

One of the problems connected with this synthesis is the decision about how high local feedback gains should be allowed. Due to the structure of the robotic system it is always possible to find local controllers which can stabilize the robot for constant parameters [59]. The problem is whether we could synthesize unique decentralized control which is robust enough to withstand all parameter variations. However, regardless of the problem of robustness of the decentralized controller, we have to answer the following question: shall we try to stabilize the robot with arbitrarily high local feedback gains, or introduce some additional feedback loops in order to satisfy the imposed control task. Namely, stabilization of the robotic system may require very high local feedback gains, because of the strong influence of coupling between subsystems which has to be compensated by a high stability degree of the subsystems. In other words, in order to satisfy conditions (2.5.19), if numbers ξ_{ij} (estimation of coupling) are large, we must increase γ_{ig} (estimation of the local subsystem stability degree). This can be achieved by higher local feedback gains. However, there are several drawbacks concerning the high feedback gains:

(a) In addition to feedback signals high feedback gains also amplify the noise that often appears in every system (noise due to chattering, vibrations, elastic modes of the robot, errors due to truncation of A/D and D/A convertors, etc.). This amplification of noise may lead to robot instability. Namely in the model of the robot considered up to now (Paragraph 2.2.1) we have not included all these effects, so the stability of the nonlinear model (2.5.1) cannot guarantee the stability of the actual robot (if these effects are too strong). For example, the model of the mechanical part of the robot (2.2.1) does not include the elastic modes of the mechanism, since we considered the mechanism as an open chain consisting of rigid links (see Chapter 1). If low gains are applied, this model is appropriate for most cases (if the links are sufficiently rigid, as is the case with most robots used in industrial practice today). However, if very high feedback gains are applied, this assumption is not valid any more: high amplification of weak elastic effects may cause high chattering of the robot and the elastic modes might become predominant. Thus, the rigid body model of robot mechanism is inappropriate in such regimes, and the above presented stability analysis is not valid.

(b) Too high local gains may lead to high suboptimality of the control with respect to a chosen criterion. We have shown in [2] how it is possible to estimate the suboptimality of the decentralized control (if nominal programmed control is applied). Here, we shall not discuss this problem in detail. We shall just remember that in [2] we have shown that by introducing additional (global) feedback loop we can reduce suboptimality of the decentralized controller. Actually, it is evident that too high local feedback gains can require too high energy consumption and that they are suboptimal from the standpoint of the required power supply.

For the above reasons, it is recommenable to avoid too high local gains. Evidently, by introducing additional feedback loops we can stabilize the robot with lower local gains. By this, we abandon the decentralized control structure and introduce global control. However, the problem is: how high local feedback gains may we allow, or, in other words, at which point should we break the above mentioned algorithm for decentralized control synthesis and introduce global control? This problem is difficult to solve in general case without considering more appropriate models of robots for high-gain regimes (i.e. the models that include elastic modes, the vibrations of robot basement, noise, etc.) and without considering control suboptimality. Both problems are out of the scope of this book, but, in our opinion, this decision should be left to the robot designer. He should, according to his own experience, decide whether he will try to find unique local controllers which satisfy the stated control task (2.2.18), or introduce additional global control.

2.6 Global Control Synthesis

If the unique decentralized controller which stabilizes the robotic system in a desired manner and which is robust enough to withstand all allowable parameter variations, cannot be determined, or if too high local feedback gains are required, we can introduce global control. The aim of the global control is to compensate for the influence of coupling among subsystems, i.e. to compensate for the dynamic interactions between joints of the robot.

The local controllers, synthesized in the previous Paragraph, stabilize local decoupled subsystems (i.e. decoupled joints and corresponding actuators). If the coupling between the subsystems (2.5.3) is weak in comparison with the local stability degree, the local controllers can withstand the influence of coupling. However, in case of strong coupling, we must introduce additional control loops to compensate for the coupling and ensure stability of the robot. In doing this we should try to preserve simplicity of the decentralized control structure, i.e. we should try to synthesize global control as simple as possible and introduce only those feedback loops which are necessary to achieve good tracking of desired trajectory.

Global control might be introduced in several ways. We shall first discuss the general form of global control law that follows from the analysis of the stability of the robotic system. The coupling term in the time derivative of function $v_i(t, x^i)$ is given by $(\text{grad}v_i)^T\tilde{f}^i(\theta^i)\cdot z^i(q_i)\bar{P}_i(d, x)$. If we want to compensate for this term in equation (2.5.4) we may introduce control in the form [2, 70, 71]:

$$\Delta u_i^G = -k_i^G[(\text{grad}v_i)^T\tilde{b}^i]^{-1}(\text{grad}v_i)^T\tilde{f}^i(\theta^i)z^i(q_i)\bar{P}_i^*(d, t, x) \quad (2.6.1)$$

where $k_i^G \in R^1$ is the global gain (which has to be chosen) and $\bar{P}_i^*(d, t, x): D\times T\times R^N \to R^1$ is the term which corresponds to term $\bar{P}_i(d, t, x)$ and the implementation of which will be discussed in the text to follow. If we consider the derivative of function v_i along the solution of the nonlinear model (2.5.1) and if we assume perfect (ideal) global control in which $P_i^*(d, t, x) = \bar{P}_i(d, t, x)$, $k_i^G = 1$, and if we neglect constraints upon the input of the subsystem we get:

$$\dot{v}_{i(\text{along}(2.5.1),\ (2.6.1))} = \dot{v}_{i(\text{along}\ (2.5.2))} + (\text{grad}v_i)^T\tilde{b}^i \Delta u_i^G +$$

$$(\text{grad}v_i)^T \cdot \tilde{f}^i(\theta^i) z^i(q_i) \cdot \bar{P}_i(t, d, x) = \dot{v}_{i(\text{along}\ (2.5.2))} \quad (2.6.2)$$

This means that, in an ideal case, by global control (2.6.1) we may achieve perfect compensation of coupling. In this case robotic system behaves as if it were perfectly decoupled to a set of independent subsystems-joints. This means that, in case of perfect compensation of coupling, the stability of the robot is proved by the stability of the decoupled subsystems.

Unfortunately, this perfect decoupling is impossible to achieve for several reasons. First, we cannot implement function $\bar{P}_i^*(d, t, x)$ such that perfectly coincides with real coupling $\bar{P}_i$, since the model of the system is not perfect and the parameters of the mechanical part of the robot d are not known in advance. Thus, in practice it is impossible to compute perfectly the driving torques in robotic joints. Second, real actuator inputs are constrained by amplitude. The control law (2.6.1) might require very high control signals: if $(\text{grad}v_i)^T\tilde{b}^i$ is a small value, the control signal might be very high and violate the amplitude limit upon the actuator input. In that case the control (2.6.1) cannot compensate for complete coupling even if $\bar{P}_i^*$ perfectly coincides with real coupling $\bar{P}_i$.

Actually, we have to limit the upper amplitude of global control, i.e. the value of $(\text{grad}v_i)^T\tilde{b}^i$ - since it might happen that $(\text{grad}v_i)^T\tilde{b}^i = 0$ which would require an infinite control signal. Thus, we may define the region $X^\varepsilon(t) \subset X^t(t)$, $\forall t \in T$, in the state space as:

$$X^\varepsilon(t) = X^{\varepsilon(1)}(t) \times X^{\varepsilon(2)}(t) \times \cdots \times X^{\varepsilon(n)}(t)$$

$$X^{\varepsilon(i)}(t) = \{x^i\colon (\text{grad}v_i)^T\tilde{b}^i < \varepsilon_i \ .\text{AND}.\ x^i \in X^{t(i)}(t)\}, \quad \forall t \in T \qquad (2.6.3)$$

where $\varepsilon_i > 0$ are real-valued numbers which have to be chosen in the process of control synthesis. Instead of control (2.6.1), we have to apply the global control in the form:

$$\Delta u_i^G(t) = \begin{cases} -k_i^G[(\mathrm{grad}v_i)^T\tilde{b}^i]^{-1}(\mathrm{grad}v_i)^T\tilde{f}^i(\theta^i)z^i(q_i)\bar{P}_i^*(t,\ d,\ x) \\ \qquad\qquad \text{for } x^i \in X^{t(i)}(t).\mathrm{AND}.x^i \notin X^{\varepsilon(i)}(t) \\ -k_i^G \cdot \varepsilon_i^{-1}(\mathrm{grad}v_i)^T\tilde{f}^i(\theta^i)z^i(q_i)\bar{P}_i^*(t,\ d,\ x),\ x^i \in X^{\varepsilon(i)}(t) \end{cases} \tag{2.6.4}$$

Evidently, the global control (2.6.4) cannot perfectly compensate for the coupling between the subsystems in the region X^ε even if $\Delta\bar{P}_i^* = \Delta\bar{P}_i$.

Although the global control cannot perfectly decouple the robotic system in reality, this control is still capable of reducing the influence of coupling upon the stability of the robot. However, the efficiency of the global control mostly depends on the choice of function $\bar{P}_i^*$. In the text to follow we shall discuss three possible implementations of the global control:

(1) Force feedback can be used as global control. The coupling among subsystems is represented by driving torques (forces) acting around joints' axes. These torques or forces can be directly measured if we introduce force transducers in the joints. In this way we can get complete information about the coupling at each subsystem (joint), and theoretically, compensate for the total influence of coupling between the subsystems [2, 70, 71]. Namely, by measuring the forces in the joints we get the information on driving torques $P_i(d,\ x)$. If we consider subsystems without introducing the aggregate model of the mechanism (2.4.5), i.e. if we assume that $P_i^* = 0$, the coupling term is defined by $\tilde{f}^i(\theta^i)z^i(q_i)P_i$. By measuring P_i we can obtain the information about the complete (actual) coupling between the joints of robot and if we implement function $P_i^*(d,\ t,\ x) = P_i(d,\ x)$ in the global control (2.6.4) we can compensate for the complete influence of the robot dynamics upon its stability (obviously, this holds only if $x \in X^t(t)$ and $x \notin X^\varepsilon(t)$, $\forall t \in T$).

The control scheme with decentralized controller and global control with force feedback is presented in Fig. 2.22 (under the assumption that global control is introduced in all the joints of the robot).

This solution has some adventages:

(a) The information control structure is simple. By introducing force feedback as global control, we increase the number of

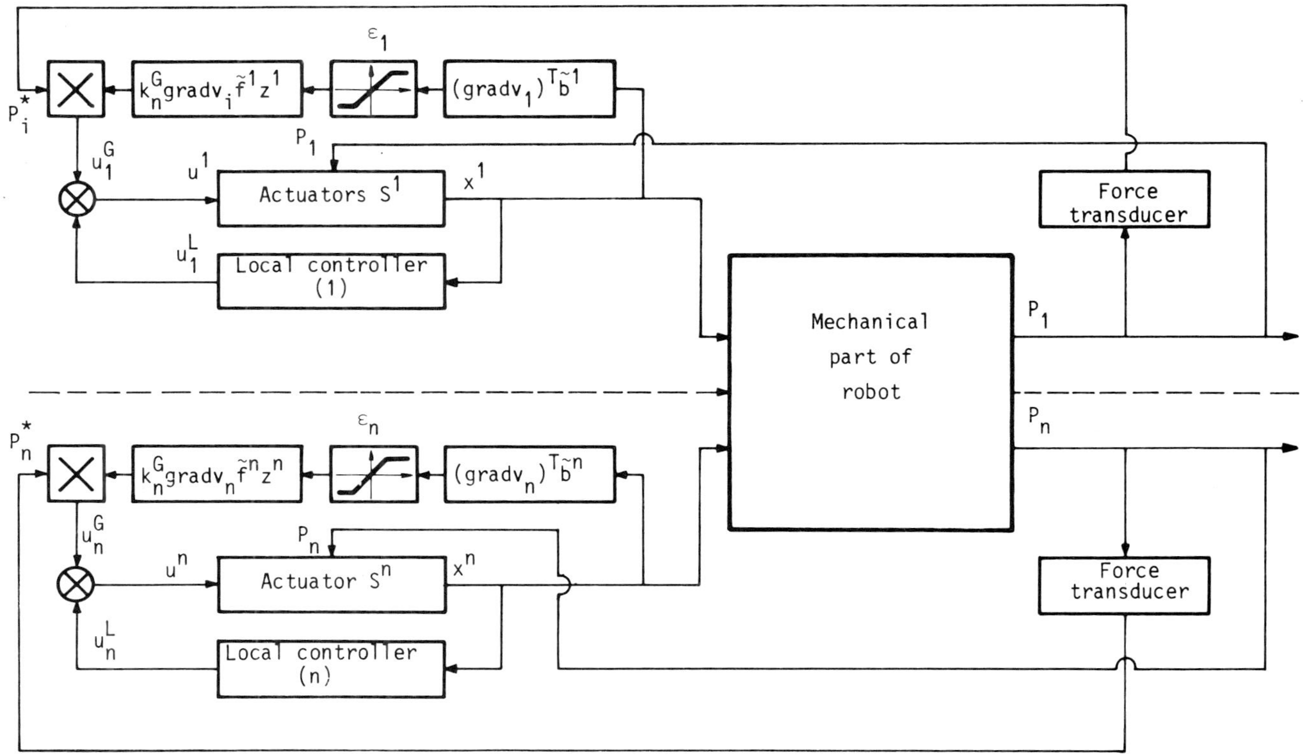

Fig. 2.22. Control scheme with global control implemented by force feedback

feedback loops by, at most, n in comparison with decentralized control.

(b) The computation of the global control with force feedback is extremely simple. We have just to compute $z^i(q_i)$ and perform no more than n multiplications to compute real coupling terms $z^i(q_i) \cdot P_i$ and $2n_i \times n$ multiplications, n division and $n \times 2(n_i-1)$ additions to compute terms $k_i^G[(\text{grad}v_i)^T\tilde{b}^i]^{-1}(\text{grad}v_i)^T\tilde{f}^i(\theta^i)$. Thus, such a global control does not require powerful microprocessor for its computation, but a simple, cheap microprocessor may be implemented.

(c) Since by measuring forces we get real information on coupling, this control is robust to variations of the mechanism d. The global control with force feedback compensates the influence of the robot dynamics regardless of the parameters d, since force transducers measure the actual value of driving torques.

Unfortunatelly, there are a few drawbacks of force feedback implementation:

(a) Force measurement by force transducers is not very reliable even if very precise transducers are used. Possible vibrations and chattering of the robot may result in "a lot of noise in the measured force". The proposed control (2.6.4) assumes direct feedback by so-measured forces, which might cause the instability of the whole system. Clearly, insertion of a filter might help to eliminate noise. However, we think that this solution with a direct feedback from transducers to actuator input might be unreliable.

(b) The mounting of force transducers on robot axes (shafts) is not simple and is connected with many problems concerning the mechanical design of the robot. As already mentioned in Paragraph 2.3.4, in [32] are discussed some technical problems in mounting the force transducers on the Stanford robot.

(c) The price of force transducers is not negligible. Since highly precise transducers are required to obtain accurate information on forces in the joints, their price may happen to be compatible with the price of microprocessors that would be needed to compute torques (if the global control is implemented by on-line computation of the robot driving torques-see the text to follow). Thus, the benefits of introducing force feedback as the global control should be carefully studied.

(2) The second possible implementation of global control is by on-line computation of the robot dynamics using its mathematical model (2.2.1). We can compute the driving torques acting in robot joints, and, in that case, function $\bar{P}_i^*(d, t, x)$ in (2.6.4) represents the computed driving torque in the ith joint. If the parameters of the robot d were precisely known in advance, and if the model of the robot dynamics were sufficiently accurrate, we should achieve the perfect compensation of coupling. However, these two assumptions are rarely satisfied in practice. On the other hand, as we have stated for several times, the driving torques are, in general case, very complex functions of all state variables. Thus such a computation of the driving torque might be very cumbersome and time consuming, and it might require a powerful multiprocessor system to achieve a sampling period compatible with the dynamics of the robot. This solution of the global control leads to a very complex information control structure, so the simplicity of the decentralized structure is lost. Actually, by this implementation of global control we face the same problems as those listed for the "inverse problem" technique (Paragraph 2.3.3):

(a) complexity of the information structure,

(b) complex computation requiring a powerful control multiprocessor system,

(c) sensitivity of the control to parameter variations.

The control scheme with global control which utilizes the computed driving torques is presented in Fig. 2.23.

We can simplify the computation of coupling by using the approximate model of the robot dynamics. In Paragraph 2.3.3 we have already mentioned several possible approximation models of the robot dynamics. Here we shall list some of the approximate models which may be used to compute global control:

(a) The simplest approximate model includes just gravity forces (moments) in the robot joints, which have to be compensated by global control in order to reduce static errors of local controllers. Thus, the functions $\bar{P}_i^*$ are given by:

$$\bar{P}_i^*(x, d^*) = g_i(q, d^*) \tag{2.6.5}$$

where $g_i(q)$ is the gravity moment (force) acting around the ith joint (for particular values of parameters d^*).

(b) A more complex model includes on-line computation of diagonal elements in the inertia matrix $H(q, d)$ (i.e. the eigen-inertia factors around each joint of the robot)*). Thus, the functions $\bar{P}_i^*$ are given by:

$$\bar{P}_i^*(x, d^*) = H_{ii}(q, d^*)\ddot{q}_i + g_i(q, d^*) \tag{2.6.6}$$

where H_{ii} is the ith diagonal element in the inertia matrix $H(q, d^*)$.

(c) The next approximate model of the robot dynamics includes the complete inertia matrix. In addition to the diagonal elements we should compute in real time cross-inertia terms, i.e. off--diagonal elements in inertia matrix $H(q, d)$. The functions $\bar{P}_i^*$ that have to be computed are given by:

$$\bar{P}_i^*(x, d^*) = H_i(q, d^*)\ddot{q} + g_i(q, d^*) \tag{2.6.7}$$

where by $H_i(q, d^*)$ we denote the ith row in the inertia matrix $H(q, d^*)$, $H_i \in R^n$.

(d) A more complex model of the robot dynamics also includes the centrifugal forces. Thus, the computed functions $\bar{P}_i^*$ include the following terms:

$$\bar{P}_i^*(x, d^*) = H_i(q, d^*)\ddot{q} + g_i(q, d^*) + \dot{q}^T C^i(q, d^*)\dot{q} \tag{2.6.8}$$

---

*) It should be noted that these inertia factors and gravity forces might be included in the model of subsystems as the aggregate model (2.4.5). However, we have assumed inertia factors $\bar{H}_{ii}$ to be constant and the gravity term linear by $x^i(\bar{h}^i q_i)$. Here, the nonlinear inertia and gravity terms are computed for global control.

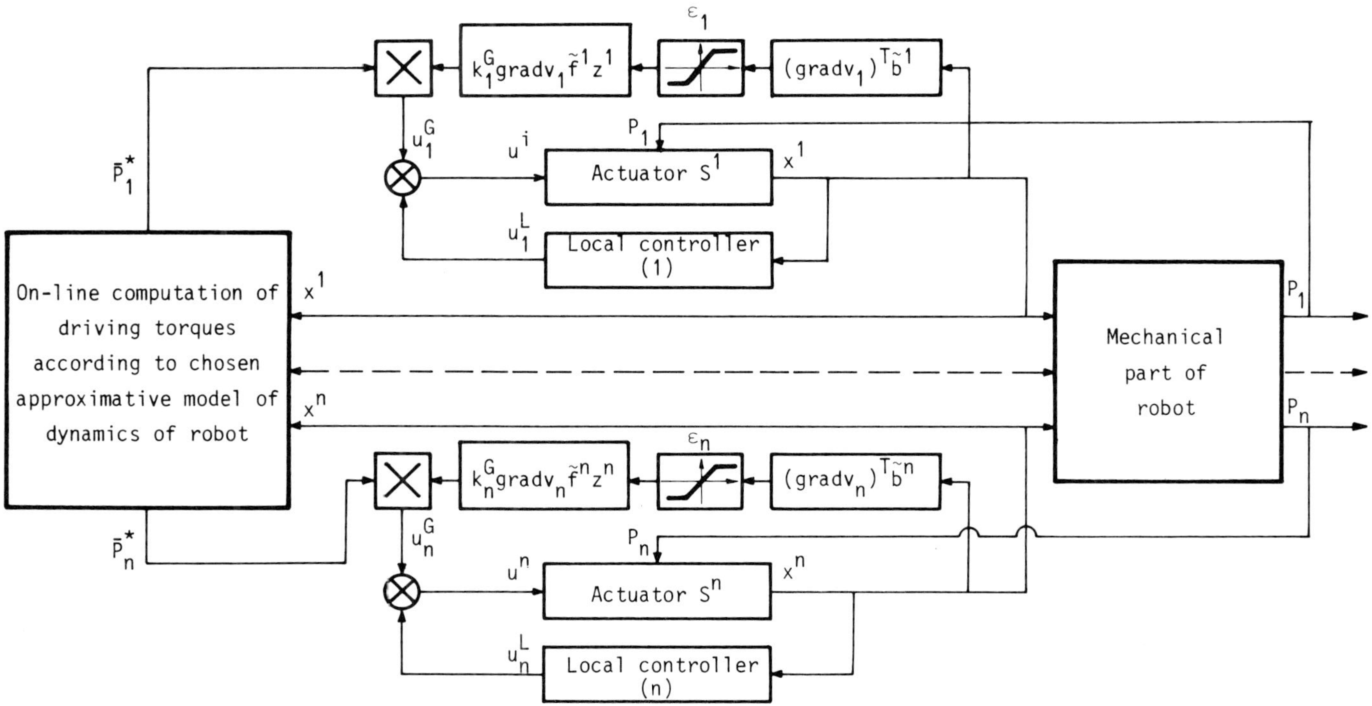

Fig. 2.23. Control scheme with global control implemented by on-line computation of the robot dynamics

where $C^i(q, d^*)$ denotes the n×n "matrix of centrifugal forces".

Various combinations of the above mentioned approximate models might be applied. For example, we might include diagonal inertia elements $H_{ii}(q, d^*)$ and centrifugal forces $\dot{q}^T C^i(q, d^*)\dot{q}$ in the computed model.

Evidently, the more complex the model applied, the more computational efforts we have to make (i.e. a larger number of multiplications and additions have to be performed in each sampling period). In Chapter 5 we shall discuss in detail the complexity of various approximate models of the robot dynamics, and present the number of multiplications and additions that are required to compute various approximate models for several types of manipulation robots. Evidently, more complex models require more complex and expensive multiprocessors to achieve a sampling period compatible with the robot dynamics. On the other hand, the application of more complex models improves robot performances. If the computer function $\bar{P}_i^*$ includes a more adequate model of the robot, global control (2.6.4) can better compensate for the coupling between the subsystems, thus providing a better tracking of the nominal trajectory (since the practical stability of the robot will be ensured if the influence of the coupling is reduced).

A trade-off between the complexity of the control equipment required and the quality of tracking should be made [10]. Actually, the robot designer's aim is to synthesize as simple a control law as possible (in order to minimize computation efforts) ensuring, at the same time, satisfaction of the imposed control task. To solve this problem we shall establish an algorithm for iterative choice of the simplest approximate dynamic model of the robot dynamics which can satisfy the imposed control task (2.2.18), if we use it to compute the global control. This algorithm will be presented in the text to follow.

We should note that the problem of variable parameters has not been solved by the above listed approximate models. Namely, we can compute these models for particular values of parameters $d^* \in D$, but the parameters may vary and, since we assume that their variations are not known in advance, we cannot compensate for these variations. However, we can still partially reduce the influence of the robot dynamics upon its performance by computing on-line approximate models even with constant parameters d^*. This will be explained in the text to follow.

(3) The third possible implementation of the global control which we shall consider here is called "robust global control" [72]. This global control has the following form:

$$\Delta u_i^G = -k_i^G[(\text{grad}v_i)^T\tilde{b}^i]^{-1}\cdot\bar{u}_m^i \tag{2.6.9}$$

where $\bar{u}_m^i$ is some constant (maximum) value, which has to be chosen in the process of control synthesis. Actually, the value $\bar{u}_m^i$ might be chosen to satisfy.

$$\bar{u}_m^i \geqslant \max|[(\text{grad}v_i)^T\tilde{f}^i(\theta^i)z^i(q^i)\bar{P}_i^*(t, x, d)]| \tag{2.6.10}$$

where the maximum is taken over region $\forall x\in X^t(t)$, $\forall t\in T$, $\forall d\in D$, $\forall\theta\in\Theta$. However, since the actuators' inputs are constrained by (2.2.4), the global control is also constrained by amplitude, so we cannot always reach the condition (2.6.10). The control scheme with the global control (2.6.9) is presented in Fig. 2.24.

The main advantage of this form of global control is its extreme simplicity and the fact that it does not depend on mechanism parameters. However, since this global control actually has the form of "bang-bang" control law, it suffers from all the disadvantages of "bang-bang" control:

(a) chattering of control signals which might produce oscillatory motion of the robot and its unstability (due to the presence of elastic modes of the robot which are not included in its model (2.2.1));

(b) high suboptimality of such control since it permanently sends the maximum signal regardless of the value of the coupling between the subsystems which may be less;

(c) the energy consumption required by this control law might be very high.

To prevent chattering of the control around the nominal trajectory, the control (2.6.9) is modified in [72]:

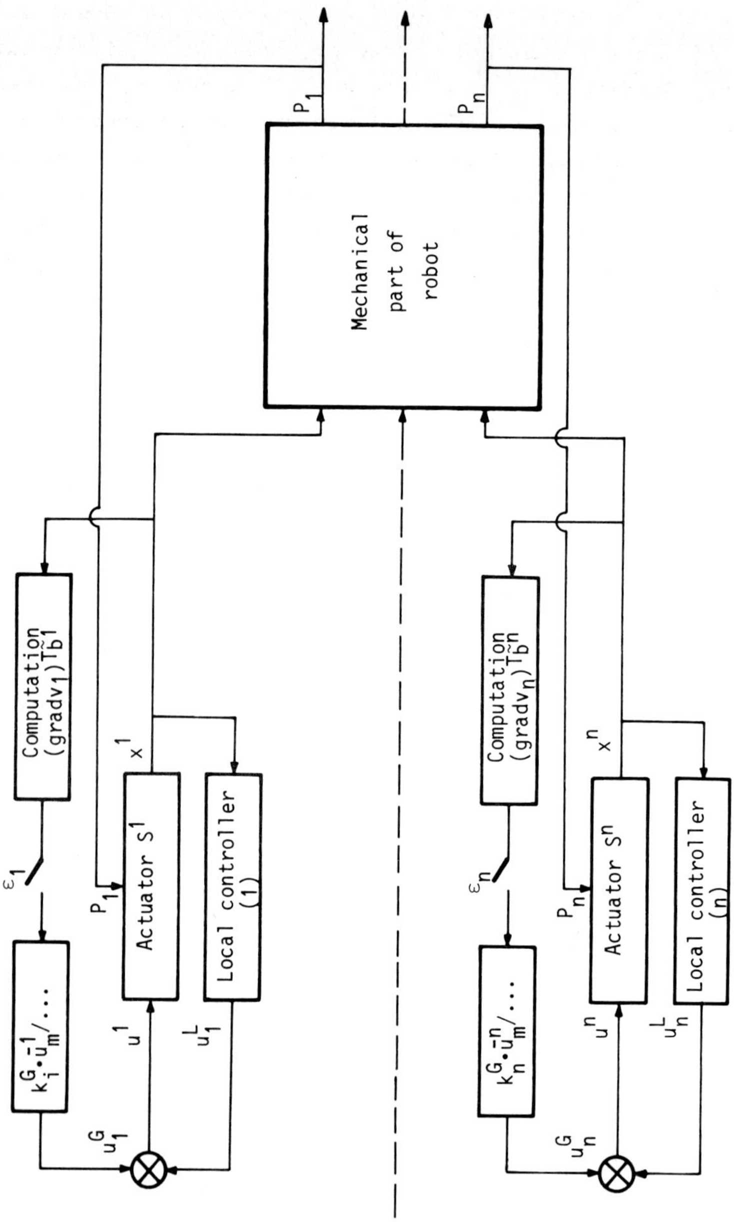

Fig. 2.24. Control scheme with "robust global control"

$$\Delta u_i^G = \begin{cases} -k_i^G[(\text{grad}v_i)^T\tilde{b}^i]^{-1}\bar{u}_m^i & \text{for} \quad (\text{grad}v_i)^T\tilde{b}^i > \varepsilon_i \\ 0 & \text{for} \quad (\text{grad}v_i)^T\tilde{b}^i < \varepsilon_i \end{cases} \tag{2.6.11}$$

where $\varepsilon_i > 0$ is an arbitrary number which should be chosen in the process of control synthesis. However, the choice of the number ε_i is very difficult; if we choose a small number ε_i, chattering of the control signal might still appear; on the other hand, if ε_i is large, it is questionable whether this control law can stabilize the robot and ensure tracking of the nominal trajectory (since the coupling between subsystems is not compensated for by the global control (2.6.11) in the region $X^\varepsilon(t)$, $\forall t \in T$). Thus, this control law, although very simple and robust to parameter variations, may result in poor tracking of nominal trajectory*).

Having presented three possible implementations of global control, we shall now describe how we can analyze the stability of the robot with various forms of global control.

Let us consider the stability of the nonlinear model of the robot (2.2.2) if the control in the form:

$$u^i = \bar{u}^{oi}(t) + \Delta u_i^L(\Delta x^i) + \Delta u_i^G(\Delta x^i, \bar{P}_i^*(d^*, t, x)) \tag{2.6.12}$$

is applied. If the local control is taken as (2.4.19) the nonlinear model with the control (2.6.12) can be written as:

$$\Delta\dot{x}^i = (\tilde{A}^i - \lambda_{ig}\tilde{b}^i r_i^{-1}\tilde{b}^i K_i)\Delta x^i + \tilde{b}^i N_G(t, \Delta x^i, \Delta u_i^G) + $$
$$+ \tilde{f}^i(\theta^i)\cdot z^i(q_i)\cdot \bar{P}_i(t, x, d) \tag{2.6.13}$$

where $N_G(t, \Delta x^i, \Delta u_i^G)$ is the constraint upon the global control amplitude:

$$N_G(t, \Delta x^i, \Delta u_i^G) = \begin{cases} u_m^{iG}(t, x^i) & \text{if} \quad \Delta u_i^G \geq u_m^{iG} \\ \Delta u_i^G & \text{if} \quad -u_m^{iG} < \Delta u_i^G < u_m^{iG} \\ -u_m^{iG} & \text{if} \quad \Delta u_i^G \leq -u_m^{iG} \end{cases} \tag{2.6.14}$$

---

*) Actually, the control (2.6.11) does not compensate for the influence of the robot dynamics along nominal trajectory, so it cannot ensure the practical stability of the robot.

with $u_m^{iG}(t, \Delta x^i)$ standing for $u_m^{iG} = \pm u_m^i - \bar{u}^{oi}(t) - \Delta u_i^L(\Delta x^i)$.

To analyze the stability of the robot, let us consider the derivative of function $v_i(t, \Delta x^i)$ (introduced in Paragraph 2.4.4) with respect to the solution of model (2.6.13). As we have written in (2.6.2), the time derivative of function $v_i(t, x^i)$ is given by:

$$\dot{v}_{i(\text{along }(2.6.13))} = \dot{v}_{i(\text{along }(2.5.2))} + \\ + (\text{grad}v_i)^T \tilde{b}^i N_G(t, \Delta x^i, \Delta u_i^G) + \\ + (\text{grad}v_i)^T \tilde{f}(\theta^i)\cdot z^i(q_i)\cdot \bar{P}_i(t, d, x) \quad (2.6.15)$$

Analogously to (2.5.9), we shall divide the function $\bar{P}_i^*(t, d, x)$ into two parts*):

$$\bar{P}_i^*(t, d^*, x) = \bar{P}_i^{o*}(t, d^*, x^o(t)) + \Delta\bar{P}_i^*(t, d^*, \Delta x) \quad (2.6.16)$$

where $\bar{P}_i^{o*}$ denotes the value of function $\bar{P}_i^*$ for nominal coordinates $x^o(t)$, $\forall t\in T$, and $\Delta\bar{P}_i^*$ the difference between $\bar{P}_i^*$ (for actual state coordinates x) and $\bar{P}_i^{o*}$ (for nominal state coordinates $x^o(t)$). Let us substitute (2.6.16) in (2.6.4), and consider coupling term and global control term in (2.6.15):

$$(\text{grad}v_i)^T \tilde{b}^i N_G(t, \Delta x^i, \Delta u_i^G) + (\text{grad}v_i)^T \tilde{f}^i(\theta^i) z^i(q_i) \bar{P}_i(t, d, x) =$$

$$= \begin{cases} (\text{grad}v_i)^T \tilde{f}^i(\theta^i) z^i(q^{oi})\cdot(\bar{P}_i^o - k_i^G \bar{P}_i^{o*}) + (\text{grad}v_i)^T \tilde{f}^i(\theta^i)\cdot[z^i(q_i)\cdot \\ (\Delta\bar{P}_i - k_i^G \Delta\bar{P}_i^*) + \Delta z^i(q_i)(\bar{P}_i^o - k_i^G \bar{P}_i^{o*})] \text{ for } x^i \in X^{t(i)}(t) \text{ .AND. } x^i \notin X^{\varepsilon(i)}(t) \\ (\text{grad}v_i)^T f^i(\theta^i) z^i(q^{oi})[P_i^o - k_i^G(\text{grad}v_i)^T \tilde{b}^i\cdot \bar{P}_i^{o*}/\varepsilon_i] + \\ + (\text{grad}v_i)^T \tilde{f}(\theta^i)\cdot\{z^i(q_i)[\Delta P_i - k_i^G(\text{grad}v_i)^T \tilde{b}^i \Delta P_i^*/\varepsilon_i] + \\ + \Delta z^i(q_i)[\bar{P}_i^o - k_i^G(\text{grad}v_i)^T \tilde{b}^i\cdot \bar{P}_i^{o*}/\varepsilon_i]\} \text{ for } x^i \in X^{\varepsilon(i)}(t) \end{cases} \quad (2.6.17)$$

where we have neglected (for simplicity) the amplitude constraint upon global control (2.6.14). Now, analogously to what we have done in Paragraph 2.5, we can consider separately two parts of coupling: "nominal

---

*) Obviously this consideration does not hold for global control (2.6.9). It is very simple to establish stability test for this global control.

coupling" and the deviation of coupling from the nominal value. This second part of coupling together with the corresponding part of global control can be estimated by:

$$\sum_{j=1}^{n} \xi_{ij}^{*} ||\Delta x^{j}|| \geq \begin{cases} (\text{grad}v_i)^T \tilde{f}^i(\theta^i)[z^i(q_i)(\Delta P_i - k_i^G \Delta P_i^*) + \\ +\Delta z^i(q_i)(\bar{P}_i^o - k_i^G \bar{P}_i^{o*})] \text{ for } x \in X^t(t) \text{.AND.} x^i \notin X^{\varepsilon(i)}(t) \\ (\text{grad}v_i) f^i(\theta^i)\{z^i(q_i)[\Delta \bar{P}_i - k_i^G (\text{grad}v_i)^T \tilde{b}^i \Delta P_i^* / \varepsilon_i] + \\ +\Delta z^i(q_i)[\bar{P}_i^* - k_i^G (\text{grad}v_i)^T \tilde{b}^i \bar{P}_i^{o*} / \varepsilon_i]\} \\ \quad \text{for } x \in X^t(t) \text{.AND.} x^i \in X^{\varepsilon(i)}(t) \end{cases} \tag{2.6.18}$$

where $\xi_{ij}^* \geq 0$ are constant numbers. The relation should hold for $\forall d \in D$ and $\forall t \in T$. It should be noted that it is always possible to determine numbers ξ_{ij}^* satisfying (2.6.18) since ΔP_i^* also satisfies (analogously to (2.5.12)):

$$\lim_{\Delta x^j \to 0, \forall j \in I} \Delta P_i^*(t, \Delta x, d^*) \to 0, \quad \forall d^* \in D, \ \forall t \in T \tag{2.6.19}$$

In (2.6.18) we have not included the nonlinear constraint upon global control (2.6.14), but, this constraint has to be taken into account when numbers ξ_{ij}^* are determined.

Considering the "nominal coupling" and its corresponding part of the global control, analogously to (2.5.18) we may write:

$$\rho_i^*(t, d) \geq \begin{cases} \int_o^t z^i(q^{oi})[\bar{P}_i^o(t', d) - k_i^G \bar{P}_i^{o*}(t', d)]dt', \\ \int_o^t z^i(q^{oi})[\bar{P}_i^o(t', d) - k_i^G \min\limits_{x^i \in X^{\varepsilon(i)}(t)} (\text{grad}v_i) ||\tilde{b}^i|| \cdot \bar{P}_i^{o*}(t', d^*)/\varepsilon_i]dt' \end{cases} \tag{2.6.20}$$

We have to determine the function $\rho_i^*(t, d)$ so that both inequalities (2.6.20) are satisfied.

Now, if we substitute (2.6.17) and (2.6.18) into (2.6.15) we get:

$$\dot{v}_{i(\text{along } (2.6.13))} \leq \dot{v}_{i(\text{along } (2.5.2))} + \sum_{j=1}^{n} \xi_{ij}^* ||\Delta x^j|| + \\ + (\text{grad}v_i)^T \tilde{f}^i(\theta^i) z^i(q^{oi}) \cdot (\bar{P}^o - k_i^G \cdot \bar{P}_i^{o*}) \tag{2.6.21}$$

If we consider the state of the robot satisfying $x \in \partial X^t(t)$, and take function $v_i(t, x^i)$ in the form (2.5.43), (2.5.47) we get:

$$\dot{v}_{i(\text{along }(2.6.13))} < -\gamma_{ig}\lambda_m^{1/2}(H_i)\bar{X}^{I(i)}\exp(-\alpha^{(i)}t) + \sum_{j=1}^{n}\xi_{ij}^{*}\bar{X}^{t(j)}$$

$$\cdot\exp(-\alpha^{(j)}t) + (\text{grad}v_i)^T\tilde{f}^i(\theta^i)z^i(q^{oi})(\bar{P}^o - k_i^G\cdot\bar{P}_i^{o*}) = \psi_i'(t, d),$$

$$\forall x \in \tilde{X}^t(t),\ \forall t\in T,\ \forall d\in D \qquad (2.6.22)$$

To test the practical stability of the robot, we have to test condition (2.5.16) for all subsystems. Thus, taking into account relation (2.6.20), we get the stability test in the form:

$$-\gamma_{ig}\lambda_m^{1/2}(H_i)\bar{X}^{I(i)}(1-\exp(-\alpha^{(i)}t))/\alpha^{(i)} + \sum_{j=1}^{n}\xi_{ij}^{*}\cdot\frac{\bar{X}^{t(j)}}{\alpha^{(j)}}[1-\exp(-\alpha^{(j)})] +$$

$$+\frac{\lambda_M(H_i)}{\lambda_m^{1/2}(H_i)}||\tilde{f}^i(\theta^i)||\rho_i^*(t,d) < \lambda_m^{1/2}(H_i)\bar{X}^{t(i)}\exp(-\alpha^{(i)}t) -$$

$$-\lambda_M^{1/2}(H_i)\bar{X}^{I(i)},\quad \forall t\in T,\ \forall i\in I,\ \forall d\in D \qquad (2.6.23)$$

If conditions (2.6.23) are satisfied, we can guarantee that the robot is practically stable with control (2.6.12) i.e. that it satisfies control task (2.2.18). Clearly, conditions (2.6.23) are sufficient but not necessary for practical stability. We may use tests (2.6.23) for the synthesis of global control which can stabilize the robot. Actually we can determine the simplest global control which should satisfy the stated control task. To achieve this, we shall establish an iterative algorithm for the synthesis of global control. Before establishing this algorithm we have to make several remarks concerning the global control and analysis of the practical stability of the robot if global control is applied.

(I) Tests (2.6.23 correspond to static local controllers (2.4.19). By analogy, we can write these conditions for dynamic local controllers (2.4.56), if we choose functions $v_i(t, \hat{x}^i)$ in form (2.4.57). Of course, control (2.6.1) must take into account the fact that the subsystems are extended and that functions v_i also include an additional state coordinate.

(II) In conditions (2.6.23) we have introduced numbers γ_{ig} which take into account amplitude constraints upon the actuators inputs. If conditions (2.4.67) are not satisfied for some of the subsystems, it means that a local control signal reaches the amplitude constraint in the region $X^t(t)$. This, in turn, means that $u_m^{iG}(t, \Delta x^i)$ in (2.6.14) may be zero, and this leads to conclusion that there is "no room left for global control signal". Nevertheless, even in this case application of global control may be effective. Since the coupling is a nonlinear function of robot state, it may happen that the maximum coupling is not achieved for $x \in \partial X^t(t)$ but for some other $x \in X^t(t)$, $\forall t \in T$. Thus, although the global control signal may be zero for $x^i \in \partial X^{t(i)}(t)$, it may compensate for the maximum coupling and, thus, reduce numbers ξ^*_{ij} in (2.6.18) in comparison to ξ_{ij} in (2.5.13). It must be remembered that in determining numbers ξ^*_{ij} in (2.6.18) and functions $\rho^*_i(t, d)$ in (2.6.20) we must take into account amplitude constraints upon global control (2.6.14), i.e. for each $x \in X^t(t)$, $\forall t \in T$ we must test (2.6.14) and then determine the global control signal and, for this control signal, determine numbers ξ^*_{ij} and functions $\rho^*_i(t, d)$.

(III) The robustness of the above presented global control laws has to be carefully studied. The control law (1) with force feedback is obviously robust to parameter variations since it uses information on actual forces measured on the robot. However, problems arise when we want to apply global control law (2) to a robot with variable parameters. As we have already noted, we may compute the driving torques using the approximate model of the robot dynamics for some particular values of parameters $d^* \in D$. Evidently, if the parameters of the robot differ from d^*, the global control cannot compensate for influence of the corresponding forces (which are taken into account in the robot model). However, we may still expect that such global control can reduce the influence of coupling for all allowable parameter variations. For example, let us assume that a parameter d_1 varies between d_{1min} and d_{1max} and that the cross-inertia element H_{ij} linearly depends on d_1 (H_{ij} = $k'd_1+k''$). If we compute this factor for the parameter value d_{1min} ($H^*_{ij}=k'd_{1min}+k''$) and use the so-computed term for global control ($\Delta u^G_i \sim H^*_{ij}\ddot{q}_j$), then we may reduce the influence of this term (it reduces to $[k'(d_1 - d_{1\,min})]$ for all allowable parameter values, although we compute it for only one particular value. It should be noted that the values of parameters should be chosen so as to minimize the numbers ξ^*_{ij} in (2.6.18) and functions ρ^*_i in (2.6.20). The algorithm which we shall present below iteratively searches for such d^* for which the global control can accommodate all parameter variations and ensure practical stability for $\forall d \in D$.

(IV) The form of global control (2.6.1) follows from equation (2.6.2), i.e. from the condition to reduce the influence of the coupling in the time derivative of functions $v_i(t, x^i)$ (or, in other words, to reduce numbers ξ^*_{ij}). This form actually takes into account the dynamics of the actuators through the factor $[(\text{grad}v_i)^T\tilde{b}^i]^{-1}(\text{grad}v_i)^T\cdot\tilde{f}^i(\theta^i)$. The computation of this terms is not complex but it requires $2n_i$ multiplications, $2(n_i-1)$ additions and one division. We may desire to use a simpler form of the global control and to neglect the dynamics of the actuators in the global control. If we consider the second-order model of actuator $n_i=2$ (see Appendix 2.A), then vectors $\tilde{b}^i$ and $\tilde{f}^i$ have forms $\tilde{b}^i=(0, \tilde{\tilde{b}}^i)^T$ and $\tilde{f}^i=(0,\tilde{\tilde{f}}^i)^T$. If functions $v_i(t, x^i)$ are of form (2.4.43), then $\text{grad}v_i=(H^i_{11}x_1+H^i_{12}x_2, H^i_{21}x_1+H^i_{22}x_2)^T/2v_i$ where H^i_{kj} are the elements of the matrix H^i. In this case the global control (2.6.4) reduces to:

$$\Delta u^G_i = -k^G_i\tilde{\tilde{f}}^i(\theta^i)/\tilde{\tilde{b}}^i\cdot z^i(q_i)\bar{P}^*_i(t, x, d^*) \qquad (2.6.24)$$

and there is no need to introduce the region $X^\varepsilon(t)$, $\forall t\in T$. Obviously, control (2.6.24) is much simpler than (2.6.4). Actually, in (2.6.24) we have neglected the delays in actuators which always exist (although they may be negligible if we consider relatively slow motions - see Appendix 2.A).

A compromise between control laws (2.6.4) and (2.6.24) has been suggested in [2] as:

$$\Delta u^G_i = k^G_i\tilde{\tilde{f}}^i(\theta^i)/\tilde{\tilde{b}}^i \ \text{sgn}(\Delta u^L_i(\Delta x^i))\cdot z^i(q_i)\cdot\bar{P}^*_i(t, x, d^*) \qquad (2.6.25)$$

where $\text{sgn}(\Delta u^L_i(\Delta x^i))$ denotes the sign of the corresponding value, i.e.

$$\text{sgn}(\Delta u^L_i(\Delta x^i)) = \begin{cases} +1 & \text{if} \quad \Delta u^L_i(\Delta x^i)\geqslant 0 \\ -1 & \text{if} \quad \Delta u^L_i(\Delta x^i)<0 \end{cases}$$

In [2] we have explained why we have chosen the global control of form (2.6.25).

Obviously, both forms of global control, (2.6.24) and (2.6.25), can be tested for practical stability using the same procedure as we have described above for control (2.6.4).

(V) We can estimate the suboptimality of control (2.6.12) with respect to a chosen criterion, as we have done in [2] for the case when nominal programmed control, synthesized with the complete centralized model of the robot, is applied. However, this is out of the scope of this book, but it is evident that the introduction of global control may reduce the suboptimality of the control compared with the suboptimality of the local controllers.

We shall now briefly establish the algorithm for synthesizing the global control which stabilizes the robotic system around the given nominal trajectory for all allowable parameters of the mechanism and actuators. The algorithm is intended to determine the simplest global control which satisfies stability test (2.6.23). Actually, we have to choose the form of global control; the form of approximate model $\bar{P}_i^*(t, x, d^*)$ (or to choose control laws (1) or (3)) and to determine the global gains k_i^G and numbers ε_i. A brief flow-chart of the algorithm is presented in Fig. 2.25. It includes the algorithm for the synthesis of decentralized controller presented in Fig. 2.21. Actually, this algorithm represents one step in the algorithm for global control synthesis.

The algorithm can be performed according to the following steps:

(1) The synthesis of decentralized control is performed acc. to the algorithm presented in Fig. 2.21. If it is possible to determine a unique decentralized controller which satisfies control task (2.2.18) and if the obtained local feedback gains are not too high, there is no need to synthesize global control. If it is impossible to synthesize adequate local control we should introduce global control. Let us assume that for the jth subsystem (joint) it is impossible to find adequate local controller, so we want to introduce the global control for this subsystem, while for all subsystems $i<j$ we have found local controller which satisfy practical stability of the system.

(2) We have to decide which form of the global control we want to implement. If the algorithm should make this decision automatically, then we may start with the simplest form of global control (2.6.24). If this form has already been tested, the algorithm should choose form (2.6.25). If this form has already tested, the algorithm should choose the most complex form (2.6.4).

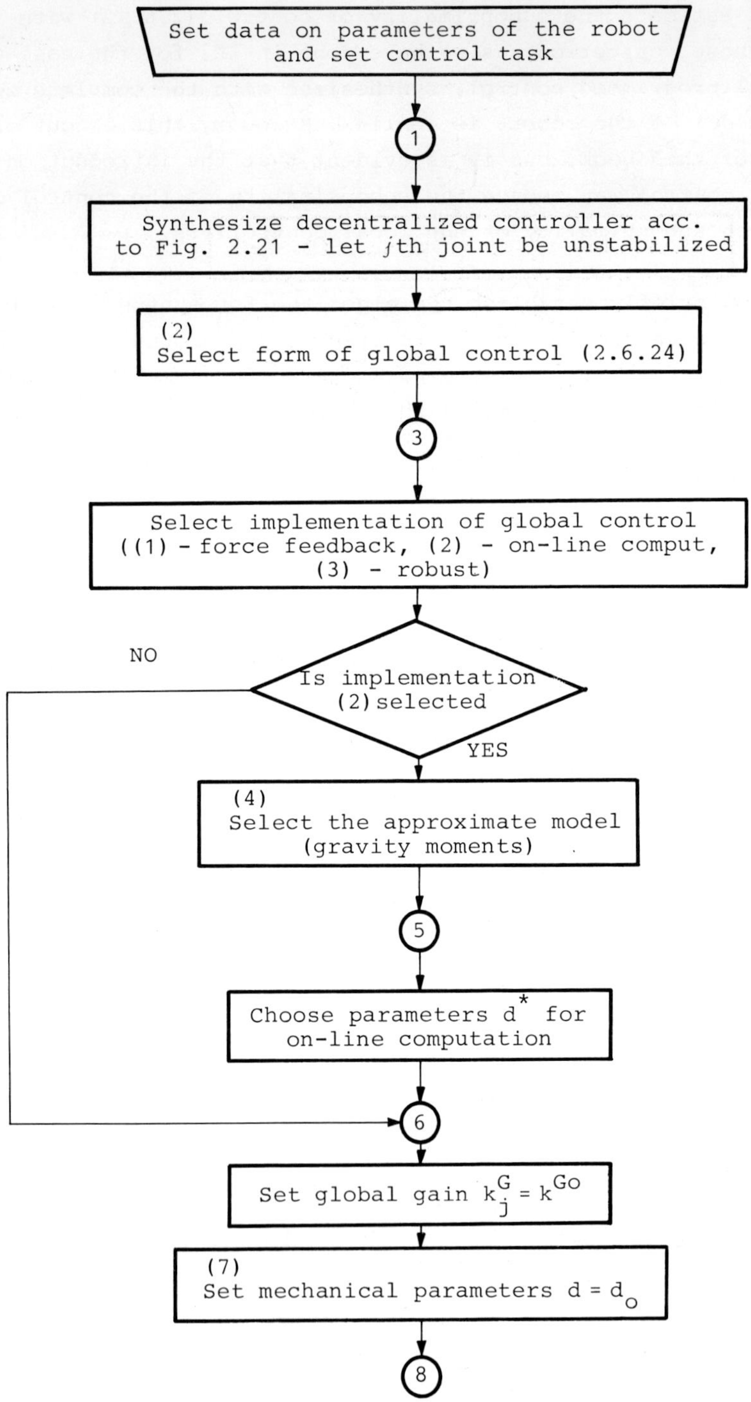

Fig. 2.25. Flow-chart of the algorithm for global control synthesis

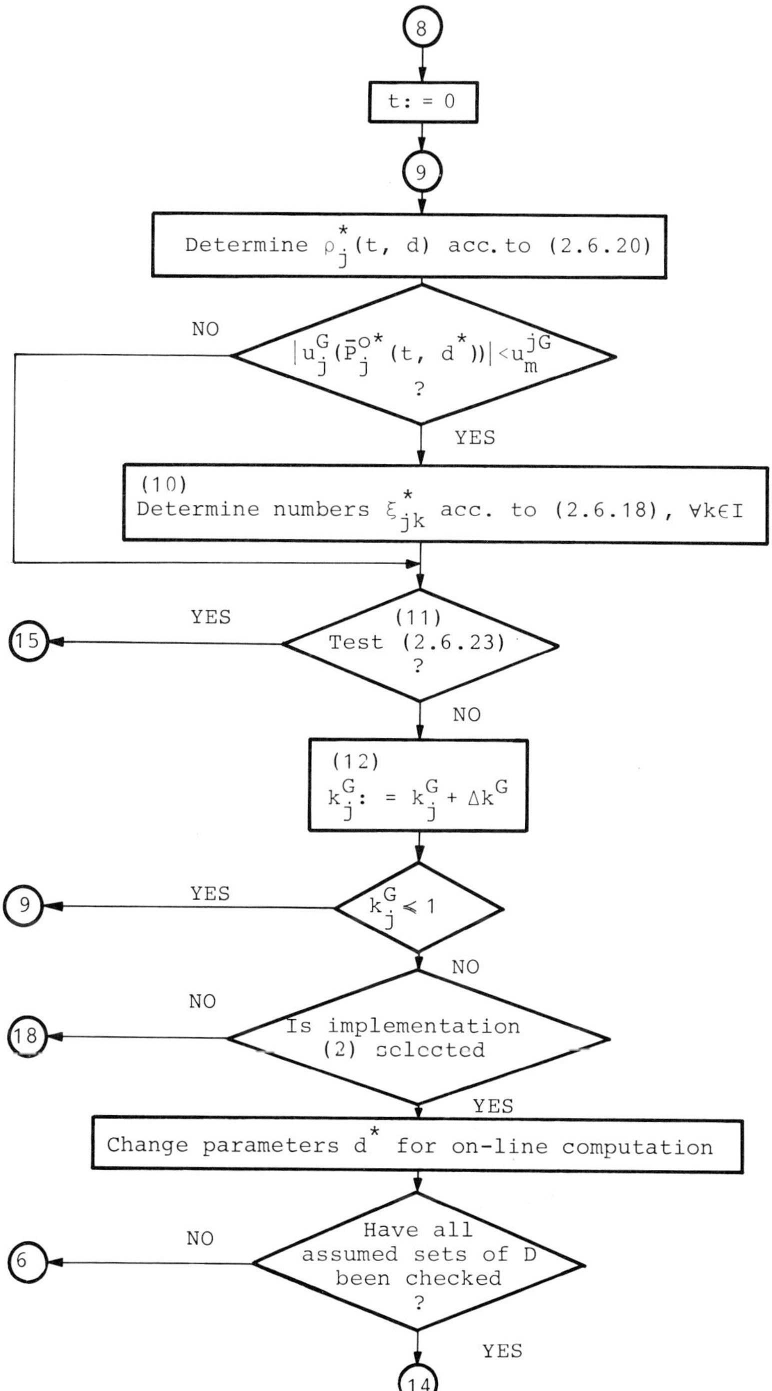
8
t: = 0
9
Determine $\rho_j^*(t, d)$ acc. to (2.6.20)
NO
$|u_j^G(\bar{P}_j^{o*}(t, d^*))| < u_m^{jG}$
?
YES
(10)
Determine numbers ξ_{jk}^* acc. to (2.6.18), $\forall k \in I$
YES
(11)
Test (2.6.23)
?
15
NO
(12)
$k_j^G: = k_j^G + \Delta k^G$
YES
$k_j^G \leqslant 1$
9
NO
NO
Is implementation
(2) selected
18
YES
Change parameters d^* for on-line computation
NO
Have all
assumed sets of D
been checked
?
6
YES
14

Fig. 2.25. (Cont.)

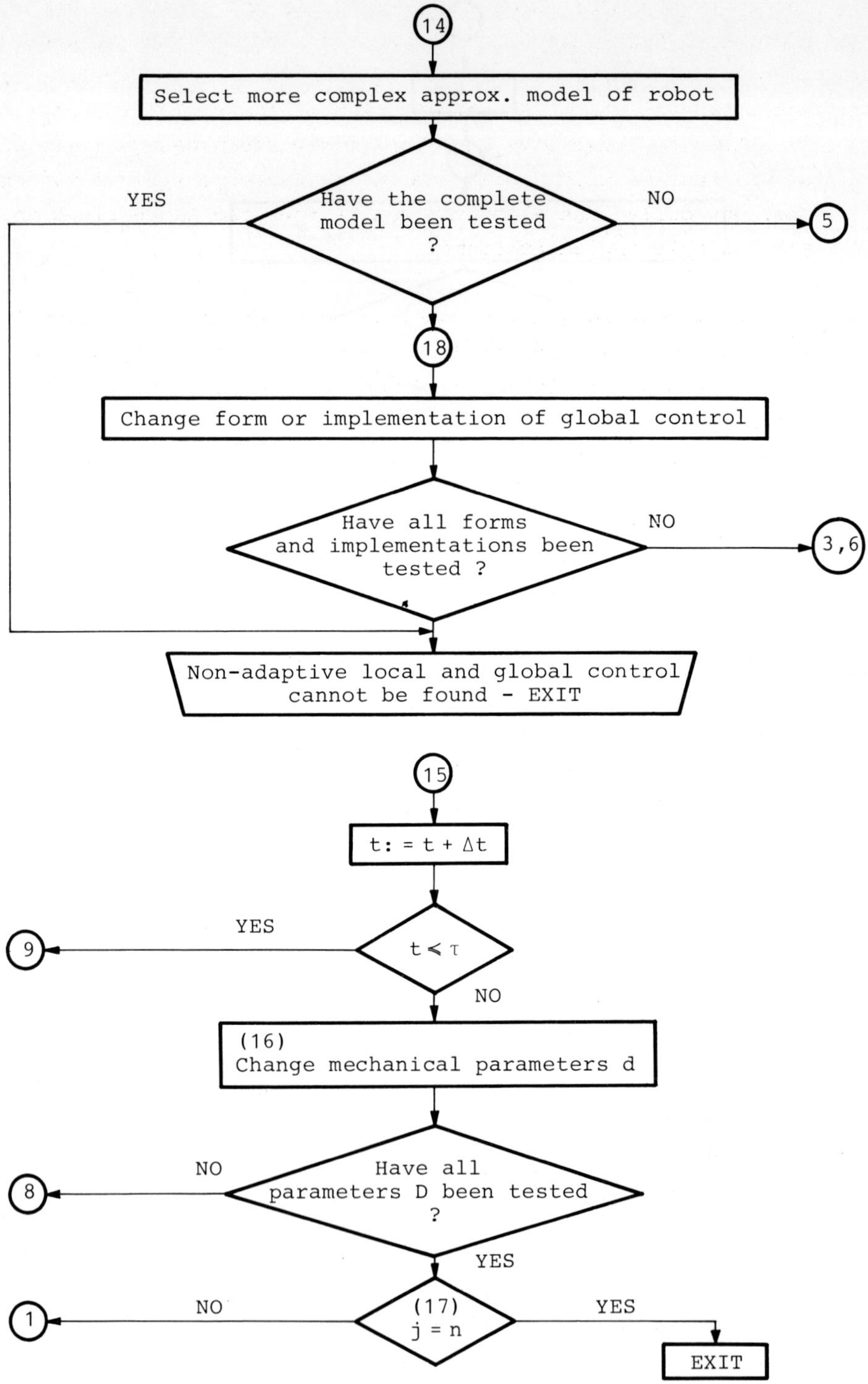
14
Select more complex approx. model of robot
YES
Have the complete model been tested ?
NO
5
18
Change form or implementation of global control
Have all forms and implementations been tested ?
NO
3,6
Non-adaptive local and global control cannot be found - EXIT
15
t: = t + Δt
YES
9
t ≤ τ
NO
(16) Change mechanical parameters d
NO
8
Have all parameters D been tested ?
YES
NO
1
(17) j = n
YES
EXIT

Fig. 2.25. (Cont.)

(3) Choose the form of function $\bar{P}_j^*$: actually the algorithm has to decide which among the three above presented implementations of the global control will be chosen. In principle, the algorithm can automatically check all three implementations, but in practice it is better to leave this decision to the robot designer. We shall assume that we have chosen one of the three possible implementations. Obviously, the algorithm can be repeated three times to check all possible implementations.

(4) If implementation (2) with on-line computation of the coupling $\bar{P}_i^*(x, d^*)$ has been selected, the algorithm has to choose the approximate model of the robot dynamics which will be used to compute global control. The algorithm starts with the simplest approximate model (2.6.5) which includes just gravitational moments.

(5) Choice of the values of parameters $d^* \in D$ has to be made. For these values, the approximate model $\bar{P}_j^*(x, d^*)$ will be computed in the global control (if type (2) implementation has been selected).

(6) The global gain k_j^G is set to an initial value k_j^{Go}.

(7) From this step we start the stability analysis of the nonlinear model of robot with a decentralized controller and global control. Set the initial values of the parameters of the mechanical part of robot $d=d_o \in D$.

(8) Set the initial time instant (t=0.).

(9) Determine the number $\rho_j^*(t, d)$ which satisfies (2.6.20). In doing this, it should be checked whether the global control with "nominal coupling" $\bar{P}_j^{o*}(t, d^*)$ exceeds the constraint upon the amplitude of global control (2.6.14). If yes, it means that just "nominal coupling" exhausts the allowable amplitude of global control signal and then there is no need to determine the numbers ξ_{jk}^* (but ξ_{jk} may be needed for the stability test).

(10) Determine numbers ξ_{jk}^* that satisfy inequality (2.6.18) for the jth subsystem and for all $k \in I$. We must again take care about the constraint upon the global control amplitude (2.6.14), and determine the number ε_j and region $X^{\varepsilon(j)}$ as defined by (2.6.3), in order to ensure that the global control signal (2.6.4) is beyond the amplitude constraint (2.6.14).

(11) Test the condition (2.6.23) for the jth joint. If the condition is not satisfied we have to change the global control by changing either global gain k_i^G or function $\bar{P}_j^*$. So, if the condition (2.6.23) is not met, the algorithm goes to step (12). If the condition (2.6.23) is satisfied, go to step (15).

(12) Increase the global gain k_j^G by a chosen increment Δk^G. Test if $k_j^G<1$; if yes, go back to step (9); if not, we should not increase global gain k_j^G over 1; thus, the algorithm passes to the next step.

(13) Change parameter values d^* for which the approximate model $\bar{P}_j^*(x, d^*)$ is computed. We should depict several sets of parameter values $d^* \in D$ for which we should try to compute efficient global control (since there is no sense in exploring an infinite number of possible parameter values in the allowable parameter region D). If all predicted sets of parameters have already been examined, the algorithm passes to the next step; if not, it goes back to step (6) (of course, this holds if type (2) implementation has been selected).

(14) Change the approximate model $P_j^*(x, d^*)$ for global control; the algorithm takes a more complex model than in the previous iteration according to the listed possible approximate models (2.6.5)-(2.6.8). If the algorithm has already examined all possible approximate models and tried to stabilize the robot even with the complete dynamic robot model and the stability test has not been satisfied, this means that we cannot determine unique non-adaptive local and global control which can stabilize the robot along the desired trajectory and for all allowable parameters of the mechanism. In this case we have to stop this algorithm and implement adaptive control (see Chapter 3). Obviously, if the choice of the global control form and implementation is left to the algorithm, it should select the next possible global control form or the next possible implementation and go back to step (3) or step (6). If the algorithm has not examined all assumed approximate models yet, it has to select the next more complex model and go back to step (5).

(15) Increase time instant by Δt ($t: = t+\Delta t$); test if $t<\tau$; if yes, go back to step (9); if $t>\tau$, go to the next step.

(16) Change the parameters of the mechanical part of robot $d \in D$; check

whether we have analyzed the practical stability of the nonlinear model of robot for all allowable values of parameters $d \in D$; if not, go back to step (8).

(17) Check if $j=n$; if yes, it means that the algorithm has already synthesized either local control or local and global control for all robot joints and, thus, the imposed control task (2.2.18) is solved; if $j<n$, the algorithm goes back to step (1) and tries to synthesize local controller for the next $(j+1)$th joint.

It should be noted that in this algorithm we have not considered the variation of actuator parameters θ. As we have already explained, this problem can be treated in two ways. The first is to try to synthesize unique decentralized and global controller which can withstand all allowable parameter variations. In this case we should introduce an additional loop to test stability of the robot with selected global control for various actuator parameters. The alternative approach is to synthesize a few decentralized and global controls for various actuator parameters and change control parameters whenever actuator parameters are changed. The latter approach is essentially adaptive control. However, since we have assumed that the variation of actuator parameters is slow, there is no need to implement such adaptive control on--line; it is sufficient to change control parameters in some time periods.

Although the above algorithm is prepared to be implemented on a digital computer and to be run fully automatically, it is obvious that the interaction with the robot designer might significantly speed up the solution (convergence of the algorithm). Actually, designers experience might be of great help especially in the synthesis of global control, since the designer might directly select the most appropriate approximate model knowing the robot performance. This interaction with the user will be discussed in Chapter 4.

2.7 Example

In order to illustrate the efficiency of the control synthesis presented, we shall consider an example of a particular industrial manipulator. Manipulator UMS-3B of semianthropomorphic type, with 6 d.o.f., is presented in Fig. 2.26. Data on nominal parameters (masses, moments of inertia, lengths of links) are given in Table 2.1. All manipulator joints (d.o.f.) are powered by hydraulic actuators (linear for the first three d.o.f. - "Knapp" - 2.9. 40125 - and vane actuators - "Knapp" ROTAC. D-10-250-1. for the three d.o.f. of gripper), the models of which are given by (2.2.2). The orders of subsystems S^i are $n_i = 3$ and subsystems state vectors are $x^i = (\ell^i, \dot{\ell}^i, p^i)$ where ℓ^i is the piston position, $\dot{\ell}^i$ is the piston speed, p^i is the pressure. The connections among the angles of manipulator q_i and the piston positions ℓ^i are linear $q_i = c_i \ell^i$ for all d.o.f. except for the i=2 where $\ell^2 = (a+b\cos(q_2+c))^{1/2}+d$. The elements of the actuator model matrices $A^i = [a_{ij}]$ and $b^i = [b_j]$ are given in Table 2.2.

The control task is imposed as follows: $L_1=3$ nominal trajectories $x_\ell^o(t)$ are given, for $\tau_1 = \tau_2 = 1$ s and $\tau_3 = 0.6$ s. Nominal trajectories are presented in Fig. 2.27. They correspond to manipulator motion during a particular task of transfer of various objects. We assume that all parameters of the manipulation system are constant except for the mass m_p and moment of inertia J_p of the objects which are transferred by the manipulator. Thus, the regions of the allowable values of variable parameters are given by: $D_1 = \{d: m_p=0, J_p = 0.\}$ $D_2 = \{d: m_p \in (0., 20 \text{ kg}), J_p \in (0., 0.006 \text{ kgm}^2)\}$, $D_3 = \{d: m_p = 10 \text{ kg}, J_p = 0.003 \text{ kgm}^2\}$. The parameters of practical stability regions are given by: $\bar{X}_\ell^{I(i)} = 0.01$, i=1, 2,3, $\bar{X}_\ell^{I(i)} = 0.1$, i=4,5,6, $\bar{X}_\ell^{t(1)} = 0.02$, $\bar{X}_\ell^{t(i)} = 0.04$, i=2,3, $\bar{X}_\ell^{t(i)} = 0.4$, i=4,5,6, $\alpha_\ell^{(i)} = 3.$, $\forall i \in I$, $\forall \ell \in L$. Nominal trajectories $x_\ell^o(t)$, $\forall t \in T$, are so imposed that $u_\ell^o(t)$ can be calculated which satisfy (2.4.14).

The synthesis of local controllers is performed by an iterative procedure set on a digital computer. We have synthesized the local optimal regulators. In this procedure the weighting matrices in (2.4.16) are varied and the conditions are tested for all imposed nominal trajectories and all allowed parameter values. The algorithm cannot find unique weighting matrices for which local controllers can accommodate all three nominal trajectories and all assumed parameter variations. Thus, we had to introduce global control in the form of force feedback (2.6.4). The algorithm iteratively increases global gains k_i^G, $\forall i \in I$ and tests the conditions of practical stability, in order to find out the smallest global gains for which the system S is practically stable round all $x_\ell^o(t)$ and for all $d \in D_\ell$, $\forall \ell \in L$. The synthesized local and global gains are presented in Table 2.3. In order to verify the results obtained we have simulated dynamics of the manipulator during tracking of all three nominal trajectories for the perturbations of the initial conditions type. We choose initial conditions on the bounds of the regions $X_\ell^{I(i)}$.

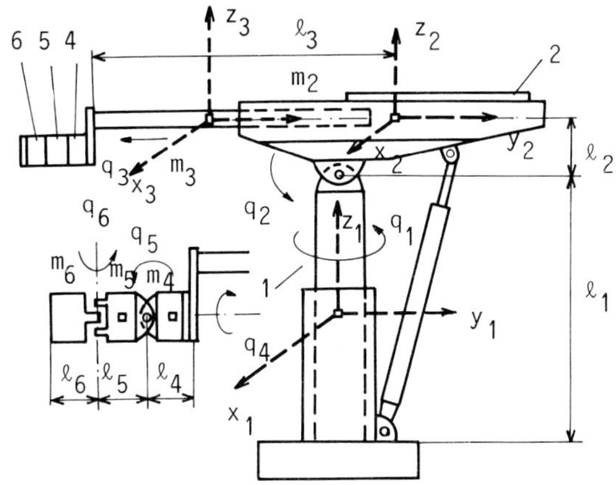

Fig. 2.26. Manipulator UMS-3B with 6 d.o.f.

Link	1	2	3	4, 5	6
Mass m_i (kg)	-	27.8	22.8	2.33	0.33
Length ℓ_i (m)	1.2	0.142	1.14	0.14	0.26
J_{ix} (kgm^2)	-	2.98	1.21	0.004	0.0007
J_{iy} (kgm^2)	-	-	-	0.004	0.0009
J_{iz} (kgm^2)	0.322	3.701	1.21	0.004	0.0009

Table 2.1. Parameters of the mechanism

i	a^i_{22}	a^i_{23}	a^i_{32}	a^i_{33}	b^i_3	f^i_2
1,2	-11.32	47.5	-1122.	-66.18	74.9	-1.885
3	-9.77	41.04	-850.	-49.6	56.1	-1.885
4,5,6	-166.	800.	-300.	-80.	800.	-330.

Table 2.2. Parameters of actuators

We simulate tracking by only local controllers and by local and global control. The results of simulations are presented in Fig. 2.28.

i	1	2	3	4	5	6
k_1^i [mA/m]	1707.	986.5	309.	71.5	73.8	71.2
k_2^i [mA/m/s]	196.2	113.3	5.	1.36	2.97	1.15
k_3^i [mA/bar]	0.46	0.42	0.4	0.33	0.29	0.35
k_i^G	0.10	0.2	0.2	0.0	0.2	0.1

Table 2.3. Local $k^i = (k_1^i, k_2^i, k_3^i)^T$, global k_i^G gains

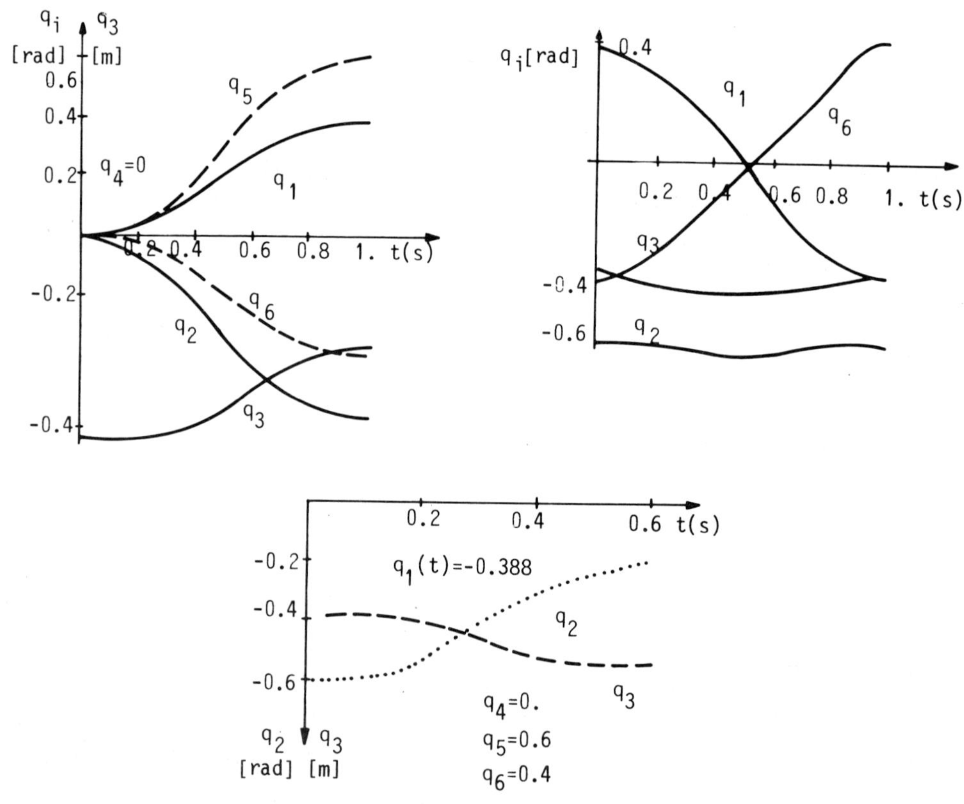

Fig. 2.27. Nominal trajectories

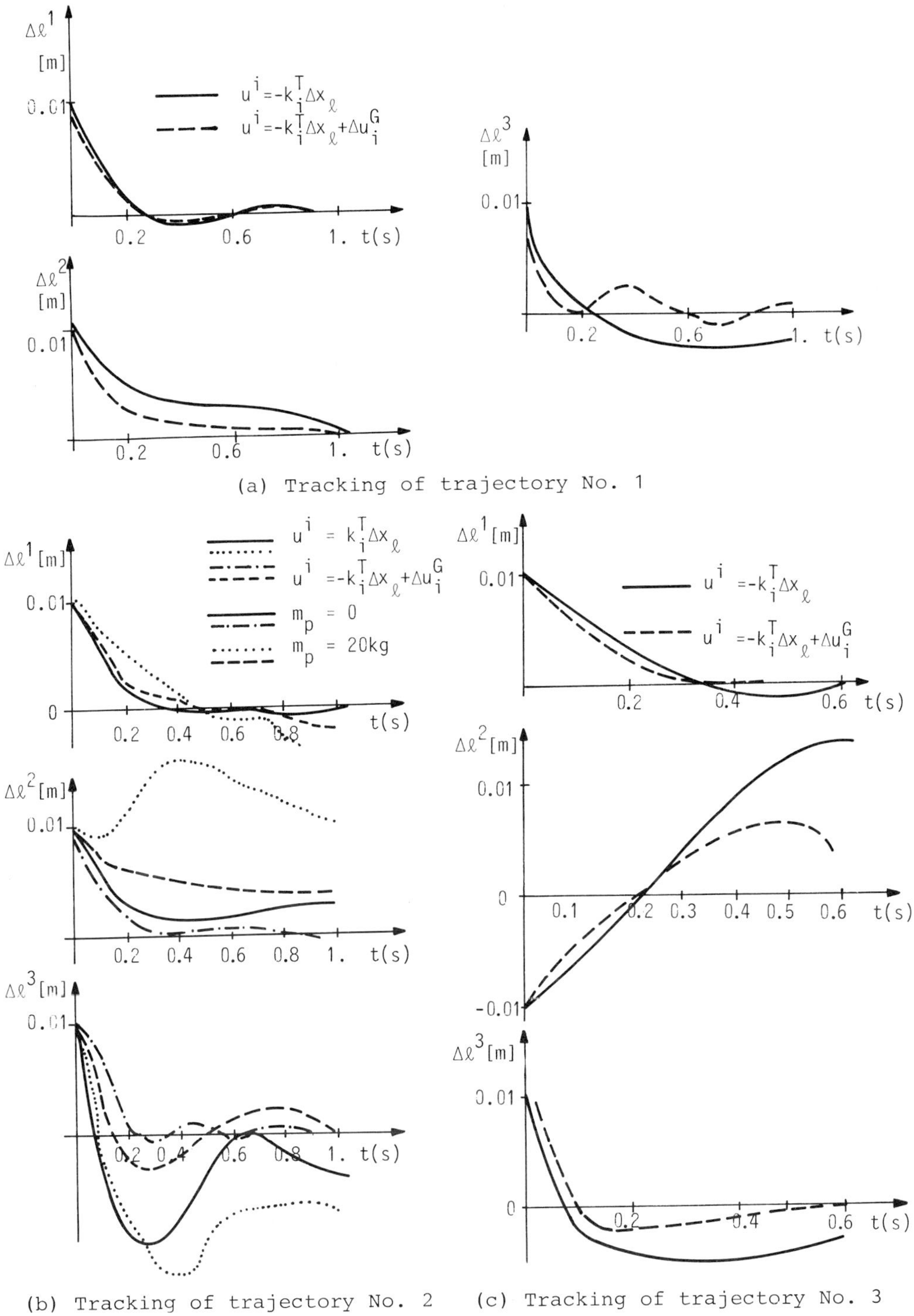

(a) Tracking of trajectory No. 1

(b) Tracking of trajectory No. 2

(c) Tracking of trajectory No. 3

Fig. 2.28. Tracking of nominal trajectories

Conclusion

In this chapter we have presented the synthesis of non-adaptive control for manipulation robots. We have restricted our attention to considering the executive control level of robotic system. Although we have given a brief survey of various approaches to control synthesis for manipulation robots, our prime aim has been to consider the synthesis of decentralized control for robotic systems. We have established the algorithm for iterative synthesis of decentralized control which can stabilize the robotic system with variable parameters. We have developed the procedure for testing the practical stability of the robotic system around a given nominal trajectory. Using this procedure we can verify whether the robotic system is practically stable with the chosen decentralized controller for all allowable variations of parameters.

We have described the synthesis of various static and dynamic decentralized controllers, and analyzed the robustness of these controllers to parameter variations. We have also proposed various global control laws which can be added to the decentralized controller in order to compensate for the influence of coupling between the joints of the robot. Thus, our intention has been to synthesize a unique robust controller which is capable of withstanding all nonlinearities and parameter variations. However, we have underlined that it is not always possible to find a convenient unique controller which is able to accommodate all requirements set in a particular control task. In that case it is necessary to implement adaptive control, which will be considered in the following chapter.

The presented algorithm for synthesis of decentralized and global control has served as a basis for developing a software package for computer-aided synthesis of control for manipulation robots. We shall describe this package in Chapter 4. We have presented a brief example of control synthesis for a particular robotic system. In Chapter 4 we shall give a more extensive example of control synthesis using the described algorithm.

It should be noted that we have considered a few particular forms of decentralized controller and global controller. However, the elaborated approach to control synthesis via analysis of the practical stability of the complete robotic system, can also be used for the synthesis of various other control laws (e.g., nonlinear decentralized control, etc.).

References

[1] Vukobratović K.M., Kirćanski V.M., Scientific Fundamentals of Robotics, 3, Kinematics and Trajectory Planning, Monograph, Springer-Verlag, Berlin, 1985.

[2] Vukobratović K.M., Stokić M.D., Scientific Fundamentals of Robotics, 2, Control of Manipulation Robots: Theory and Application, Monograph, Springer-Verlag, Berlin, 1982, also in Russian, Nauka, 1985.

[3] Vukobratović K.M., Potkonjak V., Scientific Fundamentals of Robotics, 1, Dynamics of Manipulation Robots: Theory and Application, Monograph, Springer-Verlag, Berlin, 1982.

[4] Vukobratović K.M., Legged Locomotion Robots and Anthropomorphic Mechanisms, Monograph, Institute Mihailo Pupin, Beograd, (in English), 1974, also publish by Mir, Moscow, (in Russian), 1976.

[5] Albus S.J., McLean R.C., Barbera J.A., Fitzgerald L.M., "Hierarchical Control for Robots in an Automated Factory", Proc. of the 13th ISIR, pp. 13-29 ÷ 13-43, Chicago, April, 1983.

[6] Luh Y.S.J., "An Anatomy of Industrial Robots and Their Controls", IEEE Trans. on Automatic Control, Vol. AC-28, No 2, pp. 133-152, February, 1983.

[7] Vukobratović K.M., Stokić M.D., "Simplified Control Procedure of Strongly Coupled Complex Nonlinear Mechanical Systems", (in Russian), Avtomatika and Telemechanika, No 11, also in English, Automatica and Remote Control, Vol. 39, No 11, 1978.

[8] Vukobratović K.M., Stokić M.D., "Contribution to the Decoupled Control of Large-Scale Mechanical Systems", Automatica, No 1, January, 1980.

[9] Vukobratović K.M., Stokić M.D., "One Engineering Concept of Dynamic Control of Manipulators", Journal of Dynamic Systems, Measurement and Control, Trans. of the ASME, Vol. 103, No 2, pp. 108--118, June 1981.

[10] Vukobratović K.M., Stokić M.D., "Is Dynamic Control Needed in Robotic Systems, and if so to What Extent?", International Journal of Robotic Research, Vol. 102, June, 1982.

[11] Kahn M.E. Roth B., "The Near Minimum Time Control of Open Loop Articulated Kinematic Chains", Trans. of the ASME, Journal of Dynamic Systems, Measurement and Control, September, pp. 164-172, 1971.

[12] Young D.K.K., "Control and Optimization of Robot Arm Trajectories", Proc. IEEE Milwaukee Symp. on Automatic Computation and Control, pp. 175-178, April, 1976.

[13] Vukobratović K.M., Kirćanski V.M., "A Method for Optimal Synthesis of Manipulation Robot Trajectories", Trans. of ASME, Journal of Dynamic Systems, Measurement, and Control, Vol. 104, No 2, pp. 188-193, 1982.

[14] Vukobratović K.M., Stokić M.D., "Engineering Approach to Dynamic Control of Industrial Manipulators, Part I: Synthesis of Dynamic Nominal Regimes", (in Russian), Mashinovedeniya AN USSR, No 3, 1981.

[15] Popov E.P., Vereschagin A.F., Ivkin A.M., Leskov A.S., Medvedov V.S., "Synthesis of Control System of Robots Using Dynamic Models of Manipulation Mechanisms", Proc. of VI IFAC Symp. on Automatic Control in Space, Erevan, USSR, 1974.

[16] Vukobratović K.M., Kirćanski M.N., "Computer-Oriented Method for Linearization of Dynamic Model of Active Spatial Mechanisms", Journal of Mechanism and Machine Theory, Vol. 17, No. 1, 1982.

[17] Popov E.P., Vereschagin A.F., Filaretov F.V. "Synthesis of Quasi-optimal Nonlinear Feedback Control System of Manipulator", (in Russian) Teknicheskaya kibernetika, No 6, pp. 91-101, 1976.

[18] Takegaki M., Arimoto S., "A New Feedback Method for Dynamic Control of Manipulators" Trans. of the ASME, Journal of Dynamic Systems, Measurement, and Control, June, pp. 113-125, Vol. 102, 1981.

[19] Paul R.C., Modeling, Trajectory Calculation and Servoing of a Computer Controlled Arm, A.I. Memo 177, Stanford Artificial Intelligence Laboratory, Stanford University, Sept. 1972, also in Russian, Nauka, Moscow, 1976.

[20] Bejczy K.A., Robot Arm Dynamics and Control, Technical Memorandum 33-669, Jet Propulsion Laboratory, February, 1974.

[21] Pavlov A.V., Timofeyev V.A., "Calculation and Stabilization of Programmed Motion of a Moving Robot-Manipulator", (in Russian), Teknicheskaya kibernetika, No 6, pp. 91-101, 1976.

[22] Vukobratović K.M., Kirćanski M.N., Scientific Fundamentals of Robotics, 4, Real-time Dynamics of Manipulation Robots, Monograph, Springer-Verlag, Berlin, 1984.

[23] Hollerbach M.J., "A Recursive Langrangian Formulation of Manipulator Dynamics and a Cooperative Study of Dynamics Formulation Complexity", IEEE Trans. on Systems, Man and Cybernetics, Vol. SMC-10, pp. 730-736, November, 1980.

[24] Bejczy K.A., Paul P.R., "Simplified Robot Arm Dynamics for Control", Proc. of IEEE Conf. on Automatic Control, pp. 261-262, 1981.

[25] Raibert H.M., Horn P.K.B., "Manipulator Control Using the Configuration Space Method", The Industrial Robot, Vol. 5, No 2, pp. 69-73, June, 1978.

[26] Timofeyev V.A., Ekalo V.Yu., "Stability and Stabilization of Programmed Motion of Robots-Manipulators", (in Russian), Avtomatika and Telemechanika, No 10, pp. 148-156, 1976.

[27] Saridis N.G., Lee G.S.C., "An Approximation Theory of Optimal Control for Trianable Manipulators", IEEE Trans. on Systems, Man, and Cybernetics, Vol. SMC-9, No 3, March, 1979.

[28] Vukobratović K.M., Stokić M.D., Hristić S.D., "A New Control Concept of Anthropomorphic Manipulators", Proc. of Second Conference of Remotely Manned Systems, Los Angeles, June, 1975.

[29] Vukobratović K.M., Hristić S.D., Stokić M.D., "Algorithmic Control of Anthropomorphic Manipulators", Proc. of V Inter. Symp. on Industrial Robots, Chicago, Illinois, September, 1975.

[30] Hewit R.J., Burdess S.J., "Fast Dynamic Decoupled Control for

Robotics, Using Active Force Control", Mechanism and Machine Theory, Vol. 16, No. 5, pp. 535-542, 1981.

[31] Wu H.C., Paul P.R., "Manipulator Compliance Based on Joint Torque Control", Proc. 19th IEEE Conf. Decision Control, Albuquerque, NM, Vol. 1, pp. 88-94, December, 1980.

[32] Luh Y.S.J., Fisher D.W., Paul C.P.R., "Joint Torque Control by a Direct Feedback for Industrial Robots", IEEE Trans. on Automatic Control, Vol. AC-28, No 2, February, 1983.

[33] Nevins L.J., Whithey E.D., "The Force Vector Assembler Concept", Proc. of the I International Conf. on Robots Manipulator Systems, Udine, Italy, September, 1973.

[34] Paul P.C.R., Shimano E.B., "Complience and Control" Proc. of Joint Automatic Control, San Francisko, CA, pp. 694-699, 1976.

[35] Raibert H.M., Craig J.J., "Hybrid Position/Force Control of Manipulators", Journal of Dynamic Systems, Measurement and Control, Trans. of the ASME, Vol. 103, No 2, pp. 126-133, 1981.

[36] Vukobratović K.M., Stokić M.D., "Dynamic Control of Manipulators via Load-Feedback", Journal of Mechanism and Machine Theory, Vol. 17, No 2, pp. 107-118, 1982.

[37] Popov E.P., Vereschagin F.A., Zenkevich S.L., Manipulation Robots: Dynamics and Algorithms, (in Russian) Series "Scientific Fundamentals of Robotics", Nauka, Moscow, 1978.

[38] Wu H.C., Paul P.R., "Resolved Motion Force Control of Robot Manipulator", IEEE Trans. on Systems, Man, and Cybernetics, Vol. SMC-12, No 3, May/June, 1982.

[39] Paul P.R., "The Mathematics of Computer Controlled Manipulator", Proc. of JACC, Vol. 1, pp. 124-131, 1977.

[40] Whitney E.D., "Resolved Motion Rate Control of Manipulators and Human Prostheses", IEEE Trans. on Man-Machine Systems, Vol. MMS-10, No 2, pp. 47-53, June, 1969.

[41] Whitney E.D., "The Mathematics of Coordinated Control of Prostheses Arms and Manipulators", Trans. of ASME, Journal on Dynamics Systems, Measurement, and Control, December, 1972.

[42] Luh Y.S.J. Walker W.M., Paul P.R., "Resolved-Acceleration Control of Mechanical Manipulators", IEEE Trans. on Automatic Control, Vol. AC-25, No. 3, pp. 468-474, June, 1980.

[43] Krutko P.D., Lakota N.A., "Construction of Algorithms for Control of Motion of Manipulation Robots on the Basis of Solution of Inverse Dynamic Task", (in Russian), Tehnicheskaya kibernetika, No 1 pp. 52-58, 1981.

[44] Yuan S-C.J., "Dynamic Decoupling of a Remote Manipulator System", IEEE Trans. on Automatic Control, Vol. AC-23, No 4, pp. 713-717, 1978.

[45] Roessler J., "A Decentralized Hierarchical Control Concept for Large-Scale Systems", Proc. of the II IFAC Symp. on Large-Scale Systems, pp. 171-179, Toulouse, 1980.

[46] Freund E., "Fast Nonlinear Control with Arbitrary Pole-Placement for Industrial Robots and Manipulators", Int. Journal Robotic Research, 1(1), pp. 65-78, 1982.

[47] Medvedov B.S., Leskov G.A., Juschenko S.A., Control Systems of Manipulation Robots, (in Russian), Nauka, Moscow, 1978.

[48] Paul P.R., Robot Manipulators: Mathematics, Programming, and Control, The MIT Press, Cambridge, 1981.

[49] Arimoto S., Miyazaki F., "Stability and Robustness of PID Feedback Control for Robot Manipulators and Sensory Capability", First Int. Symp. of Robotic Research, Bretton-Woods, New Hampshire, USA, 1983.

[50] Miyazaki F., Arimoto S., Takegaki M., Maeda Y., "Sensory Feedback Based on the Artificial Potential for Robot Manipulators", Preprints of 9th IFAC World Congress, Vol. 6, pp. 27-32, Budapest, July, 1984.

[51] Albus S.J., "A New Approach to Manipulator Control: The Cerebellar Model Articulation Controller (CMAC)", Trans. of the ASME, Journal of Dynamic Systems, Measurement, and Control, pp. 220-227, September, 1975.

[52] Young K.K.D., "Controller Design for a Manipulator Using Theory of Variable-Structure Systems", IEEE Trans. on Systems, Man and Cybernetics, Vol. SMC-8, 1978.

[53] Book J.W., Maizza-Neto O., Whitney E.D., "Feedback Control of Two Beam, Two Joint Systems With Distributed Flexibility", Journal of Dynamic Systems Measurement and Control, Trans. of the ASME, Vol. 97, pp. 424-431, December, 1975.

[54] Truckenbrodt A., "Modelling and Control of Flexible Manipulator Structure", Preprints of IV CISM-IFToMM Symposium on Theory and Practice of Robots and Manipulators, pp. 110-120, Warsaw, 1981.

[55] Yushimoto K., Wakatsuki K., "Application of the Preview Tracking Control Algorithm to Servoing of Robot Manipulator" First Int. Symp. of Robotic Research, Bretton-Woods, New Hampshire, USA, 1983.

[56] Hanafusa H., Yoshihiko Y., "Autonomous Trajectory Control of Robot Manipulators", First Int. Symp. of Robotic Research, Bretton. Woods, New Hampshire, USA 1983.

[57] Sandell M.R., Varaiya D., Athans M., Safonov G.M., "Survey of Decentralized Control Methods for Large-Scale Systems", IEEE Trans. on Automatic Control AC-23, pp. 108-128, 1978.

[58] Davison J.E., "The Robust Decentralized Control of a General Servomechanism Problem", IEEE Trans. on Automatic Control AC-21, pp. 14-24, 1976.

[59] Šiljak D.D., Large-Scale Dynamic Systems: Stability and Structure, North-Holland, New York, 1978.

[60] Athans M., Falb L.D., Optimal Control, McGraw-Hill, New-York, 1966.

[61] Davison J.E., "Decentralized Robust Control of Unknown Systems Using Tuning Regulators", IEEE Trans. on Automatic Control, Vol. AC-23, No 2, pp. 276-288, 1978.

[62] Davison J.E. "The Robust Control of a Servomechanism Problem for Linear Time-Invariant Multivariable Systems", IEEE Trans. on Automatic Control, Vol. AC-21, No 1, pp. 25-34, 1976.

[63] D'Azzo J.J., Houpis C.H., Feedback Control System Analysis and Synthesis, McGraw-Hill Book Company, New York, 1966.

[64] Chen C.T., Introduction to Linear System Theory, Hort, Rinehard and Winston, New York, 1970.

[65] Johnson D.C., "Accomodation of External Disturbances in Linear Regulator and Servomechanism Problems", IEEE Trans. AC-16, pp. 552--554, 1971.

[66] Stokić M.D., Vukobratović K.M., "Decentralized Regulator and Observer for a Class of Large Scale Nonlinear Mechanical Systems", Journal of Large Scale Systems, 5, pp. 189-206, 1983.

[67] Luanberger D.G., "An Introduction to Observers", IEEE Trans. on Automatic Control, Vol. AC-16, No 6, 1971.

[68] Michel N.A., "Stability, Transient Behaviour and Trajectory Bounds of Interconnected Systems", Int. Journal of Control, Vol. 11, No 4, pp. 703-715, 1970.

[69] Stokić M.D., Vukobratović K.M., "Practical Stabilization of Robotic Systems by Decentralized Control", Automatica, Vol. 20, No 3, pp. 353-358, 1984.

[70] Vukobratović K.M., Stokić M.D., "Significance of the Force Feedback in Realizing Movements of Extremities", IEEE Trans. on Biomedical Engineering, December, 1980.

[71] Vukobratović K.M., Stokić M.D., "Contribution to Suboptimal Control of Manipulation Robots", IEEE Trans. on Automatic Control, June, 1983.

[72] Cvetković V., Vukobratović K.M., "One Robust Dynamic Control Algorithm for Manipulation Systems", The Int. Journal of Robotic Research, Vol. 1, No 4, pp. 15-28, 1982.

[73] Vukobratović K.M., Stokić M.D., "Choice of Decoupled Control Law of Large Scale Mechanical Systems", Journal of Large Scale Systems, 3, 1981.

[74] Vukobratović K.M., Stokić M.D., Kirćanski M.N., "Towards Nonadaptive and Adaptive Control of Manipulation Robots", IEEE Trans. on Avtomatic Control, Vol. AC-29, No 9, pp. 841-844, 1984.

[75] Vukobratović K.M., Stokić M.D., "Suboptimal Synthesis of Robust Decentralized Control for Large Scale Mechanical Systems", Automatica, Vol. 20, No 6, pp. 803-807, 1984.

Appendix 2A

Analysis of the Influence of Actuator Models Complexity on Manipulator Control Synthesis

In this appendix we shall briefly analyze the influence of the actuator model on manipulator control synthesis. In Paragraph 2.2.1 we have chosen the models of the actuators driving robot joints in linear form (2.2.2) (with nonlinearity upon the input of amplitude saturation type). However, the problem arising is the order of these models which should be selected so as to achieve adequate representations of a real process. On the other hand, higher-order models require a longer computing time, larger memory capacity and may lead to a larger numerical error. Thus, we have to make a trade-off between the complexity of models and their practical justifiability.

We shall discuss the control synthesis for robotic systems for various degrees of complexity of actuator models.

System behaviour with two types of actuators has been studied: electromechanical and hydraulic.

A d.c. motor with a constant magnetic field controlled by rotor current has served as an electromechanical actuator. Differential equations describing the system are [1]:

$$\begin{aligned} L_r \dot{i}_r + R_r i_r + C_E \dot{q} &= u \\ -C_M i_r + J_r \ddot{q} + B_C \dot{q} &= -M \end{aligned} \qquad (2.A.1)$$

where i_r is the rotor current (A), u is the voltage at motor poles (V), R_r is the rotor resistance (Ω), L_r is the rotor inductance (H), q is the rotational angle of motor output shaft (rad), C_M is the constant of proportionality of moments (Nm/A), C_E is the constant of proportionality of electromotive force (V/rad/sec), J_r is the inertial moment of motor (kg m^2), B_C is the coefficient of viscous friction (Nm/rad/s), M is the external torque load of motor (Nm).

If we accept $x_i = (q_i, \dot{q}_i, i_R^i)^T$ to be the state vector and present the model of motor in the form (2.2.2)

$$A^i = \begin{bmatrix} 0 & 1 & 0 \\ 0 & -(B_c/J_r) & C_M/J_r \\ 0 & -(C_E/L_r) & -(R_r/L_r) \end{bmatrix} \quad f^i = \begin{bmatrix} 0 \\ -(1/J_r) \\ 0 \end{bmatrix} \quad b^i = \begin{bmatrix} 0 \\ 0 \\ 1/L_r \end{bmatrix}. \tag{2.A.2}$$

If we neglect inductance L_r and viscous friction B_c and accept $x^i = (q_i, \dot{q}_i)^T$ to be the state vector, the system reduces to a second-order system given by

$$A^i = \begin{bmatrix} 0 & 1 \\ 0 & -\frac{C_E C_M}{J_r R_r} \end{bmatrix} \quad f^i = \begin{bmatrix} 0 \\ \frac{1}{J_r} \end{bmatrix} \quad b^i = \begin{bmatrix} 0 \\ \frac{C_M}{J_r R_r} \end{bmatrix} \tag{2.A.3}$$

A hydraulic actuator consists of servovalves and cylinder and may be described by the following system of eqns [2, 3]

$$\begin{aligned} Q &= A_c \cdot \dot{l} + C_{tc} \cdot p_d + \frac{V_u}{4} \dot{p}_d \\ F_c &= A_c p_d = m_t \cdot \ddot{l} + B_c \dot{l} + F_t \\ Q &= Q' - k_c \cdot p_d \\ k'_q \cdot i &= C_1 Q' + C_2 \dot{Q}' + \ddot{Q}' \end{aligned} \tag{2.A.4}$$

where Q' and Q are the theoretical and real flows, respectively (m^3/s), A_c is the piston area (m^2), C_{tc} is the $(C_{kc}+C_{sc})/2$, C_{kc}, C_{sc} are the coefficients of internal and external leakage, respectively ($m^3/s/N/m^2$), p_d is the pressure differential between two piston sides (N/m^2), V_u is the cylinder working capacity (m^3), β is the coefficient of compressibility of fluid $(m^2/N)^{-1}$, m_t is the piston mass (kg), l is the piston displacement (m), B_c is the coefficient of viscous friction (Ns/m), F_t is the force of external load (N), k_c is the slope of servovalve characteristic in the working point ($m^3/s/N/m^2$), k'_q is the coefficient of servovalve ($m^3/s^3/mA$), C_1 and C_2 are the characteristics of servovalves ($1/s^2$) and ($1/s$), respectively.

If we accept the state vector in the form $x^i = (l^i, \dot{l}^i, p_d^i, Q'^i, \dot{Q}'^i)^T$, matrix A^i and vectors b^i and f^i are given in the form:

$$A^i = \begin{bmatrix} 0 & 1 & 0 & 0 & 0 \\ 0 & -\frac{B_c}{m_t} & \frac{A_c}{m_t} & 0 & 0 \\ 0 & -\frac{4\beta A_c}{V_u} & -\frac{4\beta(k_c+C_{tc})}{V_u} & \frac{4\beta}{V_u} & 0 \\ 0 & 0 & 0 & 0 & 1 \\ 0 & 0 & 0 & -C_1 & -C_2 \end{bmatrix} \quad b^i = \begin{bmatrix} 0 \\ 0 \\ 0 \\ 0 \\ k_q \end{bmatrix} \quad f^i = \begin{bmatrix} 0 \\ -\frac{1}{m_t} \\ 0 \\ 0 \\ 0 \end{bmatrix} \tag{2.A.5}$$

Since servovalve bandwidth is wide enough to allow its modes to be neglected, the system may be reduced to a third-order system of the form

$$A^i = \begin{bmatrix} 0 & 1 & 0 \\ 0 & -\frac{B_c}{m_t} & \frac{A_c}{m_t} \\ 0 & -\frac{4\beta}{V_u}A_c & -\frac{4\beta}{V_u}(k_c+C_{tc}) \end{bmatrix} \quad b^i = \begin{bmatrix} 0 \\ 0 \\ \frac{4\beta k_q}{V_u C_1} \end{bmatrix} \quad f^i = \begin{bmatrix} 0 \\ -\frac{1}{m_t} \\ 0 \end{bmatrix} \tag{2.A.6}$$

if the state vector is given by $x^i = (l^i, \dot{l}^i, p_d^i)^T$.

Two mechanical configurations, UMS-1 and UMS-3B have been tested [4]. Figure 2.A.1 shows the minimal configuration (3 d.o.f) of UMS-1 manipulator, and Table 2.A.1 presents its mechanical characteristics.

Mass concentration at the very tip of the third link served for as faithful as possible imitation of load transfer by a gripper; D.C. electromotors produced by Globe Industries Division of TRW Inc., 102A200 type, whose parameters are given in Table 2.A.2 were selected for actuators.

Such a movement was considered that the tip of minimal configuration moved along a straight line between points A and B defined by internal angles according to A(0; 1.1; 0.5), B(0.08; 0.94; 1.25). Velocity time history was triangular. The absolute value of acceleration was 4.9 (m/s^2).

At the stage of nominal regimes, the trajectories of internal angles, velocities and accelerations were obtained.

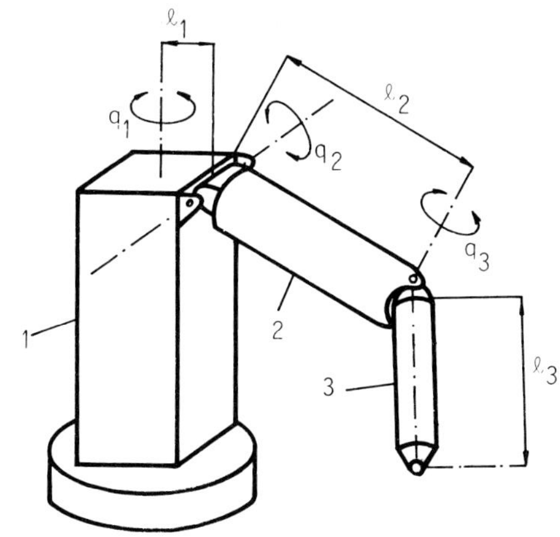

Fig. 2.A.1. UMS-1 manipulator (minimal configuration)

Link	1	2	3
Mass	15	4.7	10
Length	0.11	0.334	0.35
J_{ix} (kgm^2)	-	0.000623	0.000623
J_{iy} (kgm^2)	-	0.001246	0.00124
J_{iz} (kgm^2)	0.0092	-	-

Table 2.A.1. Parameters of mechanism (for UMS-1)

L_r (H)	$B_C(\frac{Nm}{rad/s})$	J_r (kgm^2)	$C_M(\frac{Nm}{mA})$	$C_E(\frac{V}{rad/s})$	R_r (Ω)
0.01	1.6	1.52	4.31	7	2.45

Table 2.A.2. Parameters of actuators (for UMS-1)

Local feedback gains, used at the stage of perturbed regimes, were synthesized for (according to expression (2.4.17)):

$$Q_i = \begin{bmatrix} 10 & & 0 \\ & 10 & \\ 0 & & 0.1 \end{bmatrix}, \quad r_i = 0.5, \quad \beta_i = 2.5, \quad \forall i \in I$$

and the following values were obtained

$$k_1 = 48.14; \quad k_2 = 6.042; \quad k_3 = 0.194$$

The values for Q_i, r_i and β_i were accepted on the basis of analysis described in Paragraph 2.5.

The movement being simulated was defined as follows: nominal trajectories were first generated for the prescribed conditions, and system behaviour with perturbation of initial conditions type was under observation. These perturbations were selected so as to load the actuators to a maximum. This was done to make the conditions imposed to the manipulation system as difficult as possible so that possible effects due to different complexity degrees of mathematical model of actuator would be as prominent as possible.

Two degrees of complexity of the mathematical model of electromechanical actuator were tested: second and third-order models. Two kinds of nominal trajectories were generated for a particular velocity of the tip of manipulator: for higher and lower complexity degrees of actuator models. Movement simulation was then performed on the basis of each nominal trajectory, applying first the second-order model and then the third-order one.

At the initial moment, only the deviation of angular coordinates Δq_i was accepted as the perturbation, while the remaining two coordinates in the state vector at the beginning of the movement were assumed to have nominal values. The following initial errors were assumed: $\Delta q_1(0) = 0.395$ (rad), $\Delta q_2(0) = 0.17$ (rad), $\Delta q_3(0) = -0.27$ (rad).

The length of the path of the tip of minimal configuration was 292 (mm), and the deviation of the tip from point A at the initial moment was 160 (mm).

Figure 2.A.2 presents the nominal trajectories of driving torques and voltage at electromotor poles, while Figs. 2.A.3 and 2.A.4 show the tracking of nominal trajectories when actuator models of different complexity degrees were used.

Norm of deviation was formed combining the deviations of generalized coordinate and velocity from the nominal in the form

$$C^i = \sqrt{\Delta q^{i2} + 0.1\Delta \dot{q}^{i2}}$$

It is presented as a diagram in Fig. 2.A.5(a-d).

For comparison, these diagrams also show deviation norm when the acceleration of tip is 1.5 (m/s^2) (solid lines) so that the highest velocity of the tip of minimal configuration is 0.6 (m/s). The increase in movement velocity obviously affects considerably the deviation of actual coordinates from the nominal during the movement.

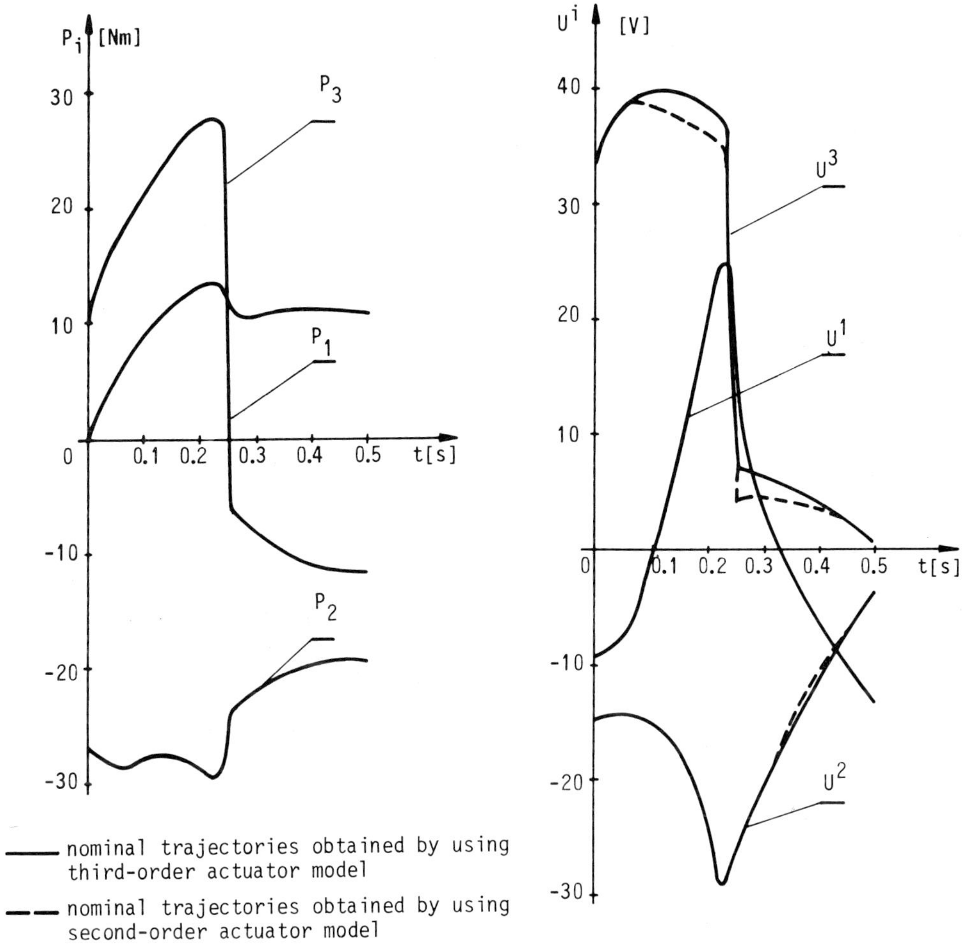

Fig. 2.A.2. Driving torques and voltage of electromotor inputs

The second set of simulation results was obtained for UMS-3B (Fig. 2.26) manipulator with minimal configuration consisting of three links connected by two rotational and one linear joint.

Mass was concentrated at the tip of the third link for more faithful simulation of the movement of a manipulator with a spraygun.

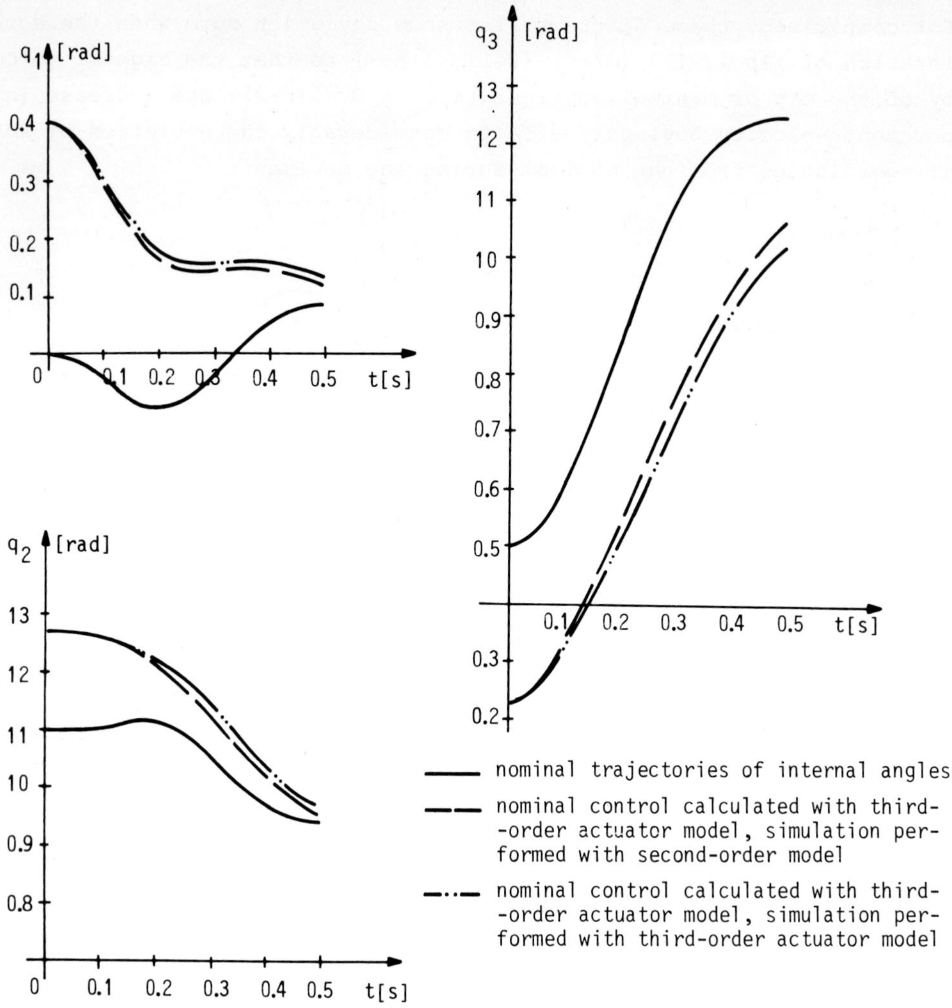

Fig. 2.A.3. Tracking of nominal trajectories

A movement similar to that of UMS-1 was considered. The tip of minimal configuration moves along a straight line between points A and B, with triangular velocity time history. Points A and B are defined by: A(0; 0; -0.68), B(0.4; -0.64; -0.4); the acceleration of tip is 4.9 (m/s^2), and its maximum velocity is 1.68 (m/s).

Feedback gains were synthesized in the same way as for UMS-1, but the following values were accepted for

$$Q_i = \begin{bmatrix} 0.1 & & 0 \\ & 100 & \\ 0 & & 0.1 \end{bmatrix}, \quad r_i = 5, \quad \beta_i = 2.5$$

so, the gains obtained were: $k_1 = 89.2$, $k_2 = 1.5$ and $k_3 = 0.11$.

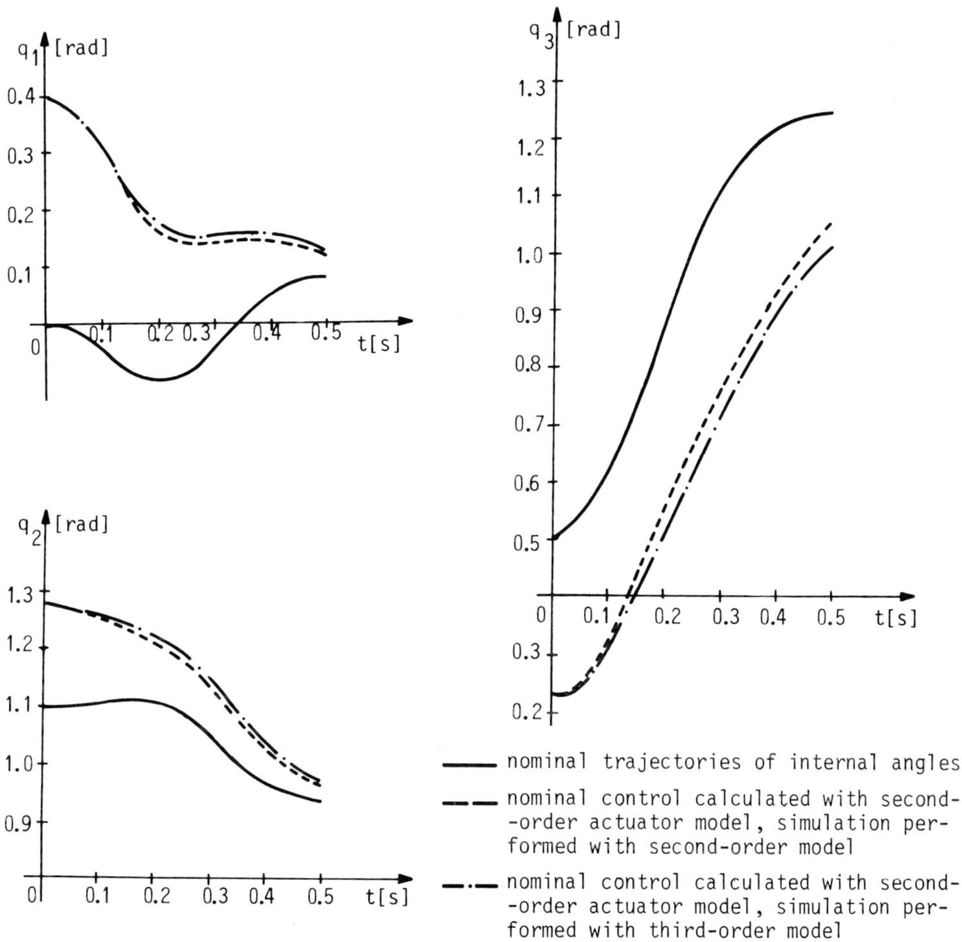

Fig. 2.A.4. Tracking of nominal trajectories

In principle, the simulation was performed in the same manner as for UMS-1. The trajectories of internal angles, velocities and accelerations were first calculated at the level of nominal regimes and then tracked in the presence of perturbation of initial conditions type.

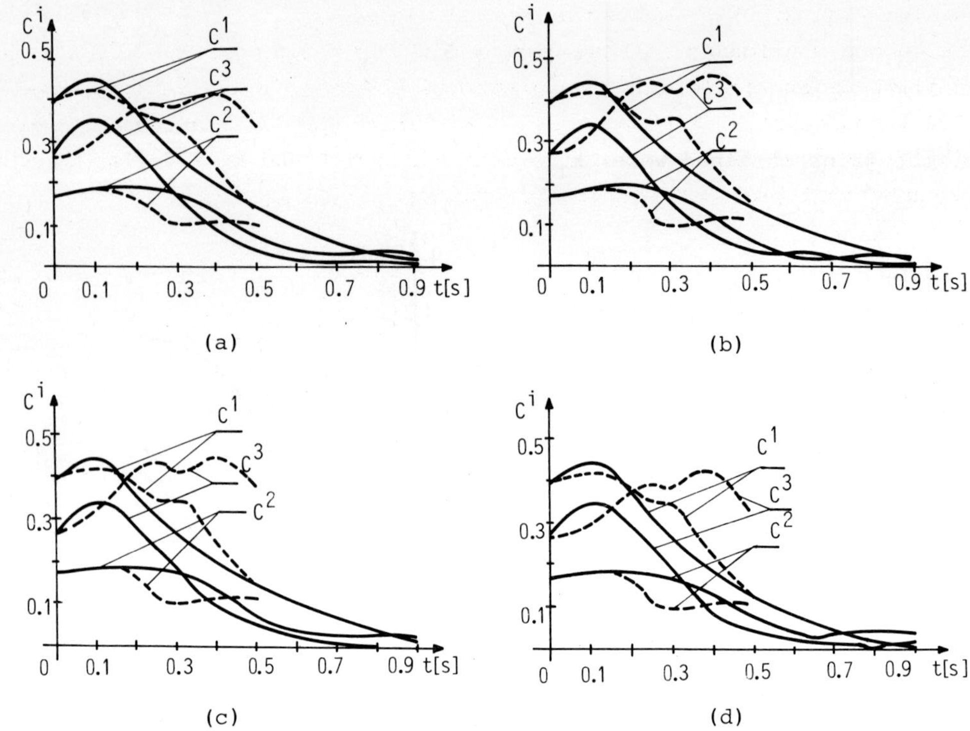

Fig. 2.A.5. Norm of combined deviation, (a) Nominal control synthesized with second-order actuator model, simulation performed with second-order model, (b) Nominal control synthesized with second-order actuator model, simulation performed with third--order model, (c) Nominal control synthesized using third--order actuator model simulation performed with third-order model, (d) Nominal control synthesized using third order actuator model, simulation performed with second-order model

Perturbations at the initial moment were accepted in the same way as in the previous case, and were: $\Delta q_1 = 0.415$ (rad), $\Delta q_2 = 0.25$ (rad), $\Delta q_3 = -0.22$ (m).

Two complexity degrees of the mathematical model of hydraulic actuator were tested: third and fifth-order model.

Nominals were synthesized using the third-order model only, while simulation was performed with both third and fifth-order models.

Figure 2.A.6 shows nominal forces and currents, while Fig. 2.A.7 presents the tracking of nominal trajectories of internal angles upon the simulation of movement with initial positional error.

Finally, Fig. 2.A.8 presents the total deviation during tracking. A point worth mentioning is that diagrams obtained by applying third - and fifth-order models coincide practically, so they are not drawn twice. Solid lines denote the error in the case when the tip of minimal configuration moved with the acceleration equal to 1.5 (m/s^2) to which a maximum velocity of 0.87 (m/s) corresponds.

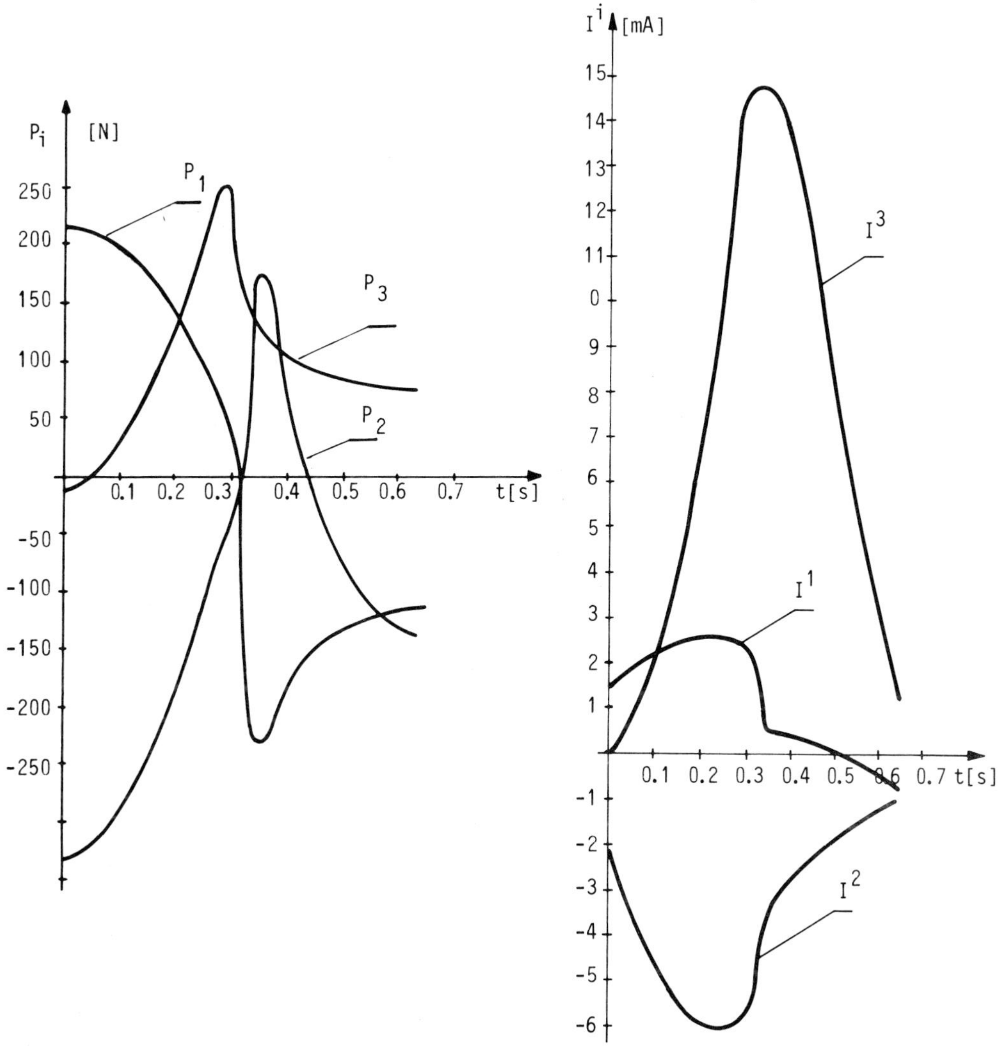

Fig. 2.A.6. Nominal forces and currents

A question of the degree of actuator model complexity which should be taken into account maintaining at the same time the quality of the tracking of nominal trajectories by manipulation robots arises in exact modelling of manipulation mechanisms. The result presented in this appendix has given a sufficiently meritorious answer to such a question and the following conclusion may be reached: as far as conventional operating speeds of manipulation robots are concerned, it is enough to consider second-order models of electromechanical actuators and third-order models of electrohydraulic actuators.

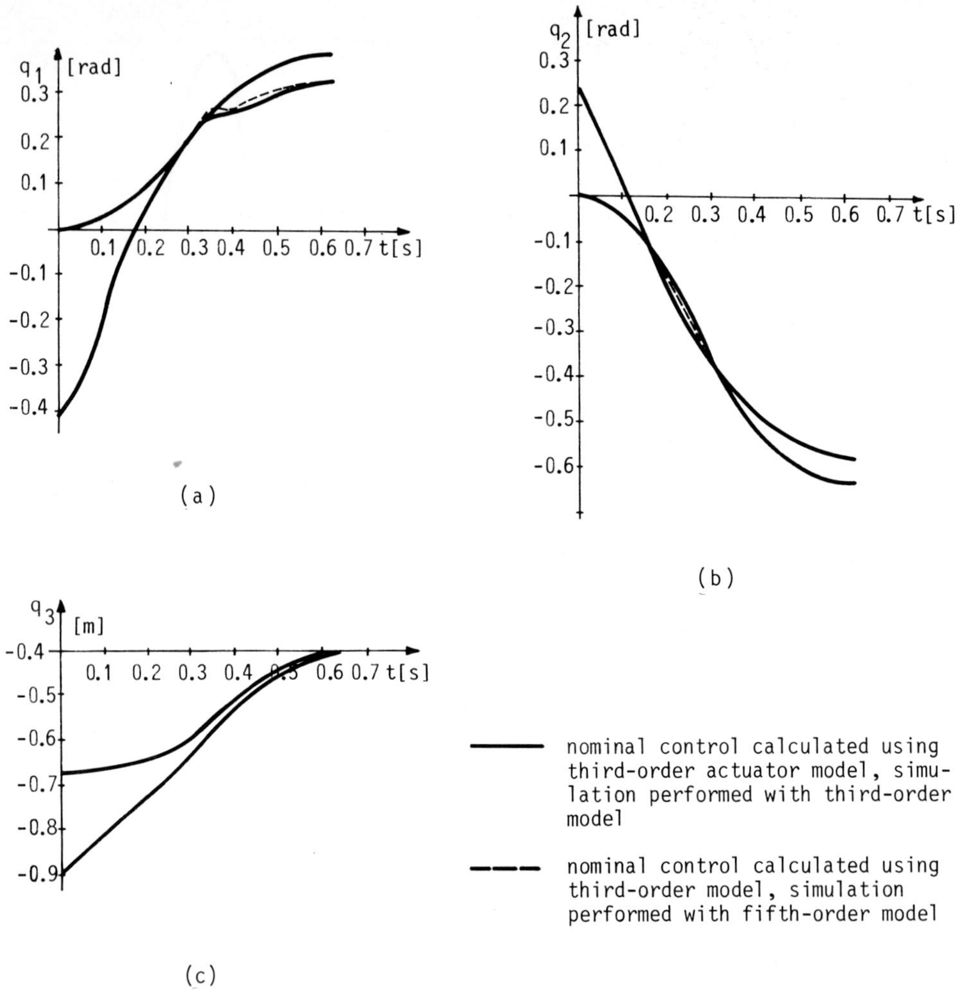

Fig. 2.A.7. Tracking of nominal trajectories of internal angles

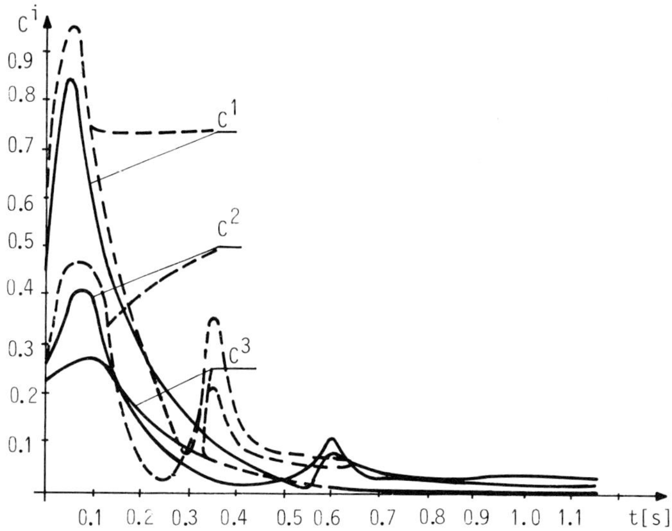

Fig. 2.A.8. Norm of combined deviation vs time

[1] Merrit E.N., Hydraulic Control Systems, Wiley, New York, 1971.

[2] Kuleshov V.S. and Lakota N.A., Dynamics of Control Systems of Manipulators, (in Russian), Energiya, Moscow, 1971.

[3] Vukobratović M., Hristić D. and Stokić D., "Dynamic Control of Industrial Manipulators and its Application", The Industrial Robot, No. 2, 1981.

[4] Borovac B., Vukobratović M., Stokić D., "Analysis of the Influence of Actuator Model Complexity on Manipulator Control Synthesis", Mechanism and Machine Theory, Vol. 18, No. 2, pp. 113-122, 1983.

Appendix 2B
Practical Stability of Manipulation Robots

In this Appendix we shall present the proof of test for the practical stability of robotic system (2.5.17). In Paragraph 2.5 we have presented how we can analyze the practical stability of the robotic system, but we have not proved the validity of this test.

The methods for stability analysis of large-scale nonlinear systems are generally based on two approaches: via Lyapunov first method and approach via input-output stability [1]. These two approaches are in essence parallel [2], but both have certain advantages and disadvantages. Here, we shall not survey various methods for stability analysis of large-scale systems. There are a number of very good surveys on various topics concerning the stability of large-scale systems (see [1-5]). We have chosen the approach via Lyapunov method for stability analysis of robotic systems. Actually, we shall consider the method for practical stability analysis of large-scale systems proposed by Michel [6].

We want to analyze the stability of the system the model of which is given as a set of subsystems:

$$\dot{x}^i = \bar{A}^i x^i + \bar{b}^i u^i + \bar{f}^i(x^i)\bar{P}_i(d,\ x), \qquad \forall i \in I \tag{2.B.1}$$

where $x^i \in R^{n_i}$ is the state vector of the ith sybsystem, $\bar{A}^i \in R^{n_i \times n_i}$, $\bar{b}^i \in R^{n_i}$, $\bar{f}^i$: $R^{n_i} \to R^{n_i}$ are corresponding matrices and vectors, $u^i \in R^1$ is the scalar input to the ith subsystem, $\bar{P}_i$: $R^{\ell} \times R^n \to R^1$ is the scalar coupling acting upon the ith subsystem which is a function of the parameters $d \in R^{\ell}$ and the state of the complete system $x(t) = (x^{1T}(t),\ x^{2T}(t), \dots$ $\dots, x^{nT}(t))^T$, $x \in R^N$; n_i is the order of the ith subsystem, n is the number of subsystems, ℓ is the order of the parameter vector d, and the order of the complete system is given by

$$N = \sum_{i=1}^{n} n_i$$

The set I is defined as I = {i: i=1,2,...,n}.

We have to analyze whether the system (2.B.1) is practically stable around the nominal trajectory $x^o(t)$, $\forall t \in T$ where T is the given time set $T = \{t: t \in (0, \tau)\}$ with τ being the defined time duration. The practical stability of the system is defined by (2.2.18). Thus, we have to explore whether the system (2.B.1) is practically stable with respect to $(X^I, X^t(t), D, T)$.

Decomposition-aggregation methods for stability analysis are always performed in two steps:

(1) first the stability of the local (decoupled) subsystem is analyzed,

(2) using the results of the first step and by estimating the coupling between the subsystems, the stability of the global system is analyzed.

First, we shall present the practical stability of local (decoupled) subsystems: then, we present the proof of the method for analysis of robotic system if coupling is taken into account; finally, we shall give an example of practical stability analysis for a particular manipulation robot.

Practical stability of local subsystems

We shall prove the test for practical stability of local subsystem. We consider local (decoupled) subsystem in which coupling is neglected:

$$\dot{x}^i = \bar{A}^i x^i + \bar{b}^i u^i(t, x^i) \qquad (2.B.2)$$

Here, we have assumed that control for the ith subsystem has already been synthesized as a function of t and subsystem state x^i. Thus, we shall analyze local stability for an arbitrary local control law.

The practical stability of local subsystem is defined by (2.4.8).

The following statement has to be proved:

If there exist a real valued, continuously differentiable function $v_i(t, x^i)$ and a real-valued function of time $\psi_i(t)$ which is integrable over time interval T, such that:

$$\dot{v}_{i(\text{along solution of (2.B.2))}} < \psi_i(t), \qquad \forall x^i \in \tilde{X}^{t(i)}(t) \qquad (2.B.3)$$

$$\int_0^t \psi_i(\lambda)d\lambda < v_{im}^{\partial X^{t(i)}(t)}(t) - v_{iM}^{\partial X^{I(i)}}(0), \qquad \forall t \in T \qquad (2.B.4)$$

then the subsystem (2.B.2) is practically stable with respect to $(X^{I(i)}, X^{t(i)}, T)$. In (2.B.4) $v_{im}^{\partial X^{t(i)}(t)}(t)$ denotes $\min_{x^i \in \partial X^{t(i)}(t)} v_i(t, x^i)$ and $v_{iM}^{\partial X^{I(i)}}(0)$ denotes $\max_{x^i \in \partial X^{I(i)}} v_i(0, x^i)$. $\partial X(t)$ denotes the boundary of the corresponding region, and $\tilde{X}^{t(i)}(t) = X^{t(i)}(t) - \bar{X}^{I(i)}\exp(-\alpha^{(i)}t)$.

The test defined above has been used in Paragraph 2.4.4 (inequalities (2.4.40), (2.4.41)).

We have to prove that if the inequalities (2.B.3), (2.B.4) hold and if $x^i(0) \in X^{I(i)}$, then the solution of the subsystem (2.B.2) must satisfy $x^i(t) \in X^{t(i)}(t)$, $\forall t \in T$. Let us prove this statement by contradiction. Let us assume that there is at least one initial state $x^i(0) \in X^{I(i)}$ such that the solution of the subsystem (2.B.2) for this initial state and chosen control $u^i(x^i)$ reaches the boundary of the region $X^{t(i)}(t)$, i.e. let us assume that $\exists x^i(0) \in X^{I(i)}$ implies $x^i(t_1, x^i(0)) \in \partial X^{t(i)}(t_1)$. Here t_1 is the first time instant for which $x^i(t)$ reaches the boundary of the region $X^{t(i)}(t)$.

We may write:

$$v_i(t, x^i(t)) = v_i(0, x^i(0)) + \int_0^t \dot{v}_i(\lambda, x^i(\lambda, x^i(0)))d\lambda \qquad (2.B.5)$$

If the inequality (2.B.3) holds, it follows:

$$v_i(t, x^i(t)) < v_i(0, x^i(0)) + \int_0^t \psi_i(\lambda)d\lambda \qquad (2.B.6)$$

If the inequality (2.B.4) is valid, from (2.B.6) if follows:

$$v_i(t, x^i(t) < v_i(0, x^i(0)) + v_{im}^{\partial X^{t(i)}(t)}(t) - v_{iM}^{\partial X^{I(i)}}(0)$$

Since $v_i(0, x^i(0)) < v_{iM}^{\partial X^{I(i)}}(0)$, it follows:

$$v_i(t, x^i(t)) < v_{im}^{\partial X^{t(i)}(t)}(t). \qquad (2.B.7)$$

The inequality (2.B.7) holds for $\forall t \in T$. If t_1 is the first time instant for which x^i reaches the boundary of the region $X^{t(i)}(t)$, then for this instant it also holds that:

$$v_i(t_1, x^i(t_1)) < v_{im}^{\partial X^{t(i)}(t_1)}(t_1) \qquad (2.B.8)$$

However, due to definition of $v_{im}^{\partial X^{t(i)}(t_1)}(t_1)$, the inequality (2.B.8) can be valid only if $x^i(t_1, x^i(0)) \notin \partial X^{t(i)}(t_1)$. Therefore, if the inequalities (2.B.3), (2.B.4) are fulfilled there is no such moment $t_1 \in T$ for which $x^i(t_1, x^i(0)) \in \partial X^{t(i)}(t_1)$. Since we assume that $X^{I(i)} \subseteq X^{t(i)}(0)$, i.e. $\forall x^i(0) \in X^{t(i)}(0)$, it follows that the state of the subsystem must belong to region $X^{t(i)}(t)$ and it cannot reach the boundary of this region $\partial X^{t(i)}(t)$ if the inequalities (2.B.3), (2.B.4) are valid. Thus, the conditions (2.B.3), (2.B.4) are sufficient to provide the practical stability of the subsystem (2.B.2) with respect to $(X^{I(i)}, X^{t(i)}(t), T)$.

For those not familiar with the notion of time derivative of function along the solution of corresponding subsystem, we have to denote that this means:

$$\dot{v}_i(t, x^i)_{(\text{along solution of (2.B.2)})} = \frac{\partial v_i}{\partial t} + (\text{grad } v_i)^T \dot{x}^i =$$

$$= \frac{\partial v_i}{\partial t} + (\text{grad } v_i)^T (\bar{A}^i x^i + \bar{b}^i u^i(t, x^i)) \qquad (2.B.9)$$

i.e. we assume the derivative of function v_i for such values of x^i which are solutions of the subsystem (2.B.2).

As we have already stated in Paragraph 2.4.4, the main problem of this test lies in the choice of the function $v_i(t, x^i)$. However, we have discussed this problem and shown that for linear time-invariant subsystems (2.B.2) we can select functions $v_i(t, x^i)$ in the form (2.4.57), where $\hat{H}_i$ satisfies (2.4.58). We have shown that this choice ensures an effective test of the practical stability of the subsystem.

Practical stability of global system

We have presented above the analysis of the practical stability of local decoupled subsystems in which the coupling between subsystems has been neglected. The second step in stability analysis of the complete interconnected system has to be presented now. We have to prove the test of the practical stability of the robotic system (2.B.1) with respect to the definition of practical stability (2.2.18). The relation between the regions of practical stability of local subsystems ($X^{I(i)}$, $X^{t(i)}(t)$) and the regions of global system (X^I, $X^t(t)$) has been established in Paragraph 2.2.2.

In order to test the practical stability of the global system, we have to estimate the coupling between subsystems. Namely, since the coupling $\bar{f}^i(x^i)\bar{P}_i(d, x)$ is a complex nonlinear function of the system state x and parameters d, we have to determine a time function which gives the boundary of the coupling. Actually, we have to determine function $\varphi_i(t, d)$ which satisfies:

$$(\text{grad } v_i)^T \bar{f}^i(x^i)\bar{P}_i(d, x) < \varphi_i(t, d), \ \forall x \in \tilde{X}^t(t), \ \forall t \in T, \ \forall d \in D \quad (2.B.10)$$

Since in robotic systems the coupling between subsystems are driving torques (forces) acting in the joints, the coupling must be limited if the system state belongs to a finite region in the state space and if parameters d are also finite values. Thus, it is always possible to determine functions $\varphi_i(t, d)$ which satisfy inequalities (2.B.10). Actually, functions $\varphi_i(t, d)$ are time estimates of the coupling.

Now, we have to prove the following statement:

Let each subsystem (2.B.2) be practically stable with respect to $(X^{I(i)}, X^{t(i)}(t), T)$ which is proved by tests (2.B.3), (2.B.4). If the following inequalities are valid for all subsystems:

$$\int_0^t \psi_i(\lambda)d\lambda + \int_0^t \varphi_i(\lambda, d)d\lambda < v_{im}^{\partial X^{t(i)}(t)}(t) - v_{iM}^{\partial X^{I(i)}}(0), \ \forall t \in T, \ \forall d \in D \quad (2.B.11)$$

where functions $\varphi_i: T \times D \to R^1$ satisfy inequalities (2.B.10), then the global complex system (2.B.1) is practically stable with respect to $(X^I, X^t(t), T, D)$.

We have to prove that if (2.B.11) holds, and if $x(0) \in X^I$, the solution of the complex system (2.B.1) must satisfy $x(t, x(0), d) \in X^t(t)$, $\forall t \in T$, $\forall d \in D$. We shall prove this statement by contradiction. Let us assume that (2.B.3), (2.B.10) and (2.B.11) are valid, but the complex system is not practically stable. If the system is not practically stable, then there is at least one initial state $x(0) \in X^I$ and $\exists d \in D$, such that the solution of (2.B.1) for this initial state and chosen control $u(x) = (u^1(x^1), u^2(x^2), \ldots, u^n(x^n))^T$ reaches the boundary of the region $X^t(t)$. Let us assume that $\exists x(0) \in X^I$ and $\exists d \in D$ imply $x(t_1, x(0), d) \in \partial X^t(t_1)$, where $t_1 \in T$ is the first moment for which $x(t)$ reaches the boundary of the region $X^t(t)$. Let us assume that the state of the jth subsystem ($j \in I$) reaches the boundary of the corresponding region $X^{t(j)}(t)$, i.e. that

$$x^j(t_1, x(0), d) \in \partial X^{t(j)}(t_1), \quad x(0) \in X^I, \quad \exists d \in D \tag{2.B.12}$$

Since we have assumed that the practical stability has been proved for each decoupled subsystem by testing inequalities (2.B.3), (2.B.4), this means that we have selected corresponding functions $v_i(t, x^i)$. For all subsystems we may write:

$$v_i(t, x^i) = v_i(0, x^i(0)) + \int_0^t \dot{v}_i(\lambda, x^i(\lambda, x(0), d))\,d\lambda, \forall i \in I \tag{2.B.13}$$

Let us consider the derivative of the function v_i along the solution of the interconnected subsystem (2.B.1):

$$\dot{v}_i(t, x^i)_{(\text{along solution of (2.B.1)})} =$$

$$= v_i(t, x^i)_{(\text{along solution of (2.B.2))}} +$$

$$+ (\text{grad } v_i)^T \cdot \bar{f}^i(x^i)\bar{P}_i(d, x), \quad \forall i \in I \tag{2.B.14}$$

Since we have assumed that for free subsystems (2.B.2) inequalities (2.B.3) are valid, and that function $\varphi_i(t, d)$ satisfying (2.B.10) may be found, we may write:

$$\dot{v}_{i(\text{along solution of (2.B.1))}} \leq \psi_i(t) + \varphi_i(t, d),$$

$$\forall i \in I, \ \forall x \in \tilde{X}^t(t), \ \forall t \in T, \ \forall d \in D \tag{2.B.15}$$

If we substitute (2.B.15) into (2.B.13), we get:

$$v_i(t, x^i) \leq v_i(0, x^i(0)) + \int_0^t \psi_i(\lambda)\,d\lambda + \int_0^t \varphi_i(\lambda, d)\,d\lambda,$$

$$\forall i \in I, \ \forall x \in \tilde{X}^t(t), \ \forall t \in T, \ \forall d \in D \tag{2.B.16}$$

Let us consider function v_j at t_1, the first moment for which the jth subsystem reaches the boundary of the region $X^{t(j)}(t)$. From (2.B.16) it follows:

$$v_j(t_1, x^j) < v_{jM}^{\partial X^{I(j)}}(0) + \int_o^{t_1} \psi_j(\lambda)d\lambda + \int_o^{t_1} \varphi_j(\lambda, d)d\lambda$$

$$\forall x \in \tilde{X}^t(t), \forall t \in T, \forall d \in D \tag{2.B.17}$$

We have assumed that for the jth subsystem inequalities (2.B.4) and (2.B.11) also hold. By substituting these inequalities into (2.B.17), we get:

$$v_j(t_1, x^j) < v_{jM}^{\partial X^{I(j)}}(0) + v_{jm}^{\partial X^{t(j)}(t_1)}(t_1) - v_{jM}^{\partial X^{I(j)}}(0) =$$

$$= v_{jm}^{\partial X^{t(j)}(t_1)}(t_1) \tag{2.B.18}$$

(2.B.18) can be valid only if $x_j(t_1, 0, x(0, d)) \notin \partial X^{t(j)}(t_1)$. However, this contradicts the original assumption (2.B.12). Therefore, it follows that $x^j(t, x(0), d)$ must belong to region $X^{t(j)}(t)$, $\forall t \in T$. Clearly, this holds for all subsystems and for all $x^i(0) \in X^{I(i)}$ and $\forall d \in D$. Thus, it follows that if the conditions (2.B.11) are valid, the complex system must be practically stable with respect to $(X^I, X^t(t), T, D)$. This completes the proof of the above statement.

Obviously, we may introduce functions $\psi_i'(t, d)$ which satisfy:

$$\psi_i(t) + \varphi_i(t, d) = \psi_i'(t, d) \tag{2.B.19}$$

and by substituting in (2.B.11) we get inequality (2.5.16) which has been used in Paragraph 2.5.

It should be mentioned that the test (2.B.11) might be very conservative, since we have estimated the coupling terms by time functions $\varphi_i(t, d)$. However, in Paragraph 2.5 we have shown that we can choose this function in the following way:

$$(\text{grad } v_i)^T \bar{f}^i(\theta^i) z^i(q_i) \bar{P}_i(x, d) < \sum_{j=1}^{n} \xi_{ij} ||\Delta x^j|| + (\text{grad } v_i)^T \bar{f}^i(\theta^i)$$

$$z^i(q^{oi}) \bar{P}_i^o(t, d) < \sum_{j=1}^{n} \xi_{ij} \bar{X}^{t(j)} \exp(-\alpha^{(j)} t) + (\text{grad } v_i)^T \bar{f}^i(\theta^i) z^i(q^{oi}) \cdot$$

$$\cdot P_i^o(t, d) = \varphi_i(t, d), \quad \forall x \in \tilde{X}^t(t), \forall t \in T, \forall d \in D \tag{2.B.20}$$

Thus, (2.B.20) determines the functions $\varphi_i(t, d)$ which estimate the coupling terms. By substituting these functions in (2.B.11) we get conditions (2.5.17) (or, (2.5.19)) which are suitable tests of the practical stability of the complete robotic system.

In the above considerations we have not assumed any particular form of local control law. Evidently, if we choose local control laws as explained in Paragraph 2.4.4, we get the test of local stability in the form (2.4.50), and the test of the complex system is obtained in the form (2.5.17) (or, (2.5.19)). However, the tests (2.B.3), (2.B.4), (2.B.11) might be used for arbitrary local control laws. Similarly, if we introduce global control (2.6.4), we can obtain the stability tests (2.6.23), starting from (2.B.3), (2.B.4) (2.B.11), as has been shown in Paragraph 2.6.

Numerical example

Finally we shall give an example of practical stability analysis for a particular manipulation robot. We have chosen a robot with very simple structure for which we can obtain the tests of practical stability in analytical form.

Let us consider the stability analysis for the "cylindrical" manipulation system UMS-2 shown in Fig. 2.B.1. Here we shall consider only the so-called minimal configuration of the manipulator (three degrees of freedom, n=3), while the gripper is not taken into account. This simplifies the stability analysis. The manipulator is powered by d.c. electromotors, the models of which are given by (2.2.2) The matrices of the models have got the form (2.A.2) ($n_i=3$, $\forall i \in I$).

The order of the complete system is N=9.

The model of the mechanical part of the system S^M (2.2.1) in this particular case has the form

$$S^M: \quad P_i = H_i \ddot{q}_i + h_i \tag{2.B.21}$$

where

$$H_1 = J_{z1} + J_{z2} + J_{z3} + m_3(q_3 + \ell^3)^2,$$

$$h_1 = 2m_3(q_3+\ell^3)\dot{q}_1\dot{q}_3$$

$$H_2 = m_2 + m_3,$$

$$h_2 = (m_2+m_3)g$$

$$H_3 = m_3,$$

$$h_3 = -m_3(q_3+\ell^3)(\dot{q}_1)^2$$

where m_i and J_{zi} are the masses and the moments of inertia of the corresponding links of the manipulator, and ℓ^3 is the length of the third link if $\ell^3 = 0$. We assume that parameters m_1, m_2, J_{z1}, J_{z2}, ℓ^3 are fixed, except for m_3 and J_{z3} which can vary between m_3^o to $\bar{m}_3$ and J_{z3}^o to $\bar{J}_{z3}$, since these parameters include mass and moment of inertia of the load carried by the manipulator, so these parameters are not known in advance. Thus, the finite region of the allowable parameter values is given by $D = \{d\colon m_1 = m_1^o,\ m_2 = m_2^o,\ J_{z1} = J_{z1}^o,\ J_{z2} = J_{z2}^o,\ \ell^3 = \ell^{3o},\ m_3 \in (m_3^o, \bar{m}_3),\ J_{z3} \in (J_{z3}^o, \bar{J}_{z3})\}$, where $d = (m_1, m_2, J_{z1}, J_{z2}, m_3, J_{z3})$ and m_1^o, m_2^o, J_{z1}^o, J_{z2}^o, ℓ^{3o} are fixed values.

Let us consider the control task stated as described in Paragraph 2.2. The nominal trajectories of the angles, velocities and accelerations of the manipulator are given by

$$q^{oi}(t) = \begin{cases} \ddot{q}^{oi}_{max}\dfrac{t^2}{2} + q^{oi}(0), & t \in (0, \tau/2) \\ \ddot{q}^{oi}_{max}\left(\tau t - \dfrac{t^2}{2} - \dfrac{\tau^2}{4}\right) + q^{oi}(0), & t \in (\tau/2, \tau) \end{cases} \tag{2.B.22}$$

where $\ddot{q}^{oi}_{max}$ are constant accelerations. The nominal trajectories of robot currents are given by

$$i_r^{oi}(t) = \begin{cases} (J_r^i\ddot{q}^{oi}_{max} - F^i\ddot{q}^{oi}_{max}t)/C_M^i, & t \in (0, \tau/2) \\ [-J_r^i\ddot{q}^{oi}_{max} - F^i\ddot{q}^{oi}_{max}(\tau - t)]/C_M^i, & t \in (\tau/2, \tau). \end{cases}$$

The nominal local programmed control $\bar{u}^{oi}(t)$ can be calculated from (2.4.14). The nominal driving torques corresponding to $q^{oi}(t)$, $\forall t \in T$ can be calculated according to (2.B.21)

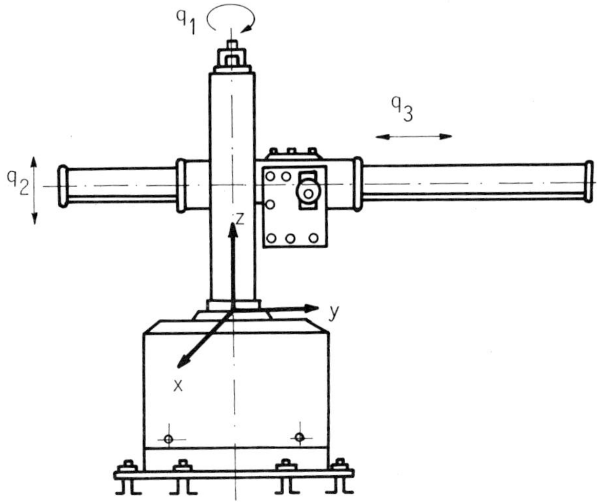

Fig. 2.B.1. Manipulator UMS-2

$$P_1^o = \begin{cases} [J_{z1} + J_{z2} + J_{z3} + m_3(\ell^3 + \ddot{q}^{o3}_{max}\frac{t^2}{2})^2]\ddot{q}^{o1}_{max} + \\ \quad + 2m_3(\ell^3 + \ddot{q}^{o3}_{max}\frac{t^2}{2})\ddot{q}^{o3}_{max}\ddot{q}^{o1}_{max}t^2, \quad t\in(0, \tau/2) \\ -\{J_{z1} + J_{z2} + J_{z3} + m_3[\ell^3 + \ddot{q}^{o3}_{max}(\tau t - \frac{t^2}{2} - \frac{\tau^2}{4})]^2\}\ddot{q}^{o1}_{max} + \\ \quad + 2m_3[\ell^3 + \ddot{q}^{o3}_{max}(\tau t - \frac{t^2}{2} - \frac{\tau^2}{4})] \times \\ \quad \times \ddot{q}^{o3}_{max}\ddot{q}^{o1}_{max}(\tau - t)^2, \quad t\in(\tau/2, \tau) \end{cases}$$

$$P_2^o = \begin{cases} (m_2+m_3)\ddot{q}^{o2}_{max} + (m_2+m_3)g, & t\in(0, \tau/2) \\ -(m_2+m_3)\ddot{q}^{o2}_{max} + (m_2+m_3)g, & t\in(\tau/2, \tau) \end{cases} \tag{2.B.23}$$

$$P_3^o = \begin{cases} m_3\ddot{q}^{o3}_{max} - m_3(\ell^3 + \ddot{q}^{o3}_{max}\frac{t^2}{2})(\ddot{q}^{o1}_{max}t)^2, & t\in(0, \tau/2) \\ -m_3\ddot{q}^{o3}_{max} - m_3[\ell^3 + \ddot{q}^{o3}_{max}(\tau t - \frac{t^2}{2} - \frac{\tau^2}{4})](\ddot{q}^{o1}_{max})^2(\tau - t)^2, & t\in(\tau/2, \tau) \end{cases}$$

First, the local control (2.4.19) for each subsystem $\tilde{S}^i$ should be synthesized. Then, the conditions (2.4.54) for practical stability of the local subsystems should be checked. After that the numbers ξ_{ij} in (2.5.13) should be determined. Starting from the model (2.B.21) and nominal values of driving torques (2.B.23), we can estimate the numbers ξ_{ij} as

$$\xi_{ij} = \frac{\lambda_M(K_i)}{\lambda_m^{1/2}(K_i)} \tilde{\xi}_{ij} ||\bar{f}^i|| \qquad (2.B.24)$$

where

$$\tilde{\xi}_{11} = \frac{[J_{z1}+J_{z2}+J_{z3}+m_3(q^{o3}_{max})^2]\sqrt{2}}{J^1_r+J_{z1}+J_{z2}+J_{z3}+m_3(q^{o3}_{max})^2} \max[(F^i+2m_3(q^{o3}\cdot\dot{q}^{o3})_{max}), C^i_M]$$

$$\tilde{\xi}_{13} = \frac{J^1_r m_3\sqrt{2}}{J^1_r+J_{z1}+J_{z2}+J_{z3}+m_3(q^{o3}_{max})^2} \max(\xi^*_{13}, \bar{\xi}_{13})$$

$$\xi^*_{13} = \frac{\max(2q^{o3}+\Delta q_3)\cdot[(J_{z1}+J_{z2}+J_{z3}+m_3(q_3)^2_{max}]}{J^1_r+J_{z1}+J_{z2}+J_{z3}+m_3(q_{3min})^2} \times$$

$$\times [(\frac{F^1}{J^1_r}(\dot{q}_1)_{max} + \frac{C^1_M}{J^1_r}(i^1_r)_{max}) + 2(q_3\dot{q}_1\dot{q}_3)_{max}] + [\max(2q^{o3}+\Delta q_3) \times$$

$$\times (\frac{F^1}{J^1_r}(\dot{q}_1)_{max} + \frac{C^1_M}{J^1_r}(i^1_r)_{max}) + 2(\dot{q}_1\dot{q}_3)_{max}]$$

$$\bar{\xi}_{13} = 2[(q^{o3}\Delta\dot{q}_1)_{max} + (q^{o3}\dot{q}^{o1})_{max}]$$

$$\tilde{\xi}_{22} = \frac{(m_2+m_3)\sqrt{2}}{m^2_r+m_2+m_3} \max(F^2, C^2_M)$$

$$\tilde{\xi}_{31} = \frac{m_3 m^3_r}{m^3_r + m_3}(q^{o3})_{max}[2(\dot{q}^{o1})_{max}+(\Delta\dot{q}_1)_{max}]$$

$$\tilde{\xi}_{33} = \frac{m_3\sqrt{2}}{m^3_r+m_3} \max(m^3_r(\dot{q}_1)^2_{max}, F^3, C^3_M)$$

where $(q^{oi})_{max}$ and $(q^{oi})_{min}$ denote maximum and minimum over the interval T, and $(q_i)_{max}$ denotes maximum of $q^{oi}(t) + \Delta q_i(t)$, for $|\Delta q_i| < \bar{X}^{t(i)}\exp(-\alpha^{(i)}t)$, $\forall t \in T$.

Now, by calculating the integrals of P^o_i over time we can calculate the functions $\rho_i(t)$ satisfying (2.5.18), e.g. for i=2

$$\rho_2(t) = \begin{cases} (m_2+m_3)\ddot{q}^{o2}_{2max}t+(m_2+m_3)gt, & t\in(0,\ \tau/2) \\ -(m_2+m_3)\ddot{q}^{o2}_{max}(t-\tau)+(m_2+m_3)gt, & t\in(\tau/2,\ \tau) \end{cases}$$

and, similarly, for $\rho_1(t)$, $\rho_3(t)$. By introducing the expressions for ξ_{ij} and $\rho_i(t)$ into (2.5.17), we obtain the connection among local controls, parameters of the manipulator and nominal trajectories and the parameters of the imposed finite regions of practical stability.

From this connection we can iteratively calculate the local controllers which should satisfy particular conditions of practical stability for the practical domain of parameters variation of this manipulator. On the other hand, if the local controllers are already chosen we can estimate either the regions of practical stability of the manipulator around the given nominal trajectory with such controllers, or the domain of parameters variation for which these controllers ensure practical stability.

Let us consider the latter case. The parameters of the manipulator UMS-2 are given in Table 2.B.1 while those of the chosen actuators (d.c. motors of IG 2315-P20 type) are given in Table 2.B.2 (elements of the matrices of actuator models). Let us consider the nominal trajectory given by (2.B.22) where $\ddot{q}^{o1}_{max}$ = 1.57 rad s^{-2}, $\ddot{q}^{o2}_{max}$ = 0.15 m s^{-2}, $\ddot{q}^{o3}_{max}$ = 0.3 m s^{-2} and τ = 2 s. The nominal trajectories are given in Fig. 2.B.2 and the computed nominal driving torques corresponding to m^o_3 and J^o_{z3} are given in Fig. 2.B.3. The regions of practical stability are imposed as described in Paragraph 2.2 where $\bar{X}^{I(1)} = 0.1$, $\bar{X}^{I(2)} = \bar{X}^{I(3)} = 0.01$ and $\alpha^{(i)} = 5$, $\forall i\in I$, and $\bar{X}^{t(i)}$ are chosen to satisfy $(\lambda_m(K_i)/\lambda_M(K_i))^{1/2} > (\bar{X}^{I(i)}/\bar{X}^{t(i)})$.

The local controllers are chosen first. Since $\alpha^{(i)} = 5$ we choose $\beta_i=10$, $\forall i\in I$ and weighting matrices in (2.4.16) as given in Table 2.B.3. Local gains are calculated and also given in Table 2.B.3. It can easily be checked whether the conditions of practical stability for decoupled subsystems are satisfied. The numbers $\tilde{\xi}_{ij}$ in (2.B.24) are evidently functions of J_{z3} and m_3, and for the values of other parameters, as given in Tables 2.B.1 and 2.B.2 those functions are

$$\tilde{\xi}_{11} = \frac{(m_3 2.37) + 0.355 + (4.23J_{z3})}{0.12 + J_{z3} + 0.5625\,m_3}$$

$$\tilde{\xi}_{13} = \frac{0.048m_3}{0.12+J_{z3}+0.2025m_3}\left[\frac{2.56J_{z3}+(1.85m_3)+2.1}{0.12+J_{z3}+m_3 0.2025} + 3.84\right]$$

$$\tilde{\xi}_{22} = \frac{29.61 + m_3 4.23}{1.028 + 0.004m_3}$$

$$\tilde{\xi}_{31} = \frac{m_3 2.325}{1 + (m_3 0.0025)}$$

$$\tilde{\xi}_{33} = \frac{m_3 4.23}{1 + (m_3 0.0025)}$$

For $m_3=5.6$ and $J_{z3}=J^o_{z3}=0.320$ these numbers get the values

$\tilde{\xi}_{11}=4.28$, $\tilde{\xi}_{13}=1.18$, $\tilde{\xi}_{22}=50.7$, $\tilde{\xi}_{31}=12.8$, $\tilde{\xi}_{33}=23.4$

Link	1	2	3
Mass (kg)	10.0	7.0	5.6
Length (m)	0.38	0.02	0.45
J_{xi} (kgm^2)	-	-	-
J_{yi} (kgm^2)	-	-	-
J_{zi} (kgm^2)	0.029	0.055	0.320

Table 2.B.1. Manipulator parameters

i	a^i_{22}	a^i_{23}	a^i_{32}	a^i_{33}	b^i_3	f^i_2
1	-2.0	20.0	-1000.0	-450.0	2000.0	-29.0
2	-3.0	0.28	-10^5	-450.0	2000.0	-0.004
3	-3.0	0.13	-10^5	-450.0	2000.0	-0.0025

Table 2.B.2. Actuator parameters

By calculating the functions $\rho_i(t)$ for these values of parameters we will see that the condition (2.5.17) is not fulfilled for the first subsystem. Therefore, we have to reselect local control for the first subsystem, or to introduce global control. In the first case, we can

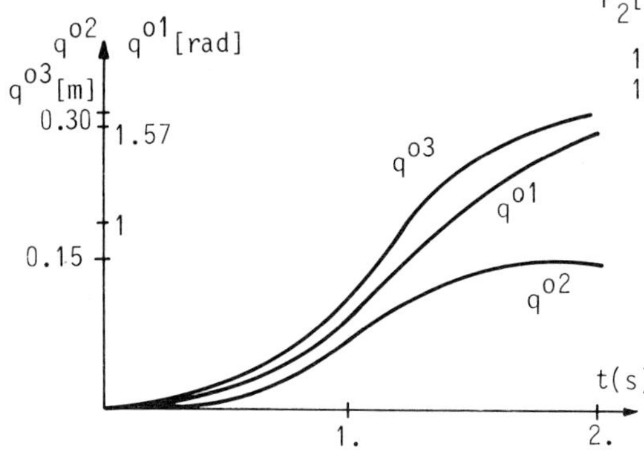

Fig. 2.B.2. Nominal trajectories

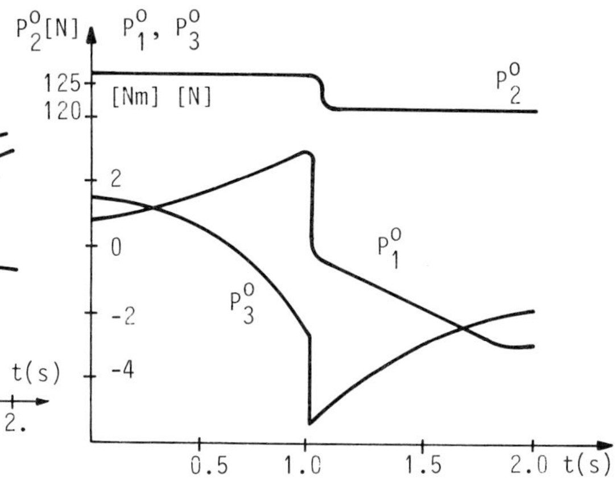

Fig. 2.B.3. Nominal driving torques

rearrange the distribution between the local subsystem $\tilde{S}^1$ and load P_1 (Paragraph 2.4). In the matrices of the local subsystem $\tilde{S}^1$ we introduce $J^1_r + \bar{H}_1$ instead of J^1_r, where $\bar{H}_1$ is chosen as $\bar{H}_1 = J_{z1} + J_{z2} + J^o_{z3} + m^o_3(\ell^3_1 + q^{o3}(0))^2 = 1.25$. Thus, the values of elements of subsystem matrices in Table 2.B.2 change (e.g. f^1_2 is now 0.778). The numbers $\tilde{\xi}_{11}$ and $\tilde{\xi}_{13}$ also change so that

$$\tilde{\xi}_{11} = \frac{(m_3 2.37) - 4.93 + 4.23 J_{z3}}{J_{z3} + 0.5625 m_3 + 0.12}$$

$$\tilde{\xi}_{13} = \frac{0.048 m_3}{J_{z3} + 0.2025 m_3 + 0.12}\left[\frac{2.56 J_{z3} + 1.85 m_3 - 1.1}{J_{z3} + 0.2025 m_3 + 0.12} + 3.84\right].$$

$$Q_i = \begin{bmatrix} 0.01 & & \\ & 100.0 & \\ & & 100.0 \end{bmatrix} \qquad r_i = 100.$$

For $\bar{H}_1 = 0.$

$$r_i^{-1}\bar{b}^{iT}K_i = (13.9,\ 1.3,\ 0.94)^T, \qquad \text{for } i=1$$

$$= (406.0,\ 76.0,\ 0.8)^T, \qquad \text{for } i=2$$

$$= (1174.0,\ 202.0,\ 0.76)^T, \qquad \text{for } i=3.$$

For $\bar{H}_1 = 1.25$

$$r_1^{-1}\bar{b}^{1T}K_1 = (23.37,\ 3.75,\ 0.91)^T$$

Table 2.B.3. Local regulator gains

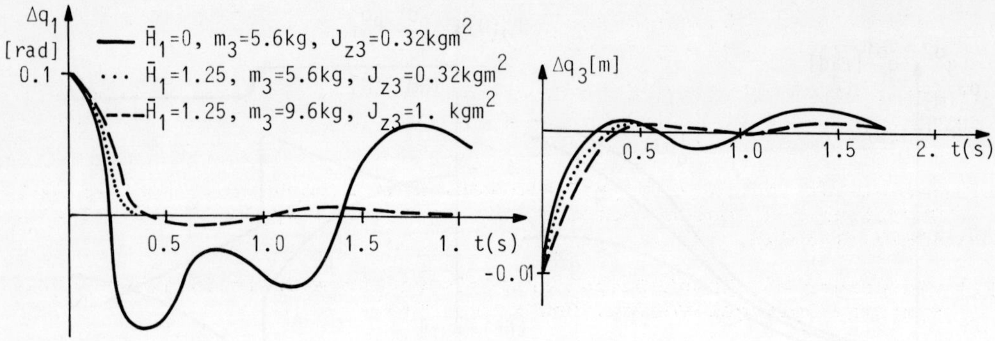

Fig. 2.B.4. Simulation of tracking for various parameters values

The local gains with reselected subsystem model are also given in Table 2.B.3. It can be shown that the system is practically stable with such local regulators for parameter variation $m_3 \in (5.6,\ 9.6)$(kg), $J_{z3} \in (0.320, 1)$ (kgm^2).

This has also been verified by digital simulation. The dynamics of the manipulation system of Fig. 2.B.1 during tracking the nominal trajectories of Fig. 2.B.2 has been simulated on a digital computer. The results of digital simulation are shown in Fig. 2.B.4. The deviations of mechanism coordinates from the nominal trajectories Δq_i (for the first and the third degree of freedom) versus time are shown for two sets of values of parameters m_3 and J_{z3}. Initial conditions are chosen on the boundary of the regions $X^{I(i)}$ i.e. $\Delta q_1(0) = 0.1$ (rad), $\Delta q_2(0) = 0.01$ (m), $\Delta q_3(0) = 0.01$ (m) and $\Delta\dot{q}_i(0) = 0$, $\Delta i_r^i(0) = 0$, $\forall i \in I$. The results show good tracking for all parameter values, i.e. the conditions of practical stability are satisfied. This means that the proposed simple control (2.4.19), consisting of local nominal control $\bar{u}^{oi}(t)$ and local regulators (2.4.17), can provide the practical stabilization of this particular manipulator for a relatively wide range of parameter variations. There is thus no need to introduce global control in this particular case.

Figure 2.B.4. also shows the results of tracking of nominal trajectories by the local control when $\bar{H}_1$ is taken to be zero (regulators gains shown in Table 2.B.3). As already stated above, in this case the system is not practically stable.

[1] Voronov A.A., "State-of-the-Art and Problems in Stability Theory", (in Russian), Avtomatika and Telemechanika, No. 1, 1983.

[2] Araki M., "Stability of Large-Scale Nonlinear Systems: Quadratic--order Theory of Composite Systems Method Using M-matrices", IEEE Transactions, AC-23, 129-142, 1978.

[3] Sandell Jr. N.R., Varaiya P., Athans M., Safonov G.M., "Survey of Decentralized Control Methods for Large Scale Systems" IEEE Trans., AC-23, 108-128, 1978.

[4] Vidyasagar M., Input-Output Analysis of Large-Scale Interconnected Systems, Springer-Verlag, New York, 1981.

[5] Michel N.A., Miller K.R., Qualitative Analysis of Large Scale Dynamic Systems, Academic Press, New York, 1977.

[6] Michel N.A., "Stability, Transient Behaviour and Trajectory Bounds of Interconnected Systems", Int. Journal of Control, Vol. 11, No. 4, pp. 703-715, 1970.

Chapter 3
Adaptive Control Algorithms

3.1 Introduction

In this chapter we shall present the theoretical background of various adaptive robot control algorithms. This subject is relatively new in robotic research. Within the past six years, only a handful of people have been actively working on this subject. Nevertheless, this area will surely be one of the most interesting in the nearest future, because the classical controllers cannot always satisfy the stability conditions, even if designed to be robust with respect to parametric and state disturbances (see Chapter 2). Thus, the adaptive control algorithm should be considered as an up-grade over the classical control approach. The adaptive control algorithms are often much more complex in numerical sense than nonadaptive laws. Also, it is more difficult to prove the stability of the overall system. But, the adaptive controllers offer more opportunities, especially when the robot works in ambient conditions which are not completely known in advance. This is the case, for example, when the payload mass is not predefined. Of course, this does not mean that the robot controllers should always be designed to be adaptive. On the contrary, in our opinion, the adaptive algorithms should be employed only when simple classical controllers cannot achieve desirable performances.

Within adaptive control theory there are two fundamental approaches. The first is Learning Model Adaptive Control (LMAC), in which an improved model of the plant is obtained by on-line parameter estimation techniques, and is then used in the feedback control. The well-known self-tuning control strategy belongs here. The second approach in adaptive control theory is called Model Referenced Adaptive Control (MRAC). The controller is adjusted so that the closed-loop behavior of a system matches that of a preselected model according to some criterion.

Within the self-tuning control theory there are many approaches based on different estimation techniques: least squares, extended least squares, maximum likelihood, etc. According to the control synthesis procedure, the self-tuning regulators can be divided into: minimum variance controllers, extended minimum variance controllers, and pole-zero placement regulators, [1-9]. The feasibility of least squares/extended minimum variance control in robotic applications is examined by Koivo and Guo [10]. The application of least squares pole zero placement in robotics is elaborated by Leininger [11, 12]. We shall discuss these results in detail in Paragraph 3.2.1 within the scope of centralized robot control strategy. PID self-tuning local regulators will be described with the scope of decentralized control of manipulators in Paragraph 3.3.1.

In this chapter we shall also discuss several basic concepts within model - reference control theory: local parametric optimization theory, methods based on Lyapunov functions and methods based on hyperstability theory [13-19]. As will be shown in the text to follow, there have been several attempts towards the application of these results to robotic devices. The local parametric optimization technique has been used by Dubowsky [20-22], and Lyapunov functions approach has been tested by Arimoto and Takegaki [23, 24]. A particularly interesting concept based on hyperstability and positivity [25-27] will be presented in Paragraph 3.2.2. It should be pointed out that these methods are applied within the centralized control strategy. The decentralized control structure based on hyperstability has been elaborated by Kirćanski and Vukobratović [28]. Let us consider now the results of Dubowsky and Arimoto in more details.

One of the first contributions in adaptive robot control is given by Dubowsky and DesForges in 1979. They developed a model-referenced-adaptive control system based on steepest descent method in order to minimize a cost function of the error between the model and the system. They defined the error between the response of the model and the system by the vector

$$e(t) = y(t) - x(t)$$

where y is the model state vector, and x is the system state vector. In this method, a scalar cost function is formulated as

$$V(e) = \frac{1}{2} e^T Q e$$

where Q is called the adaptive gain matrix. The objective of the algorithm is to drive e(t) to zero, and thereby match the system response to that of the model. Let θ be the system parameter vector including adjustable gains. Then, according to the steepest descent method this parameter vector should be varied according to the law

$$\dot{\theta} = -k\frac{\partial V}{\partial \theta} = -k\frac{\partial e^T}{\partial \theta}Qe \tag{3.1.1}$$

where k is a positive real. This formula represents the classical gradient algorithm. Since $\partial e/\partial \theta$ cannot be determined, a series of approximations must be introduced. It turns out that the sensitivity equations of reference model take part in the algorithm. Hence, using the obtained parameter vector θ, the feedback gains are calculated.

The advantages of the method are:

- it is computationally less burdensome than the methods which evaluate a nonlinear robot model,
- it has good noise rejection properties, and
- it is tested experimentally on an industrial robot [22].

There are also some undesirable properties which should be pointed out:

- the gradient algorithm built into the adaptation algorithm may negatively influence the overall system stability, and
- the criterion V(e) penalizes only the output error. As the amplitudes of input signals are not penalized, the adaptation mechanism may impose excessively large amplitudes of control signals, which cannot be realized.

Finally, it should be mentioned that the discrete - time version of this method is given in Ref. [21], and the multivariable case in Ref. [22]. Especially interesting are the experimental results obtained with a laboratory rotary positioning device (2 degrees of freedom) and with an industrial robotic manipulator, PUMA 500. The experiments show an expressive significance of adaptive robot controllers, particularly when high precision/high velocity tracking should be realized with different payloads.

Takegaki and Arimoto [23] have also considered the applicability of model reference adaptive control theory in robotics. They considered a linear centralized control law with feed-forward compensation of acceleration and gravitational effects. Then the following problem is posed: how to change the coefficients within the control law in such a way as to drive the output error to zero. For this purpose, similarly to Landau's theory [13-19], the Lyapunov function approach is applied. Thus, both the control law and the adaptation mechanism are obtained in a simple form, suitable for implementation. But, it is not quite clear how the gravity compensation affects the tracking quality, especially when weights of payloads vary over a wide range. If the payload mass (and inertias) is unknown, the gravitational forces cannot be compensated with a nonadaptive feed-forward term. Thus, the adaptation mechanism at the level of perturbation equations should take over this compensation. The same problem appears with some other adaptive algorithms [29, 30]. An attempt to improve the described control strategy was made by Arimoto and associates in 1984, [24], but the mentioned problem remained unsolved.

In Paragraph 3.2.2 we will present the robot control strategy based on hyperstability theory. Starting from the theory developed by Popov, Landau and others [13-19], Balestrino and associates applied it to robotic systems. However, before that Horowitz and Tomizuka presented an interesting attempt to apply hyperstability theory to robotics. They start from the nonlinear robot model

$$P = H(q)\ddot{q} + \dot{q}^T C(q, d)\dot{q} + g(q) \tag{3.1.2}$$

where H, C and g represent dynamic model matrices, described in Chapter 1. Then, they suggest the identification of H and C elements (they neglect gravitational effects) by using a specific adaptation mechanism based on Popov's inequality [13]. Here, the reference model is defined as a set of n linear, decoupled second-order systems (n double integrators), intended to decrease dynamic interaction between subsystems. However, in order to compensate for gravitational effect, the authors had to add an outer-loop PID controller. The evaluation of the algorithm on a 3-degree-of-freedom manipulator has been performed by Anex and Hubbard [31]. The experiments have shown that the basic algorithm must be slightly changed in order to cope with disturbances due to Coulomb friction, gravitational loading and actuator saturation. The algorithm was tested on low velocities because of the torque limitations of actuators. The experiments were not performed at higher joint rates at which adaptive control must offer significant improvements over the performance available from classical control methods.

The good features of the approach result from the fact that the adaptation mechanism is derived from the condition of overall system stability. Except for this, the control structure is based on a nonlinear dynamic robot model. Thus, the decoupled control is easy to achieve. But, the basic problem with this approach is the following: the dynamic effects are estimated without using any knowledge about the system dynamics. According to this approach it is necessary to identify H and C matrices, as if nothing about them were known ("black-box system"). Fortunately, most dynamic effects can still be calculated on line, because many parameters of a robot are mainly known and vary only slightly. For this reason, it would be more convenient to estimate only the unknown and highly variable parameters, as is the payload mass.

Another interesting attempt to introduce adaptive control strategy in robotics can be found in Lee's papers [29]. Here, the adaptive control is introduced at the level of linearized perturbation equations in the vicinity of a nominal joint trajectory. Lee suggests on-line calculation of Newton-Euler's equations for feed-forward computation of nominal control signals. A recursive least-squares identification scheme is used to identify the system parameters in the linearized perturbation equations and an optimal (one step ahead) adaptive self-tuning controller is designed to minimize the position and velocity errors along the trajectory. An extension of the idea to resolved motion rate and acceleration control is given in Ref. [30]. But, there is one problem with this approach which is still unresolved. Let us first suppose that the parameters of a robotic system are exactly known. Then, it is obvious that the identification of the linearized model is not necessary. We can use the method for linearized model calculation, which is also based on Newton-Euler's equations [32]. On the other hand, if we suppose that some parameters are unknown (for example, workpiece mass and inertias), then the nominal control cannot be calculated exactly. Thus, it is necessary to introduce an adaptive algorithm at the nominal level as well. The adaptive feed-forward compensation is not discussed in Ref. [29, 30].

Another interesting concept of adaptive robot control has been developed by Timofeev and associates [33-35]. As will be shown in Paragraph 3.2.3, this strategy is based on a quasi-gradient estimation algorithm and nonlinear feedback loops. Thus, it belongs to the class of indirect adaptive control laws. The unknown parameters are first estimated, and then fed to the controller. The controller as well as the estimator use the entire nonlinear robot model in on-line operation. For this reason, although a high quality tracking can be expected, its numerical complexity makes it very unsuitable for implementation.

Finally, we shall present an indirect decentralized adaptive control approach, which represents a trade-off between numerically burdensome but high performance and simple robust controllers suitable for implementation on low-cost microcomputers. The main difference between this and the previously described approach is in the following: here, a set of a few unknown parameters is estimated. Actually, this set includes the parameters of payload (its mass and inertias). The parameter estimation is based on an efficient quasigradient algorithm and the robot sensitivity model [36-39]. This model relates the driving torques/forces of actuators and the mass (inertias) of a payload. Vukobratović and Kirćanski [40] have shown that the sensitivity model contains much fewer numerical operations than the entire dynamic model. The estimated parameters then influence the decentralized controller gains. The feedback gains are obtained from the stability analysis of the entire system for different parameters of payload. That is, the adaptive controller increases position and velocity feedback gains when manipulating payloads with increased mass and inertias, and vice versa. This concept will be illustrated by several examples.

3.2 Centralized Adaptive Control of Robot Manipulators

In this section we shall describe several adaptive control strategies based on centralized control structure. As explained in Chapter 2, within the centralized control structure the plant to be controlled is considered as an unique multivariable system that includes the model of both robot arm and actuators. We shall pay special attention to self-tuning control, reference model-following control and a specific indirect centralized adaptive algorithm.

3.2.1. Self-tuning control strategy

There are many structures of self-tuning regulators differing in parameter estimation technique and control algorithms. But, it is usually assumed that the process to be controlled can be described by the stochastic system [1-9]

$$A(q^{-1})y(t) = B(q^{-1})u(t-t_d)+C(q^{-1})e(t) \tag{3.2.1}$$

where u and y are input and output signals, and $\{e(t)\}$ is a distrubance which represents a sequence of independent random variables. Further, $A(q^{-1})$, $B(q^{-1})$ and $C(q^{-1})$ are polynomials in the backward shift operator q^{-1}. In (3.2.1) t denotes the sampling instant, the sampling interval T is normalized to be one unit, and t_d is a positive integer specifying the time delay. Notice that $u(t-t_d)$ is equivalent to $q^{-t_d}u(t)$. Generally, the autoregressive model (3.2.1) is multivariable with u and y representing input and output vectors, respectively. Then the coefficients A and B represent the matrices whose elements are polynomials in q^{-1}.

Almost all parameter estimation algorithms for linear systems have been based on this model [1-9]. This model is linear with constant coefficients which should be estimated. On the other hand, the robot model is a highly nonlinear system. Thus, we shall not start our presentation with such an approximation, but with the nonlinear discrete model. We shall try to present a theory which is "compatible" with nonlinear robotic systems. But it should immediately be pointed out that such a theory is quite new, published in the past few years (for example, [41-43]). These results give a clear theoretical background necessary to explain various phenomena related to parameter vector dimensionality, noise influence, prediction error model design and others. We believe that these results will prove to be very applicable in the near future, regardless of enormous numerical complexity.

Nonlinear estimation

Parameter estimation techniques for nonlinear systems depend critically on the choice of model structure. The classical model structure used in nonlinear systems identification has been the functional series expansion of Volterra or Wiener [42]. Most of these expansions require a very large number of coefficients to characterize the process (several hundreds for a simple quadratic nonlinearity in cascade with a first-order linear system). The reason for this lies in the fact that these expansions map past inputs into the present output:

$$y(t) = \sum_{i=1}^{\infty} V_{k_i}(u(i),\dots,u(t)) \tag{3.2.2}$$

where V_{k_i} is a homogeneous polynomial of degree i. This expansion involves lagged inputs only and thus requires a large number of coefficients. Utilizing results from automata theory to obtain a nonlinear difference equation model, Billings and Leontaritis [41] derived the following representation

$$y(t) = F_*[y(t-1),\ldots,y(t-n_y),\ u(t-1),\ldots,u(t-n_u)] \tag{3.2.3}$$

where $F_*[\cdot]$ is some nonlinear function of u and y. Supposing that the degree of nonlinearity does not exceed ℓ, autoregressive model (3.2.3) can be represented as polynomial [42]

$$y(t) = \sum_{i=1}^{s} \theta_i V_i + \sum_{i=1}^{s}\sum_{j=1}^{s} \theta_{ij} V_i V_j + \ldots + \underbrace{\sum_{i=1}^{s}\sum_{j=1}^{s} \cdots \sum_{n=1}^{s}}_{\ell \text{ times}} \theta_{ij\ldots n} V_i V_j \cdots V_n \tag{3.2.4}$$

where $V_1 = y(t-1),\ldots,V_{n_y} = y(t-n_y)$, $V_{n_y+1} = u(t-1),\ldots,$ $V_s = u(t-n_u)$, and $s = n_y+n_u$. Introducing such polynomial expansion, we obtain the linear dependence of y(t) on parameters θ.

Assuming that the system output y(t) is corrupted by zero mean additive noise e(t) to yield the measured output signal

$$z(t) = y(t) + e(t) \tag{3.2.5}$$

and substituting in (3.2.4), we obtain

$$z(t) = F_*^{\ell'}[z(t-1),\ldots,z(t-n_y),\ u(t-1),\ldots,u(t-n_u),\ e(t-1),\ldots,e(t-n_y)] + e(t) \tag{3.2.6}$$

we see that, although the model is linear in the parameters, the inclusion of lagged outputs introduces cross-product terms between the noise and the process input-output signals. In linear identification it is usually assumed that the internal noise can be translated to be additive at the output. The superposition principle does not apply here.

Before proceeding to parameter estimation, let us return to robotic systems. The model of a robot arm powered by actuators is the nonlinear multivariable system (2.2.8). Assuming that the state-vector x is measurable, y(t) = x(t), and applying Euler's expansion, we obtain the discrete model

$$S: \quad y(t)=y(t-1)+TA_D(y(t-1),\ \theta,\ d)+TB_D(y(t-1),\ \theta,\ d)u(t-1) \quad (3.2.7)$$

where T is the sampling period, and u is the constrained input vector: $|u_i| < U_m^i$. Matrices A_D and B_C are nonlinear with respect to y(t-1). It was shown in Chapter 1 that the nonlinearities are of sine/cosine, square sine/cosine and simple quadratic type. Approximating the degree of nonlinearity by a positive integer ℓ, we obtain the model (3.2.4) with $n_y=1$, $n_u=1$. As y(t) in Eq. (3.2.7) represents an n-dimensional vector, parameters θ on the right-hand side of (3.2.4) are obviously matrices of appropriate dimensions. In order to obtain a clearer insight into the model structure, let us fix $\ell=3$ and rewrite (3.2.7) as

$$S: \quad y(t) = h+A_D^{(1)}y(t-1)+A_D^{(2)}y^2(t-1)+B_D^{(1)}y(t-1)u(t-1) + \\ + A_D^{(3)}y^3(t-1)+B_D^{(2)}y^2(t-1)u(t-1) \quad (3.2.8)$$

or concisely

$$y(t) = F_*^3[y(t-1),\ u(t-1)] \quad (3.2.8')$$

with h being a constant term due to gravity effects. Here, $A_D^{(1)}$ is an N×N dimensional matrix, A_D^2 is an N×((N(N+1))/2) dimensional matrix, etc.

Let us consider in more details the robotic system representation (3.2.8). We see that the present output y(t) depends on input/output pair at the time instant t-1. Here y involves joint coordinates and velocities (and perhaps some variables corresponding to actuators. However, accepting second-order models for actuators, the mathematical model of a robot manipulator may be written in the joint coordinate space as follows:

$$\ddot{y} = \mathcal{A}(y,\ \dot{y}) + \mathcal{B}(y)u \quad (3.2.9)$$

where $y\in R^n$ represents a joint coordinate vector, $\mathcal{A}\in R^n$ and $\mathcal{B}\in R^{n\times n}$. Following the procedure for obtaining nonlinear autoregressive model

(3.2.8), we shall see that y(t) depends on y(t-1), y(t-2) and u(t-2). Now, although the dimension of output vector is reduced ($y \in R^n$ while in (3.2.8) it was at least 2n-dimensional), the number of unknown coefficients is obviously not decreased. For a 6-degree-of-freedom robot, there are more than 2 thousand scalar parameters.

The parameter estimation of a robotic system represented by nonlinear autoregressive model (3.2.8) is a part of self-tuning control strategy. Introducing the prediction error vector $\varepsilon(t) = y(t) - \hat{y}(t)$, and formulating (3.2.8) into a prediction error model, we obtain the model:

$$y(t) = F^3[y(t-1),\ u(t-1),\ \varepsilon(t-1)] + \varepsilon(t) \tag{3.2.10}$$

which is the basis of parameter estimation algorithms. Expanding Eq. (3.2.10) and regrouping terms yields

$$y(t) = F^3_{yu}[y(t-1),\ u(t-1)]+F^3_{yu\varepsilon}[y(t-1),\ u(t-1),\ \varepsilon(t-1)]+\varepsilon(t) \tag{3.2.10a}$$

where $F^3_{yu}[\cdot]$ is a function of y and u only, and $F^3_{yu\varepsilon}[\cdot]$ represents all the cross product terms involving $\varepsilon(t)$. Separating out the unknown parameters gives

$$y(t) = \psi^{3T}_{yu}(t)\theta_{yu}(t-1)+\psi^{3T}_{yu\varepsilon}(t)\theta_{yu\varepsilon}(t-1)+\varepsilon(t) \tag{3.2.10b}$$

or simply

$$y(t) = \psi^3_{yu}(t)\theta_{yu}(t-1)+\xi(t) \tag{3.2.10c}$$

where $\zeta(t)$ is highly correlated with the elements of $\psi^3_{zu}(t)$. For the autoregressive nonlinear robot model (3.2.8), we have

$$\psi^{3T}_{yu}(t) = [y(t-1),\ u(t-1),\ y^2(t-1),\ y(t-1)u(t-1),$$

$$y^3(t-1),\ y^2(t-1)u(t-1)] \tag{3.2.11}$$

$$\theta_{yu}(t-1) = [\theta_1,\ \theta_2,\ \theta_{11},\ \theta_{12},\ \theta_{111},\ \theta_{112}]^T \tag{3.2.12}$$

Because of the high correlation of $\xi(t)$ with $\psi^3_{zu}(t)$, we cannot directly apply the least squares algorithm, since it would yield biased estimates. Thus, we must use extended least squares algorithm based on the model (3.2.10b), i.e. the prediction model:

$$y(t) = \psi^T(t)\theta(t-1)+\varepsilon(t) \tag{3.2.12}$$

with $\psi^T(t) = [\psi_{yu}^{3T}(t), \psi_{yu\varepsilon}^{3T}(t)]$ and $\theta(t) = [\theta_{yu}(t)\ \theta_{yu\varepsilon}(t)]^T$. Now, we can apply the algorithm [41]:

$$\hat{\theta}(t+1) = \hat{\theta}(t)+P(t)\psi(t+1)(\lambda(t+1) + \psi(t+1)^T P(t)\psi(t+1))^{-1}\varepsilon(t+1) \tag{3.2.13}$$

with

$$P(t+1) = (P(t) - \frac{P(t)\psi(t+1)\psi(t+1)^T P(t)}{\lambda(t+1)-\psi(t+1)^T P(t+1)\psi(t+1)})/\lambda(t+1)$$

$$\lambda(t+1) = \lambda_o\lambda(t)+(1-\lambda_o)$$

$$\varepsilon(t) = y(t+1)-\psi(t+1)^T\hat{\theta}(t)$$

where $\lambda(t)$ is a variable forgetting factor.

The main disadvantage of the extended least squares algorithm (3.2.13) is the need to include noise or prediction error terms in the estimation error. The number of entries in $\hat{\theta}(t)$ is enormously large (even several thousands) for a typical manipulator. In an attempt to limit the dimension of the vector $\hat{\theta}(t)$ an interesting suboptimal least squares algorithm has been developed in [41]. Here, instead of immeasurable noise-free output y(t), we use its estimate $\hat{y}(t)$ recursively computed from

$$\hat{y}(t) = \psi_{\hat{y}u}^T(t)\hat{\theta}_{yu}(t-1). \tag{3.2.14}$$

The parameter estimation further proceeds as in (3.2.13) except for the following meaning of $\psi(t)$ and $\theta(t)$:

$$\psi(t)^T = \psi_{\hat{y}u}^T(t)$$
$$\theta(t) = [\theta_{yu}(t)]. \tag{3.2.15}$$

Let us drow now some conclusions from the previous theory. First, least squares parameter estimation algorithms for highly nonlinear systems such as robot manipulators have been developed in the most recent literature. The estimation methods are principally the same as those for linear systems, but the problem of model structure is much more complex. It is shown that for a typical robot it is necessary to estimate more than a thousand coefficients. Although this theory is not applied in robotics yet, we have presented it because of its consistence and clearness. Apart from that, the algorithms which will be described in the text to follow, are special cases of this theory. These special cases are obtained by including certain approximations with respect to the actual plant. These approximations, however, simplify the control structures, which is important for the application on low-cost microcomputers within robot control units.

Linear model estimation

The first and a very common approximation is the model linearization about a reference trajectory [10, 29, 30]. Using the Taylor's series expansion on (2.2.8) about the nominal trajectory and assuming that the higher order terms are negligible, the associated linearized perturbation equations for this control system can be obtained

$$\delta\dot{x} = \left.\frac{\partial A_D(x,\ \theta,\ d)}{\partial x}\right|_{x=x^o} + \left.\frac{\partial B_D(x,\ \theta,\ d)}{\partial x}\right|_{x=x^o} u^o\ \delta x + \\ + B_D(x^o,\ \theta,\ d)\,\delta u \tag{3.2.16}$$

where $\delta x(t) = x(t) - x^o(t)$, $\delta u(t) = u(t) - u^o(t)$, $x^o(t)$ and $u^o(t)$ are nominal state and input vectors. Here, u^o and δu are constrained in amplitude (see (2.3.6)). Using the same notation as in (2.3.7), we obtain

$$\delta\dot{x} = \tilde{A}_D(x^o,\ \theta,\ d)\,\delta x + B_D(x^o,\ \theta,\ d)\,\delta u \tag{3.2.17}$$

where $\tilde{A}_D$ represents the Jacobian of $A_D(x,\ \theta,\ d) + B_D(x,\ \theta,\ d)u$, evaluated at the nominal states $x^o(t)$. Discretizing (3.2.17) with the sampling period T, we obtain the following discrete-time linear model

$$y(t) = F(t-1)y(t-1) + G(t-1)u(t-1) \tag{3.2.18}$$

with $u(t) \in R^n$ being a piecewise constant control input vector over the time interval between any two consecutive sampling instants, $y(t)$ is the perturbed state vector $y = \delta x$. Here we suppose that the state variables are measurable. This is quite a realistic supposition with robotic systems. Still, the estimation of the matrices F and G may be computationally awkward because of high dimensions. Supposing that 3 states for any subsystem are measured, we obtain a $(3n)\times(3n)$ dimensional F matrix, and a $(3n)\times n$ dimensional G matrix. Notice that for n=6, even 432 parameters are to be estimated. For this reason, it is convenient to follow the model (3.2.9) given in joint coordinate space. The corresponding linearized model becomes

$$\delta\ddot{y} = A_1(t)\delta\dot{y}+A_2(t)\delta y+B(y^o(t))\delta u \tag{3.2.19}$$

where $A_1 \in R^{n\times n}$ and $A_2 \in R^{n\times n}$ are the Jacobians of the right-hand side of (3.2.9) with respect to y and $\dot{y}$, evaluated at the nominal states. Here, $y \in R^n$ is the joint coordinate vector of manipulator, and n - the number of joints. Now, we can apply Euler's method and discretize the linearized model (3.2.19)

$$y(t) = A_1 y(t-1) + A_2 y(t-2) + B_2 u(t-2) + e(t) \tag{3.2.20}$$

For simplicity, the symbol $y(t)$ is used instead of $\delta y(t)$. Matrices A_1, A_2 and B_2 are $n\times n$ - dimensional, with unknown elements, and $e(\cdot)$ represents the modeling error.

Let us note that the model (3.2.20) represents a special case of autoregressive model (3.2.1), on which almost all the parameter estimation algorithms for linear systems [1-9] are based. Introducing the linear backward shift operator q^{-1}, model (3.2.20) becomes

$$[I-A_1q^{-1}-A_2q^{-2}]y(t) = B_2q^{-2}u(t) + e(t) \tag{3.2.21}$$

This is a special case of the model (3.2.1).

Model (3.2.20) can be rewritten in the form:

$$y(t) = [A_1 \quad A_2 \quad B_2]\begin{bmatrix} y(t-1) \\ \hline y(t-2) \\ \hline u(t-2) \end{bmatrix} + e(t) \tag{3.2.22a}$$

or, simply, as

$$y(t) = \theta_{yu}^T \psi(t-1) + e(t) \tag{3.2.22b}$$

with $\theta_{yu}^T = [A_1 \mid A_2 \mid B_2]$ and $\psi(t-1)^T = [y(t-1)^T, y(t-2)^T, u(t-2)]^T$. We see that θ_{yu} comprises $3n^2$ parameters. For a 6-degree-of-freedom robot it corresponds to 108 parameters. This is 1 or even 2 orders of magnitude less than for a nonlinear autoregressive robot model.

Let us point out an essential difference between the nonlinear autoregressive model (3.2.6) and the linear autoregressive model (3.2.20) as far as noise is concerned. The cross-product terms between the noise and the process input-output signals are characteristic for the nonlinear model. The superposition principle is applicable here and results in the additive character of noise. But it does not mean that a classical least squares method is directly applicable. This is a good and simple method, but its main drawback is that it gives biased estimates unless the output y(t) is directly measurable and noise-free. Otherwise, the extended least squares algorithm must be used. But the extended prediction model results in an extended parameter vector, similar to θ in (3.2.12). In most robot manipulators the measurement of joint coordinates can be considered as noise-free. Joint coordinate sensors are most often high-precision and high-quality digital encoders. Thus, we shall make the following assumption: measurement noise is negligible. We can now apply the well known classical least squares scheme to the model (3.2.22b). Thus, the recursive scheme follows

$$\hat{\theta}_i(t+1) = \hat{\theta}_i(t) + P(t+1)\psi(t)[y_i(t+1) - \hat{\theta}_i^T(t)\psi(t)] \tag{3.2.23}$$

$$P(t+1) = \left(P(t) - \frac{P(t)\psi(t)\psi^T(t)P(t)}{\lambda + \psi^T(t)P(t)\psi(t)}\right)/\lambda \tag{3.2.24}$$

where λ denotes the forgetting factor ($0<\lambda\leqslant 1$), $\hat{\theta}_i(t)$ - the estimate of the ith row of matrix $[A_1 \mid A_2 \mid B_2]$, and y_i is the output of the ith robot subsystem. Here, for simplicity, λ is taken to be a constant. It could also be time-varying as in (3.2.13). The use of forgetting factor λ is very common when tracking slowly time-varying system parameters. Now, we must impose the second assumption: variation speed of parameters is lower than the adaptation speed. This assumption is made to simplify the identification scheme and it needs to be verified experimentally. Theoretically, it cannot be proved.

Extended minimum variance self-tuning control

In the preceding text we pointed out that the parameter estimation represents the basis for self-tuning regulators. A block diagram of the entire control structure is presented in Fig. 3.1. The methods belonging to the self-tuning structures can be classified into 3 categories

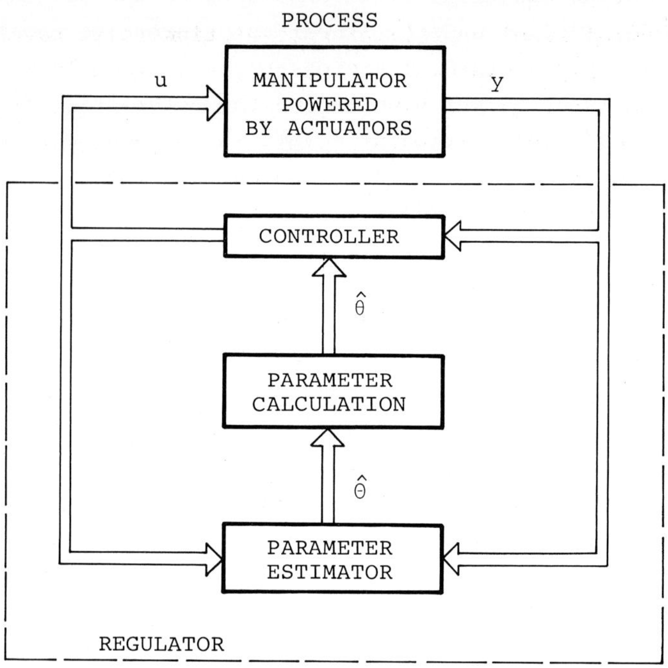

Fig. 3.1. Block diagram of the self-tuning regulator

[1 - 12]:

1) Minimum variance control
2) Extended minimum variance control
3) Pole-zero placement control.

Generally, any of these controllers is of the form [1]:

$$u(t) = \frac{G(q^{-1})}{F(q^{-1})} y(t) \tag{3.2.25}$$

where

$$F(q^{-1}) = 1 + f_1q^{-1} + \cdots + f_\ell q^{-\ell}$$
$$G(q^{-1}) = g_1 + g_2q^{-1} + \cdots + g_mq^{-m+1}. \tag{3.2.26}$$

The regulator parameters

$$v = [g_1 \cdots g_m \; f_1 \cdots f_\ell]^T \tag{3.2.27}$$

should be computed from the estimated prediction model parameters. Assuming that the prediction model (3.2.22b) is reasonably good, a regulator can be obtained by applying any design method to (3.2.22b). That would yield from a simple linear to complicated nonlinear expressions whose evaluation is far more time consuming than the updating of the parameter estimates [1].

Minimum variance controllers (MVC) as well as extended MVC are based on Linear quadratic Gaussian theory. They involve solving the Riccati equation following from the minimization of the performance index

$$V = \lim_{N\to\infty} E\{\frac{1}{N} \sum_{i=t}^{N} [y^2(i) + \zeta u^2(i)]\} \tag{3.2.28}$$

This is obviously computationally awkward and difficult for real-time implementation. There are, however, special cases when the calculations can be considerably simplified. When there is no penalty on the control, $\zeta=0$, the criterion (3.2.28) reduces to the variance of the output. The corresponding controllers are called minimum variance controllers. This procedure is essentially a pole placing algorithm where the controller places poles to cancel system zeros. The resulting control is, however, erratic and often with excessively large amplitudes.

Extended MVC includes penalization of both error and input norms (see papers of Clarke and Gawthrop [44, 45]). This approach was also followed by Koivo [10]. For multivariable linear prediction model (3.2.22):

$$y(t) = \hat{\theta}_{yu}^T \psi(t-1) + e(t) \tag{3.2.29}$$

we introduce the performance index

$$J(u) = E\{\Delta y(t+1)^T Q \Delta y(t+1) + u(t)^T R u(t)\} \tag{3.2.30}$$

where $\Delta y(t) = y(t) - y^o(t)$, $y^o(t)$ - joint coordinate vector at nominal trajectory, $E\{\cdot\}$ - mathematical expectation conditioned on the available measurements at time $\leq t-1$, $Q>0$ is a symmetric $n\times n$ matrix, and $R>0$ is a control weighting matrix. Notice that the main difference between (3.2.28) and (3.2.30) lies in the fact that (3.2.30) includes only one step evaluation, while (3.2.28) takes into account input/output variance accurring at time t, t+1,...,N for $N\to\infty$. This simplification leads to the simple control law [10]

$$u(t) = -R^{-1}B_o^TQ(\hat{y}(t+1|t-1)-y^o(t+1)) \tag{3.2.31}$$

with recursive relations

$$\hat{y}(t+1|t-1) = \hat{\theta}_{yu}^T(t-1)\hat{\psi}(t|t-1)$$

and $\hat{\psi}(t, t-1) = [\hat{y}(t|t-1), y(t-1), y(t-2), u(t-2)]$.

Here, $\hat{y}(t+k|t-1)$ means the k step ahead predicted value of y. In practical applications the control law (3.2.31) could result in a closed--loop unstable operation if unknown robot parameters are far from the initial guess and/or are fast time varying compared to adaptation speed. The following remark may be made which applies to optimal control approach in general: many iterations in selection of Q and R matrices compound the design problem. According to the simulation results of Stanford manipulator [10], there exists also the following serious problem: estimated parameters change erratically during the first part of the trajectory (in about 0.3s) producing abrupt variations in control signals (input voltages to DC motors) and erratic movements of robot arm (Fig. 3.2). Thereby, is should be kept in mind that the time evaluations shown in Fig. 3.2 are obtained after some trial-and-error iterations concerning initial guess and performance index weighting factors. Interference between links flexibility and erratic control is the second problem, that generally burdens the application of EMV self--tuning controllers.

The self-tuning approach based on pole-zero placement is presented in many papers devoted to adaptive control of linear systems (see, for example, [46]). Leininger [11, 12] examined their applicability to robotic systems. But, the conclusions drawn for extended minimum variance approach practically hold for pole-zero placement control as well. The difference is in that we impose the closed-loop poles instead of weighting matrices. On the other hand, this procedure is described for single-input, single-output linear systems only. This procedure will be the basis for PID local controllers within the scope of decentralized adaptive control strategy, which will be described in Paragraph 3.3.

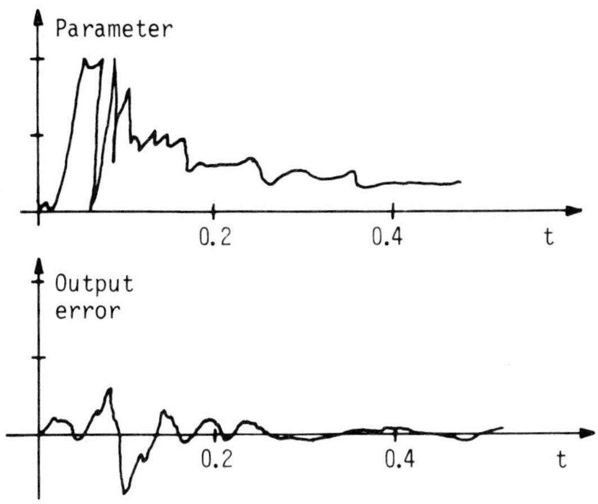

Fig. 3.2. Example of self-tuning action

3.2.2. Centralized reference model following control

There are 3 basic approaches to the design of model reference adaptive systems [13]:

1) Local parametric optimization theory
2) Lyapunov functions
3) Hyperstability and positivity concepts.

The first approach is explained in more details in the introduction to this chapter. This is chronologically the first approach developed within the scope of linear system theory. It includes the well known gradient method - the steepest descent algorithm. An interesting application of this method in robotics is given by Dubowsky and DesForges [20]. The obtained results appear to be very simple and practical. But these methods have a general drawback: nothing is said about the stability of the resulting adaptive system. This approach as well as the approach by Lyapunov functions [47, 48] will not be considered in this book. According to the results from general system theory, one can say that the stability-based approach, and in particular the use of hyperstability and positivity concepts, is the most successful approach to the design of model reference adaptive systems [13]. The application of the theory to centralized robot models is given in Ref. [26, 49].

Before proceeding to the theory of hyperstable robot controllers, let us point out an essential difference between self-tuning and hyperstable regulators. The self-tuning approach is based on the nonlinear polynomial model (3.2.4) or the autoregressive linear model (3.2.1). As the coefficients of these models are unknown, they should be estimated by minimizing an error criterion, or by placing zeros or poles of the closed-loop system. However, many dynamic effects can be computed on-line by using Newton-Euler's, Lagrange's or some other method. There are still some effects which can not be calculated due to their stochastic nature. These are measuring noise, vibrations due to elasticity of links and joints, inertial and gravitational effects due to the influence of different workpieces etc. Thus, it is quite natural to compensate the deterministic effects by "on-line" model computation, and to compensate the stochastic effects by introducing the adaptive control loops, too. Such a control structure can be obtained by combining the hyperstability theory and model reference adaptive control. First, we shall derive a suitable robot reference model in the sequel. Then, the structure of both nonadaptive and adaptive control terms will be presented.

In order to obtain a suitable centralized reference model, let us consider the models of subsystems. Each subsystem is assigned to a manipulator degree-of-freedom, describing the dynamical behaviour of the corresponding actuator and the transmission between it and the joint. Thus, let us consider the subsystem model S^i $(i=1,\ldots,n)$, defined by (2.2.2). Denoting by x^{io} the n_i - dimensional state vector of the ith subsystem on the nominal trajectory, and by $u_i^o \in R$ the nominal imput signal, we obtain

$$S^i: \quad \dot{x}^{io} = A^i(\theta^i)x^{io}+b^i(\theta^i)u_i^o+f^i(\theta^i)P_i^o$$

where P_i^o represents the nominal driving torque (or force) of the ith actuator. As P_i^o depends on x^{io}, we can extract the corresponding terms from the manipulator model

$$S^M: \quad P^o = H(q^o,\ d)\ddot{q}^o+h(q^o,\ \dot{q}^o,\ d)$$

and adjoin them to subsystem models. Thus, the diagonal elements of matrix $H(q^o, d)$ should be presented as

$$H_{ii}(q^o, d) = \bar{H}_{ii}(d) + \varepsilon_H(q^o, d)$$

with $\bar{H}_{ii}(d)>0$, $\varepsilon_H=H_{ii}-\bar{H}_{ii}$. The ith element of vector $h(q^o, \dot{q}^o, d)$ can be presented as

$$h_i(q^o, \dot{q}^o, d) = \bar{h}_i(d)q_i^o + \varepsilon_h(q^o, \dot{q}^o, d)$$

with $\varepsilon_h = h_i-\bar{h}_iq_i^o$. Subsystem model S^i now becomes

$$S^i: \quad \dot{x}^{io} = \bar{A}^i(\theta^i, d)x^{io} + \bar{b}^i(\theta^i, d)u_i^o+\bar{f}^i(\theta^i, d)\Delta P_i \qquad (3.2.32)$$

where $\bar{A}^i(\theta^i, d) = \bar{C}^i(\theta^i, d)(A^i(\theta^i) + f^i(\theta^i)\bar{h}(d)T_1^i)$

$$\bar{C}^i(\theta^i, d) = (I_{n_i}-f^i(\theta^i)\bar{H}_{ii}(d)T^i)^{-1},$$

$$\bar{b}^i(\theta^i, d) = \bar{C}^i(\theta^i, d)b^i(\theta^i),$$

$$\bar{f}^i(\theta^i, d) = \bar{C}^i(\theta^i, d)f^i(\theta^i),$$

with $T_1^i \in R^{1\times n_i}$ and $T^i \in R^{1\times n_i}$ being the transformation matrices defined as

$$T_1^ix^i = q_i \quad \text{and} \quad T^i\dot{x}^i = \ddot{q}_i.$$

Here, ΔP_i represents the difference between the actual and the approximated value of driving torque (force) of the ith actuator

$$\Delta P_i = P_i-\bar{H}_{ii}T^i\dot{x}^i-\bar{h}_iT_1^ix^i.$$

The model (3.2.32) includes both the actuator parameters and inertial and gravitational loads. These loads depend on manipulator configuration and are time-varying. Denoting by $\bar{d}$ an average value of parameter vector, the nominal control can be approximately evaluated from the free subsystem S^i (3.2.32):

$$\dot{x}^{io} = \bar{A}^i(\theta^i, \bar{d})x^{io} + \bar{b}^i(\theta^i, \bar{d})u_i^o \qquad (3.2.33)$$

with $\bar{A}^i$ and $\bar{b}^i$ being time invariant matrix and vector, respectively.

The set of subsystems (3.2.33) represents actually n independent linear time-invariant second- or third-order systems. These subsystems are decoupled and thus very useful for control synthesis. If we wish to control the nonlinearly coupled system so that it behaves as a set of linear, decoupled subsystems, it is quite natural to accept (3.2.33) for the reference model. To underline the introduction of reference model, we shall use the following notations

$$A_M^i = \bar{A}^i(\theta^i, \bar{d})$$
$$b_M^i = \bar{b}^i(\theta^i, \bar{d}) \tag{3.2.34}$$

or

$$S_M^i: \quad \dot{x}^{io} = A_M^i x^{io} + b_M^i u_i^o. \tag{3.2.35}$$

In the case of second-order subsystems, (3.2.35) becomes

$$\ddot{q}_i^o = -a_i^* \dot{q}_i^o - f_i^* q_i^o + b_i^* u_i^o \tag{3.2.36}$$

with $a_i^* = a_i/\mathcal{L}_i$, $f_i^* = f_i \bar{h}_i/\mathcal{L}_i$ and $b_i^* = b_i/\mathcal{L}_i$, where $\mathcal{L}_i = 1 + f_i \bar{H}_{ii}$. Here, the coefficients a_i, f_i and b_i are determined from the second-order model of the actuator without adjoint inertial and gravitational load: $\ddot{q}_i = a_i \dot{q}_i + b_i u_i + f_i P_i$. Substituting $P_i = \bar{H}_{ii} \ddot{q}_i + \bar{h}_i q_i$ into it, we get (3.2.36). The matrices of the reference model (3.2.35) are now

$$A_M^i = \begin{bmatrix} 0 & 1 \\ -f_i^* & -a_i^* \end{bmatrix}, \qquad b_M^i = \begin{bmatrix} 0 \\ b_i^* \end{bmatrix} \tag{3.2.37}$$

If we join together the subsystems (3.2.35), we obtain the centralized reference model

$$S_M: \quad \dot{x}^o = A_M x^o + B_M u^o \tag{3.2.38}$$

For second-order subsystems, we have

$$A_M = \left[\begin{array}{c|c} 0 & I_n \\ \hline A_{21}^M & A_{22}^M \end{array}\right] \quad \text{and} \quad B_M = \left[\begin{array}{c} 0 \\ \hline B_2^M \end{array}\right] \tag{3.2.39}$$

with $A_{21}^M = -\mathrm{diag}(f_i^*)$, $A_{22}^M = -\mathrm{diag}(a_i^*)$ and $B_2^M = \mathrm{diag}(b_i^*)$.

The entire manipulator model that includes second-order subsystems can be represented by

$$S: \quad u = \hat{H}(q, \theta)\ddot{q} + \hat{h}(q, \dot{q}, \theta) \tag{3.2.40}$$

with $\hat{H}$ and $\hat{h}$ obtained by using simple transformations upon H and h. This model can also be presented in canonical form

$$\dot{\xi} = A(\xi, \theta)\xi + B(\xi, \theta)u \tag{3.2.41}$$

with $\xi^T = [q^T \mid \dot{q}^T]$.

Let us first suppose that the parameter vector θ is constant and known. Then, we can apply the centralized control

$$u = \hat{H}(q, \theta)[\ddot{q}^o + D(\xi - \xi^o)] + \hat{h}(\xi, \theta) \tag{3.2.42}$$

with $D \in R^{2n \times 2n}$ being the feedback gain matrix. Further, we shall suppose that this matrix is equal to A_M in (3.2.39). The model (3.2.42) now becomes

$$u = [0 \mid \hat{H}(\xi, \theta)](B_M u^o + A_M \xi) + \hat{h}(\xi, \theta) \tag{3.2.43}$$

where u^o satisfies the reference model (3.2.38). It is easy to show that $\hat{h}(\xi, \theta)$ can be presented as

$$\hat{h}(\xi, \theta) = [\hat{Q}_2(\xi, \theta) \mid \hat{Q}_1(\xi, \theta)]\xi \tag{3.2.44}$$

where $\hat{Q}_1$ and $\hat{Q}_2$ are appropriate $n \times n$ matrices. Thus

$$\hat{h}(\xi, \theta) = \hat{Q}_{21}(\xi, \theta)\xi, \qquad \hat{Q}_{21} \in R^{n \times 2n}. \tag{3.2.45}$$

Substituting this in (3.2.43), we get

$$u = ([0 \mid \hat{H}(\xi, \theta)]A_M + \hat{Q}_{21}(\xi, \theta))\xi + [0 \quad \hat{H}(\xi, \theta)]B_M u^o$$

or simply

$$u = K_p(\xi, \theta)\xi + K_u(\xi, \theta)u^o \tag{3.2.46}$$

with $K_p(\xi, \theta)$ and $K_u(\xi, \theta)$ being determined by $\hat{H}$ and $\hat{Q}_{21}$ and by reference model matrices A_M and B_M. The control law (3.2.46) ideally compensates for system nonlinearity and makes it decoupled. The system model in the space of $e=x-x^o$ becomes

$$\dot{e} = A_M e \tag{3.2.47}$$

where A_M is a diagonal matrix.

Suppose now that θ is unknown. Replacing the actual value of θ by a constant "average" quantity $\bar{\theta}$, we obtain the control law

$$u = K_p(\xi, \bar{\theta})\xi + K_u(\xi, \bar{\theta})u^o \tag{3.2.48}$$

It is apparent that the asymptotic stability (3.2.47) does not hold now. Thus, we have to modify (3.2.48) by introducing some adaptive components. Supposing that $\mathcal{L}_p(\xi, e)$ and $\mathcal{L}_u(\xi, e)$ represent the matrices governed by an adaptive mechanism, we can write

$$u = [K_p(\xi, \bar{\theta})+\mathcal{L}_p(\xi, \theta)]\xi+[K_u(\xi, \theta)+\mathcal{L}_u(\xi, e)]u^o \tag{3.2.49}$$

This is the general form of direct, centralized adaptive control law. Here, we are concerned with 2 basic components:

1) a deterministic, dynamic component expressed in terms of matrices $K_p(\xi, \bar{\theta})$ and $K_u(\xi, \bar{\theta})$, and

2) an adaptive component expressed in terms of functions $\mathcal{L}_p(\xi, e)$ and $\mathcal{L}_u(\xi, e)$.

For the computation of the dynamic component, it is very useful to apply the numeric-symbolic method, described in Chapter 1. The adaptive component will be determined using the hyperstability theory. The block diagram of the complete system including reference model (3.2.38), control law (3.2.49) and robot model (3.2.41), is shown in Fig. 3.3. A linear static compensator D is also introduced in this scheme:

$$v = De \tag{3.2.50}$$

where v is the generalized error which will be explained in the text to follow.

Let us now derive the differential equation which satisfies the error

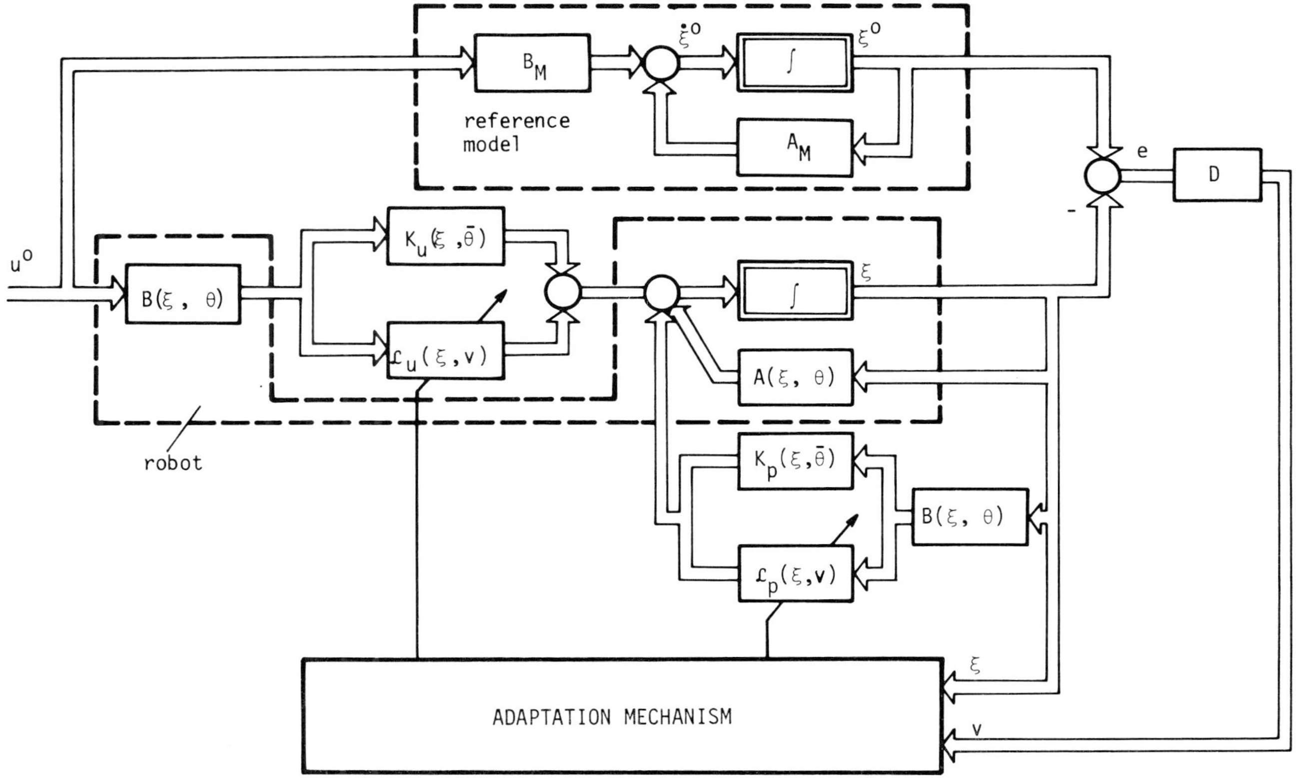

Fig. 3.3. Centrlaized adaptive control scheme

vector e. Substituting (3.2.49) into robot model (3.2.41), we get

$$\dot{\xi} = (A+BK_p+B\mathcal{L}_p)\xi + (BK_u+B\mathcal{L}_u)u^o \tag{3.2.51}$$

In a perfectly compensated system

$$A + BK_p \to A_M, \qquad BK_u \to B_M, \tag{3.2.52}$$

and $\mathcal{L}_p \to 0$, $\mathcal{L}_u \to 0$. Thus, let us introduce the notations

$$\begin{aligned} \varepsilon_A &= A + BK_p - A_M, \\ \varepsilon_B &= BK_u - B_M \end{aligned} \tag{3.2.53}$$

so that $\varepsilon_A \to 0$ and $\varepsilon_B \to 0$. Putting (3.2.53) into (3.2.51), we get

$$\dot{\xi} = (A_M+\varepsilon_A+B\mathcal{L}_p)\xi + (B_M+\varepsilon_B+B\mathcal{L}_u)u^o. \tag{3.2.54}$$

Subtracting (3.2.54) from (3.2.38), we see that $e=\xi-x^o=\xi-\xi^o$ is governed by the differential equation

$$\dot{e} = A_M e - w \tag{3.2.55}$$

where

$$-w = (\varepsilon_A+B\mathcal{L}_p)\xi + (\varepsilon_B+B\mathcal{L}_u)u^o. \tag{3.2.56}$$

The equations (3.2.55) and (3.2.56) are presented in Fig. 3.4. The system shown in this figure consists of a linear feedforward block, and a nonlinear feedback block. Then the hyperstability conditions [13] become:

1) positivity of linear feedforward block, and

2) the feedback block belongs to the family of nonlinear systems satisfying an input/output relation of the form

$$\int_0^{t_1} v^T w \, dt \geq -\gamma_o^2 \quad \text{for} \quad \forall t_1 \geq 0 \tag{3.2.57}$$

which is known as "Popov's inequality".

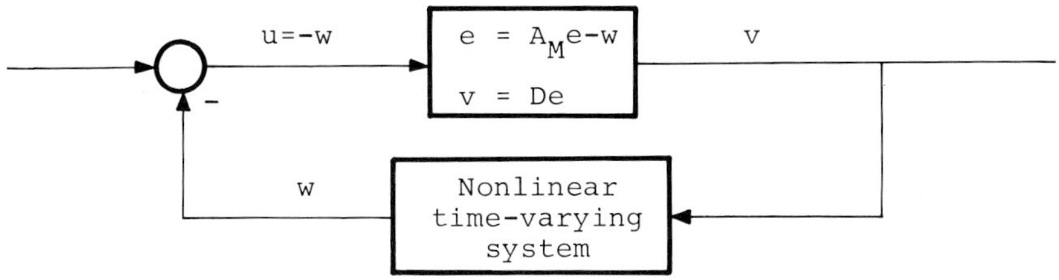

Fig. 3.4. Nonlinear control system

The static compensator D is intended to preserve the positivity of the feedforward block. The condition of positivity is fulfilled if D satisfies the Lyapunov equation

$$A_M^T D + DA_M = -Q \tag{3.2.58}$$

with $Q>0$.

From the condition 2), one obtains the adaptive components $\mathcal{L}_p(\xi, v)$ and $\mathcal{L}_u(\xi, v)$ as

$$\begin{aligned} \mathcal{L}_p &= k_p(t) v \xi^T \\ \mathcal{L}_u &= k_u(t) v u^{oT} \end{aligned} \tag{3.2.59}$$

where $k_p(t)$: $T \to R^{n\times n}$ and $k_u(t)$: $T \to R^{n\times n}$ are positive real functions. It is apparent that by choosing different forms of functions $k_p(t)$ and $k_u(t)$, we obtain a family of hyperstable adaptive controllers. For example, by choosing $k_p(t)v = k_p \operatorname{sgn}(v)$ and $k_u(t)v = k_u \operatorname{sgn}(v)$, we obtain

$$\mathcal{L}_p(t) = k_p(\operatorname{sgn}(v))\xi^T$$

$$\mathcal{L}_u(t) = k_u(\operatorname{sgn}(v))u^{oT}$$

where k_p and k_u are positive real constants. The adaption mechanism (3.2.59) can also be accepted as

$$\begin{aligned} \mathcal{L}_p &= k_p \frac{v}{\|v\|}(\operatorname{sgn} \xi)^T \\ \mathcal{L}_u &= k_u \frac{v}{\|v\|}(\operatorname{sgn} u^o)^T \end{aligned} \tag{3.2.60}$$

which is known as unit vector adaptation law [13, 50]. Obviously, the Popov's inequality cannot be satisfied for any k_p and k_v. These constants should be positive and large enough, thus making the influence of noncompensated system part reflected through ε_A and ε_B less effective than BL_p and BL_u:

$$k_p > c||(B^TB)^{-1}B^T\varepsilon_A|| \qquad\qquad (3.2.61)$$
$$k_v > c||(B^TB)^{-1}B^T\varepsilon_B||$$

where the constant c depends on the adaptation mechanism (3.2.59) - (3.2.60).

Finally, let us point out some excellences and shortcomings of the described control strategy. Here, the manipulator model is not considered as an unknown plant, but as a plant whose dynamics is partially known and computable in real-time. Consequently, this strategy offers better transient behaviour compared to that with self-tuning regulators. According to the simulation results [26], "erratic" variations do not occur in control signals and robot movements when starting the learning process, but a smooth adaptation of gains during the time, as shown in Fig. 3.5. The second major advantage of this approach follows from the fact that the adaptation mechanism is based on stability of the entire system. However, the main drawback of this approach is its extremely high numerical complexity.

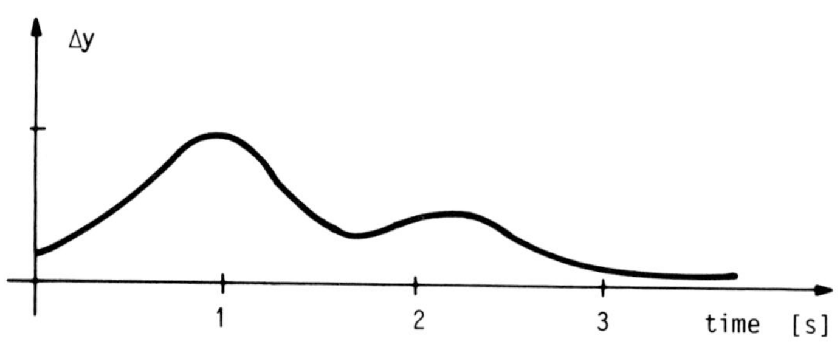

Fig. 3.5. Output error

It originates from the centralized control approach. The complete robot model included in gain matrices $K_p(\xi, \bar{\theta})$ and $K_u(\xi, \bar{\theta})$ must be computed on-line. As a 6-degree-of-freedom robot model computation requires several hundred floating-point multiplications [51, 52], it is difficult to be achieved with a low-cost microcomputer.

3.2.3. Indirect centralized adaptive control

Now we shall examine the indirect adaptive robot control, which consists of two blocks: parameter estimation block and adaptive controller. The adaptive controller is governed by the parameter estimates and output states. A specific indirect adaptive controller for robotic purposes is developed in Ref. [33-35]. It will be shortly presented in the text to follow.

Consider a centralized robot model that includes the model of both mechanical arm and actuators:

$$S: \quad u = \hat{H}(q, d, \theta_a)\ddot{q} + \hat{h}(q, \dot{q}, d, \theta_a) \tag{3.2.62}$$

where d represents the vector of mechanical parameters of robot arm, and θ_a the vector of actuator parameters.

Neglecting the dynamics, of actuators, this model becomes

$$S: \quad u = D_a H(q, d)\ddot{q} + D_a h(q, \dot{q}, d) \tag{3.2.63}$$

where D_a is a diagonal matrix. Let us separate the elements of the vector d into two subvectors

$$d = \begin{bmatrix} \theta_k \\ \hdashline \theta_d \end{bmatrix} \tag{3.2.64}$$

where θ_k is assigned to kinematical (geometrical) parameters, while θ_d represents the vector of dynamical parameters of links (masses and inertias). Vector θ_d can be represented as

$$\theta_d = [m_1 \ \cdots \ m_n \ \underline{J}_1 \ \cdots \ \underline{J}_n]^T \tag{3.2.65}$$

where m_i and J_i refer to the mass and the tensor of inertia of the ith segment, respectively.

As shown in Ref. [40, 53], dynamic model matrices H and h depend linearly on θ_d. Thus, the model (3.2.63) can be written as

$$S: \quad u = D_a \phi(q, \dot{q}, \ddot{q}, \theta_k)\theta_d \tag{3.2.66}$$

where D_a and ϕ are $n \times n$ and $n \times n_d$ matrices, respectively.

In the case when we know exactly the parameter vector d, we can apply the centralized nonadaptive control law

$$u = D_a \phi(q, \dot{q}, \ddot{q}^o + D_p(q - q^o) + D_v(\dot{q} - \dot{q}^o), \theta_k)\theta_d \tag{3.2.67}$$

where D_p and D_v are diagonal matrices.

However, θ_d includes the parameters of payload. In this case θ_d is un unknown function of time. Thus, the control law (3.2.67) is applicable only if the parameter vector is estimated on-line. Let us divide the interval $\tau = [t_o, t_f]$ into segments $[t_k, t_{k+1}]$, $k = 1, 2, \ldots, k_\tau$, and denote the estimate of θ_d at time instant t_k by $\bar{\theta}_d(t_k)$. The estimator is accepted in the form

$$\bar{\theta}_d(t_{k+1}) = \psi(\bar{\theta}_d(t_k)) \tag{3.2.68}$$

then we define a generalized error $v \in R^n$

$$v = v(q - q^o, \dot{q} - \dot{q}^o) \tag{3.2.69}$$

which refers to the state deviation from the reference trajectory. According to Ref. [33-35], the estimator (3.2.68) can be defined as

$$\bar{\theta}_d(t_{k+1}) = \bar{\theta}_d(t_k) + \phi^T(t_k')[\phi(t_k')\phi^T(t_k')]^{-1} D_a^{-1} v(t_k') \tag{3.2.70}$$

where $t_k' \in [t_k, t_{k+1})$ is defined by the condition

$$||v(t_k')|| \geqslant \delta \tag{3.2.71}$$

where δ is a given positive real number.

The preceding equations show that the centralized control law should be calculated at any time instant t_k, using the parameter estimate from the time interval $[t_{k-1}, t_k)$. Then we have to test the condition (3.2.71) at least once. Thus, we obtain the time instant t_k'. In the time interval (t_k', t_{k+1}) we have to form the estimate $\bar{\theta}_d(t_{k+1})$ according to (3.2.70). If the condition (3.2.71) is not satisfied for any t_k', then we accept $\bar{\theta}_d(t_{k+1}) = \bar{\theta}_d(t_k)$.

In conclusion, it is interesting to note that the presented concept differs from self-tuning approach because it is based on the dynamic robot model. This method differs from model-following adaptive control, too. Namely, it does not use a reference model to express the control objective. The estimation problem is here solved by using a "quasi-gradient" algorithm. Based on this estimate, the control signals are calculated following the "inverse-control" approach. The numerical complexity of such a procedure is very high. Also, it is obvious that it is unnecessary to estimate such a large number of parameters (3.2.65), because many of them are well-known and constant.

3.3 Decentralized Adaptive Control Strategy for Mechanical Manipulators

The purpose of this section is to provide an overview of decentralized adaptive control algorithms. PID self-tuning local controllers, decentralized reference model-following controllers and a specific indirect adaptive structure will be described. As will be shown, such controllers are very convenient for driving the up-to-date industrial robots. These algorithms are not numerically involved and difficult for real-time applications. Apart from that, they provide for system adaptivity with respect to different payloads and parameter variations.

3.3.1. Self-tuning local controllers

An overwhelming majority of the regulators which are used to drive industrial robots are digital or analog PID, because when properly tuned they generally achieve satisfactory performance. This is undoubtedly true when driving only one joint (or more physically decoupled joints) at a time. As pointed out in Chapter 2, coupling effects and interaction forces between links moving simultaneously may decrease the performance of the overall system and increase the tracking error during following the reference trajectory. The disturbances of unknown magnitude (due to grasping or releasing a payload with unknown mass and moments of inertia, interaction due to flexibility of links etc.) also reduce the quality of tracking.

In this section we shall deal with a possible solution for applications, because of more stringent robot performance requirement and/or the existence of larger modeling uncertainty levels. Supposing that the fixed parameter controllers are inadequate, we shall discuss self-tuning local regulators. Moreover, we shall restrict our attention to PID control structures with the aim to bridge traditional and modern adaptive control concepts.

Let us start with a classical PID regulator (Fig. 3.6)

$$U(s) = K\left[1 + \frac{1}{T_i s} + \frac{T_d s}{1+T_d s/N}\right]E(s) \tag{3.3.1}$$

where K is the gain, T_i and T_d are integral and derivative time constants, N is a positive constant. Notice that the shown PID drives only one robot actuator. Thus, we have one PID controller for each degree of freedom. Off-line tuning of PID parameters may be done utilizing optimization procedures with the simulated subsystem model (actuator with an average inertial load) or using some tuning rules. The main drawback of the first approach lies in uncertainty of model parameters. The simulation might also be time-consuming and numerically involved. The second approach refers to experiments with a real robot. Then, we can apply step-response-based methods [54, 55] or Ragazzini's direct method [55, 56]. However, a designer is forced to tune the regulator parameters at an operating point. In order to cope with variations in subsystem dynamics at different operating points (see Chapter 2) and to satisfy increasingly stringent control objectives, we can apply PID self-tuning local regulators. At the end of this section we shall discuss major advantages as well as some drawbacks of such a control strategy.

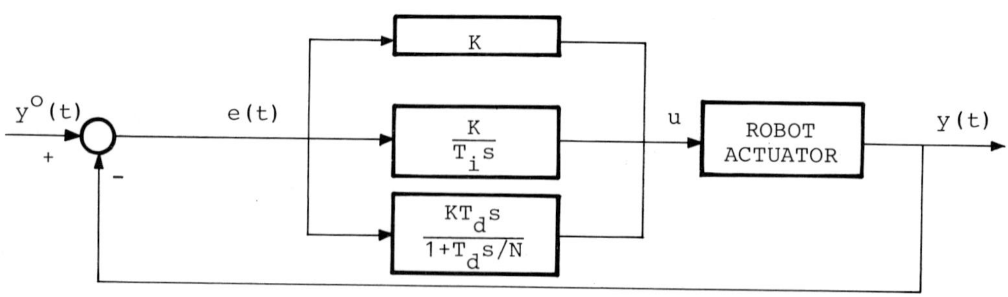

Fig. 3.6. PID controller for one degree of freedom

Differential equation corresponding to classical PID regulator (3.3.1) is

$$\frac{d^2u}{dt^2} + \frac{N}{T_d}\frac{du}{dt} = K(N+1)\frac{d^2e}{dt^2} + \frac{K}{T_iT_d}(NT_i+T_d)\frac{de}{dt} + \frac{KN}{T_iT_d}e \qquad (3.3.2)$$

using Euler's derivative approximation, we obtain the difference equation

$$u(t)+(\gamma-1)u(t-1)-\gamma u(t-2) = r_oe(t)+r_1e(t-1) + r_2e(t-2) \qquad (3.3.3)$$

where the meaning of u(t), u(t-1),... is the same as in Paragraph 3.2.1. Constants γ, r_o, r_1 and r_2 are

$$\gamma = \frac{NT}{T_d} - 1, \qquad r_o = K(\xi(\gamma+1)+N+1)$$

$$r_1 = K(\xi+\gamma-2N-1), \qquad r_2 = r_1 + K(N+1)$$

where $\xi = T/T_i$ (T - sampling period). Equations (3.3.3) can be presented in the following form

$$S(q^{-1})u(t) = R(q^{-1})e(t) \qquad (3.3.4)$$

with

$$S(q-1) = (1-q^{-1})(1+\gamma q^{-1})$$

$$R(q^{-1}) = r_o+r_1q^{-1} + r_2q^{-2}$$

where $e(t) = q^o(t) - q(t)$ represents the tracking error and q^{-1} is the backward-shift operator.

As the plant to be controlled by PID/ST regulator represents a subsystem of a manipulator (one actuator with transmission and an inertial load), it can be described by the following linear time-invariant model:

$$A(q^{-1})y(t) = B(q^{-1})u(t-1) + e(t) \qquad (3.3.5)$$

with

$$A(q^{-1}) = 1-a_1q^{-1}-a_2q^{-2}$$
$$B(q^{-1}) = b_o+b_1q^{-1} \qquad (3.3.6)$$

where a_i's and b_i's are assumed unknown scalar constants. The second order model of the subsystem is quite compatible with overall robot model (3.2.21).

Denoting by $\hat{A}(q^{-1})$ and $\hat{B}(q^{-1})$ the estimates of real polynomials (3.3.6) according to one of least squares algorithms described in Paragraph 3.2.1, and by $\hat{S}(q^{-1})$ and $\hat{R}(q^{-1})$ the corresponding polynomials of the self-tuning regulator (3.3.4), we obtain the closed-loop system scheme shown in Fig. 3.7. After estimating polynomials $\hat{A}(q^{-1})$ and $\hat{B}(q^{-1})$ we

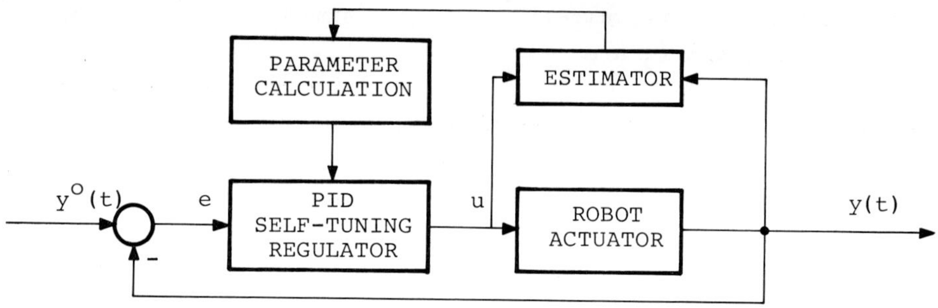

Fig. 3.7. PID self-tuning controller for one robot joint

have to calculate $\hat{S}(q^{-1})$ and $\hat{R}(q^{-1})$. For this purpose, we shall apply the well-known *certainty equivalence* assumption, i.e. the true parameter values will be replaced by their estimated when determining PID parameters. This fundamental ad-hoc hypothesis is implicitly used in Paragraph 3.2.1. The transfer function of the overall system: PID/ST regulator $u(t) = [\hat{R}(q^{-1})/\hat{S}(q^{-1})]e(t)$ together with predicted plant model $y(t) = [\hat{B}(q^{-1})/\hat{A}(q^{-1})]u(t)$ is

$$\frac{y(t)}{y^o(t)} = \frac{\hat{R}(q^{-1})\hat{B}(q^{-1})}{\hat{S}(q^{-1})\hat{A}(q^{-1})+\hat{R}(q^{-1})\hat{B}(q^{-1})} \tag{3.3.7}$$

Now, we can apply pole-placement method in order to calculate $\hat{S}(q^{-1})$ and $\hat{R}(q^{-1})$. By equating the characteristic polynomial of (3.3.7) with a desired polynomial $C_R(q^{-1})$, we obtain

$$\hat{S}(q^{-1})\hat{A}(q^{-1})+\hat{R}(q^{-1})\hat{B}(q^{-1}) = C_R(q^{-1}) \tag{3.3.8}$$

where

$$C_R(q^{-1}) = 1 + \sum_{i=1}^{n_c} c_i q^{-1}, \qquad n_c \leqslant 4.$$

The polynomial identity (3.3.8) defines a set of 4 linear algebraic equations with 4 unknowns - the regulator parameters. The number of poles to be located in closed-loop is defined by n_c, being the remaining $4-n_c$ poles placed at the origin. A sensible choice for $C_R(q^{-1})$ is

$$C_R(q^{-1}) = 1-2e^{-\rho}\cos\theta q^{-1}+e^{-2\sigma}q^{-2}$$

with

$$\theta = \omega_n T\sqrt{1-\xi^2} \quad \text{and} \quad \rho = \theta\xi/\sqrt{1-\xi^2}$$

where ω_n is the undamped natural frequency and ξ the damping ratio of the corresponding continuous time second-order characteristic polynomial. Finally, the complete control strategy is summarized in Fig. 3.8.

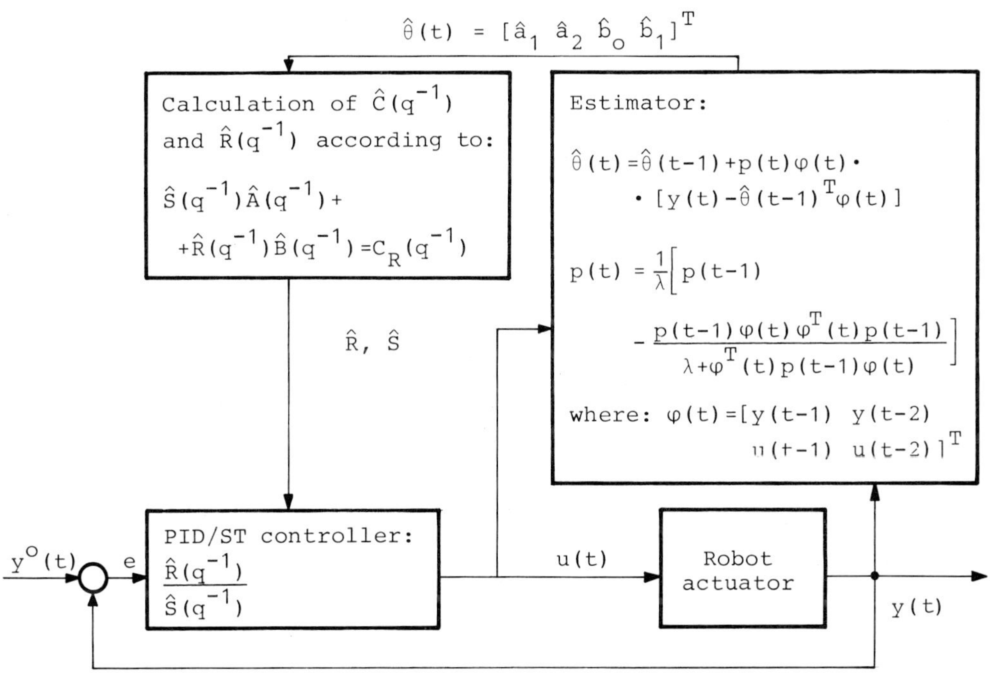

Fig. 3.8. PID self-tuning controller

The simulation of Stanford manipulator controlled by local PID/ST regulators may be found in Koivo's and Leininger's papers [10, 11]. As one could expect, the behavior of the manipulator when starting the learning process depends mostly on the accuracy in initialization of the prediction model parameters. This accuracy determines the degree of "erratic" motion at task start-up. The second problem is related to convergence of estimates. The convergence of parameter estimates and controller gains may not be achieved during the finite time over which the motion takes place. Also, when parameters change abruptly (for example, when grasping a large mass workpiece), the stability of the overall system may fail. On the other hand, the major advantage of PID self-tuning regulators is the attractive "learning mechanism" which could be very useful in repetitive tasks. Then, the last estimates of parameters from the previous run can be used as the initial estimates. Thus, the controller of the robot is "trainable" in repetitive tasks, resulting in a more accurate performance. Fig. 3.9 shows the effect of learning mechanism in "circle drawing" task.

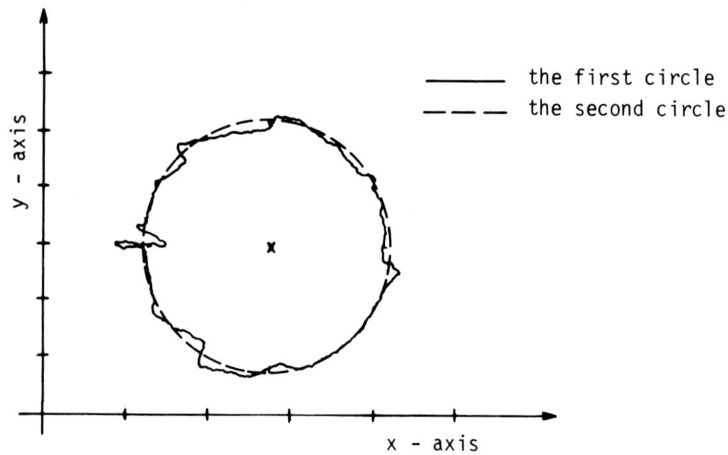

Fig. 3.9. Learning mechanism based on PID/ST regulator

3.3.2. Decentralized reference model-following control

In this section we examine the design of local adaptive robot controllers based on hyperstability and positivity concepts in model reference adaptive systems. Instead of the centralized approach described in Paragraph 3.2.2, here we shall base the control strategy on subsystems assigned to robot degrees of freedom. We shall present a pragmatic approach without formal generalizations and detailed mathematical derivations. In our opinion the algorithm, which will be described in this section, will find its place in robotics applications.

The structure of an adaptive model-following controller for a subsystem S^i is presented in Fig. 3.10. The plant (actuator, reducer and an average inertial and gravitational load) together with a variable-gain controller on one hand, and a reference model on the other hand, work in a parallel manner. Therefore, this system belongs to the family of parallel model-following control systems.

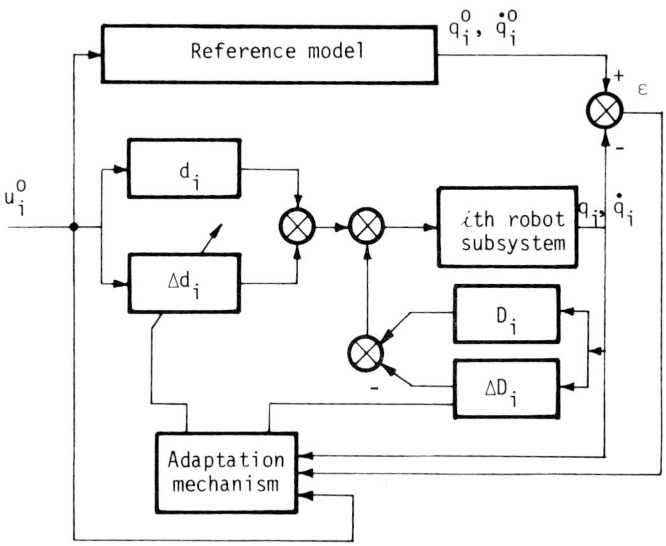

Fig. 3.10. Parallel model-following local controller for driving one degree of freedom of a robot

First, we have to define the structure of the reference model. For this purpose, we can follow the algorithm described in Paragraph 3.2.2. The above method is based on the reference model, whose structure is quite similar to the plant to be controlled. Since, the subsystem model S^i is approximately representable by (3.2.33)

$$S^i: \quad \dot{x}^i = \bar{A}^i(\theta^i, \bar{d})x^i + \bar{b}^i(\theta^i, \bar{d})u_i \tag{3.3.9}$$

it is convenient to introduce the reference model (3.2.35)

$$S_M^i: \quad \dot{x}^{io} = A_M^i x^{io} + b_M^i u_i^o \tag{3.3.10}$$

where A_M^i and b_M^i determine a desired dynamical behaviour of ith robot subsystem. Substituting $x^i = [q_i \ \dot{q}_i]^T$ in the reference model (3.3.10), we obtain the scalar form $\ddot{q}_i^o = -a_i^*\dot{q}_i^o - f_i^*q_i^o + b_i^*u_i^o$ (Equation (3.3.36)). The nominal joint coordinates, velocities and acceleration (q_i^o, $\dot{q}_i^o$ and $\ddot{q}_i^o$, respectively) are usually calculated at the kinematical level.

Thus, the nominal control u_i^o can be computed from the reference model.

Let us introduce the control in an additive form [13]:

$$u_i = u_i^{NA} + u_i^A \tag{3.3.11}$$

where u_i^{NA} is a non-adaptive linear control signal computed for the nominal dynamic parameters, and u_i^A is the adaptive component. The non--adaptive component should be very simple and convenient for application. Thus, we arrive at

$$u_i^{NA} = -D_i x^i + d_i u_i^o \tag{3.3.12}$$

where $D_i = [d_{i1} \ d_{i2}]$ is a constant feedback gain matrix of dimension 1×2, while the parameter d_i represents the feedforward gain. The expression (3.3.12) actually represents the classical local control structure, which is described in Chapter 2. We shall therefore concentrate on obtaining the adaptive component u_i^A. It will be accepted in a similar form

$$u_i^A = \Delta D_i(e^i, t)x^i + \Delta d_i(e^i, t)u_i^o \tag{3.3.13}$$

where $\Delta D_i(e^i, t): R^2\times T\to R^{1\times 2}$ is an adaptive feedback gain matrix, and $\Delta d_i(e^i, t)$ a scalar function representing a variable feedforward gain. Let us introduce the error vector $e^i = x^{io}-x^i$, i.e.

$$e^i = \begin{bmatrix} q_i^o - q_i \\ \dot{q}_i^o - \dot{q}_i \end{bmatrix} \tag{3.3.14}$$

The objective of the adaptation mechanism which will generate the time--varying matrix $\Delta D_i(e^i, t)$ and scalar $\Delta d_i(e^i, t)$ is to assure that the state error e goes to zero under certain conditions.

Substituting (3.3.12) and (3.3.13) in (3.3.11), we obtain

$$u_i = (-D_i+\Delta D_i)x^i + (d_i+\Delta d_i)u_i^o \tag{3.3.15}$$

Putting it into (3.3.9), we arrive to

$$\dot{x}^i = [\bar{A}^i-\bar{b}^i D_i+\bar{b}^i\Delta D_i]x^i+\bar{b}^i(d_i+\Delta d_i)u_i^o \tag{3.3.16}$$

This model represents the system parallel to the reference model (see Fig. 3.10). This system is called "parallel adjustable system". Denoting by $A_s(e^i, t)$ and $B_s(e^i, t)$ the matrices of the above model, we obtain

$$\dot{x}^i = A_s(e^i, t)x^i + B_s(e^i, t)u_i^o \tag{3.3.17}$$

with $A_s \in R^{2\times 2}$ and $B_s \in R^2$. Subtracting the reference model (3.3.10) from (3.3.16), we get

$$\dot{e}^i = A_M^i e^i + w^i \tag{3.3.18}$$

with

$$w^i = -[A_M^i - \bar{A}^i + \bar{b}^i(D_i - \Delta D_i)]x^i - [b_M^i - \bar{b}^i(d_i + \Delta d_i)]u_i^o$$

Let D_i^o and d_i^o satisfy the equations

$$A_M^i - \bar{A}^i = -\bar{b}^i D_i^o, \quad b_M^i = \bar{b}^i d_i^o \tag{3.3.19}$$

where $D_i^o \in R^{1\times 2}$ and $d_i^o \in R$. Thus, the vector w^i can be represented as

$$w^i = -\bar{b}^i[(D_i - D_i^o - \Delta D_i)x^i - (d_i^o - d_i - \Delta d_i)u_i^o] \tag{3.3.20}$$

Now, we have to accept some general form of the adaptation law $\Delta D_i(e^i, t)$ and $\Delta d_i(e^i, t)$. After imposing a desired stability condition, we will obtain the particular forms of adaptation mechanisms. In order to avoid a steady state error, let us introduce a proportional - integral adaptation mechanism. To ensure the asymptotic system stability, we ought to establish a linear compensator in the form [13]:

$$v^i = De^i \tag{3.3.21}$$

where D represents a 2×2 matrix which will be determined later. The adaptation mechanism will be introduced with respect to the compensator output v^i:

$$\begin{aligned} \Delta D_i(v^i, t) &= \int_0^t \phi_i(v^i, t, \tau)d\tau + \phi_i(v^i, t) + \Delta D_i(0) \\ \Delta d_i(v^i, t) &= \int_0^t \psi_i(v^i, t, \tau)d\tau + \psi_i(v^i, t) + \Delta d_i(0) \end{aligned} \tag{3.3.22}$$

Now, we can separate the system into 2 parts (as shown in Fig. 3.4):

1) a linear feedforward block comprising the error model (3.3.18) and the linear compensator (3.3.21), and

2) a non-linear feedback block consisting of the model (3.3.20) and the adaptation law (3.3.22).

Next, we can directly apply the hyperstability conditions, positivity of the feedforward block and Popov's integral inequality to the feedback block. From the first condition, it follows that the static compensator must satisfy the condition

$$D(sI-A_M^i)\bar{b}^i = \text{a strictly positive real transfer matrix.}$$

The second condition yields the adaptation mechanism

$$\Delta D_i(v^i,\ t) = \int_0^t (Fv^i)(Gx^i)^T d\tau + (Fv^i)(Gx^i)^T + \Delta D_i(0) \tag{3.3.23}$$

$$\Delta d_i(v^i,\ t) = \int_0^t (Mv^i)(Nu_i^o)^T d\tau + (Mv^i)(Nu_i^o)^T + \Delta d_i(0)$$

where $F\in R^{1\times 2}$, $G\in R^{2\times 2}$, $M\in R^{1\times 2}$ and $N\in R$ are positive.

Integral terms in (3.3.23) provide memory of the adaptation mechanism, while the proportional terms are introduced in order to accelerate the reduction of the generalized error at the beginning of the adaptation process. The discrete version of the proposed adaptive model-following scheme can be found in Ref. [13].

The action of the proposed controller will be illustrated on an example with direct-current motor speed regulation. First, the motor is regulated by an usual PI controller, and, after some time, the adaptive controller is activated [13]. The action is shown in Fig. 3.11. Some interesting examples are also presented in Ref. [26].

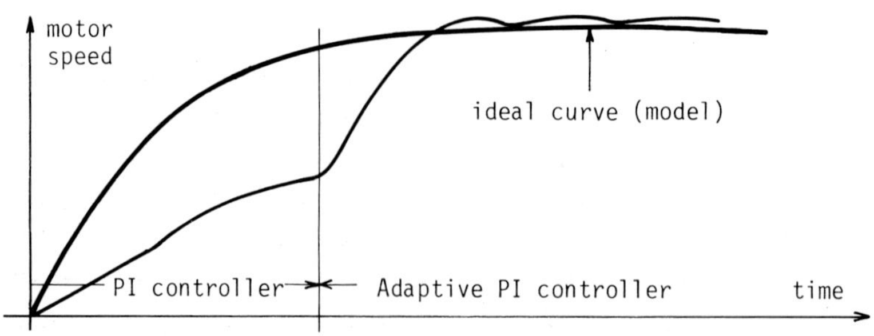

Fig. 3.11. Comparison between classical PI controller action and model-following adaptive controller

3.3.3. Indirect decentralized adaptive control

Having presented adaptive techniques based on various concepts, we will complete the picture by introducing a new adaptive scheme relying on the decoupled control strategy, described in Chapter 2. The goal is to obtain an algorithm which is both simple and stable when manipulating different payloads. From that point of view, the decoupled control strategy is very suitable because of the systematic selection of feedback loops. Note that the set of PID regulators controlling robot degrees of freedom corresponds to local regulators described in Chapter 2. Such a set of regulators represents the simplest control structure, which is applied in most of today's industrial robots. The stability theory related to this as well as to more complex control structures is presented in Chapter 2, too. Thereby, the stability analysis is applicable to any given set of parameters, or to a given range of parameters. However, for some range of parameters, it may appear that there is no fixed-gain-controller which can provide the imposed stability conditions. For instance, if the gains of local controllers are tuned to stabilize the robot carrying a 30 kg workpiece, it may happen that the same stability conditions are not satisfied without a payload or, possibly, with a 60 kg payload. If we still want to stabilize the robot motion with simple controllers, we have to introduce the force--feedback. But the employment of force/torque sensors may be difficult and expensive. The problem can be solved in the following way. First, we determine the gains of local regulators in terms of payload mass and/or inertias.

Then, if we add some recursive parameter estimator, the output of which is the payload mass/inertias, it would not be necessary to introduce the force or torque measurements. The parameters of local regulators would change in accordance with estimates.

We have seen that some algorithms of centralized adaptive robot control involve the estimation of a large number of parameters [33 - 35]. Obviously, this is not necessary, because most of them are well known. There is also a series of quasi-stationary parameters (such as temperature-dependent viscous frictions, static frictions, etc). Thus, we have to separate such parameters which change (in a very wide range) and cannot be prespecified. These are primarily the parameters of workpieces.

In this section we shall describe an indirect adaptive scheme.

Within the indirect adaptive control the procedure is devided into two steps. First, the unknown parameter estimates are generated by using some of the on-line identification algorithms. Thereafter, the obtained estimates are used as inputs to adaptive regulator, which adjusts its coefficients according to the estimates.

Consider the model of a robotic mechanism with the vector of parameters $\theta_p = \bar{\theta}_p + \Delta\theta_p$ in the form

$$P = [0 \mid H(\xi, \bar{\theta}_p + \Delta\theta_p)]\dot{\xi} + h(\xi, \bar{\theta}_p + \Delta\theta_p)$$

Taking into account the parameter deviation $\Delta\theta_p$, centralized law of nonadaptive control (2.3.20) becomes

$$u = [0 \mid \hat{H}(\xi, \bar{\theta}_p + \Delta\theta_p)](\dot{\xi}^o + D(\xi - \xi^o)) + \hat{h}(\xi, \bar{\theta}_p + \Delta\theta_p) \qquad (3.3.24)$$

The state vector ξ can be presented as $\xi = \xi^o + \Delta\xi$, where ξ^o denotes the state vector on the desired trajectory. Denoting by $u^o(\theta_p)$ the nominal control vector obtained from (3.3.24) at the conditions $\xi = \xi^o$ and $\theta_p = \bar{\theta}_p$, the control law (3.3.24) can be presented by

$$u = u^o(\bar{\theta}_p) + [0 \mid \hat{H}(\xi, \bar{\theta}_p + \Delta\theta_p)]D(\xi - \xi^o) + \Delta u^N \qquad (3.3.25)$$

where Δu^N is the nonlinear control component.

It is shown [40, 53] that the dynamic model matrices are linear with respect to dynamic parameters (masses and inertias of the links). This is a very important property which is suitable for indirect adaptive control synthesis. The control law (3.3.25) now becomes

$$u = u^o(\bar{\theta}_p) + [0 \mid \hat{H}(\xi, \bar{\theta}_p) + \hat{S}_H(\xi)\Delta\theta_p]D(\xi - \xi^o) + \Delta u^N \qquad (3.3.26)$$

with $\hat{S}(\xi)$ being the sensitivity function valid for both small and large parameter deviations. The computer-aided construction of the sensitivity function for an arbitrary manipulator structure is given in [40]. Following the idea of control decentralization, matrices $\hat{H}$ and $\hat{S}$ can be redistributed to constant parts and nonlinear parts which can be added to Δu^N (3.3.26). Thus, the control law can be significantly simplified.

$$u_i = u_i^o(\bar{\theta}_p) + (\bar{H}_{ii}(\bar{\theta}_p) + \bar{S}_{ii}\Delta\theta_p)D_i(\xi^i - \xi^{io}) + \Delta\tilde{u}_i^N$$

where $\bar{S}_{ii} \in R^{1\times n_p}$ and $D_i = [d_i^p \; d_i^v]$. Here we clearly see the nonadaptive component of local control

$$\Delta u_i^{LNA} = \bar{H}_{ii}(\bar{\theta}_p)[d_i^p \; d_i^v](\xi^i - \xi^{io}) \tag{3.3.27}$$

and the adaptive component

$$\Delta u_i^{LA} = \bar{S}_{ii}\Delta\theta_p[d_i^p \; d_i^v](\xi^i - \xi^{io}) \tag{3.3.28}$$

where $\Delta\theta_p$ is not known. For this reason, instead of real value we have to use the estimate $\Delta\theta_p^e$:

$$\Delta u_i^{LA} = \bar{S}_{ii}\Delta\theta_p^e[d_i^p \; d_i^v](\xi^i - \xi^{io}) \tag{3.3.29}$$

When only one parameter is not known, local control can be presented in the form

$$\Delta u_i^L = [D_{i(min)} + \frac{\Delta\theta_p^e}{|\Delta\theta_p|_{(max)}}(D_{i(max)} - D_{i(min)})](\xi^i - \xi^{io}) \tag{3.3.30}$$

where $|\Delta\theta_p|_{(max)}$ represents the largest possible parameter variation, $D_{i(min)}$ and $D_{i(max)}$ - gains of local regulators in the cases of minimal and maximal values of estimated parameters. The values of $D_{i(min)}$ and $D_{i(max)}$ can be obtained by using the stability analysis when the unknown parameter reaches its boundary values (see Chapter 2).

The second part of the control algorithm represents the parameter estimation of $\Delta\theta_p$. For this purpose, the time should be discretized in time intervals $[t_k, t_{k+1}]$. Let us denote by P_k^* the driving torques/ /forces of actuators (k indicates the time instant t_k) which even when the parameter vector changes its value, force the system to follow the nominal trajectory. It is shown [40, 53] that the vector P_k^* can be expressed as

$$P_k^* = [0 \mid H(\xi_k^o, \bar{\theta}_p) + S_H(\xi_k^o)\Delta\theta_p]\dot{\xi}_k^o + h(\xi_k^o, \bar{\theta}_p) + S_h(\xi_k^o)\Delta\theta_p \tag{3.3.31}$$

where S_H and S_h do not depend on θ_p. However, in reality the system is out of the nominal trajectory due to parametric or state disturbances. For this reason, real driving torques vector P_k differs from P_k^*. Vector P_k is given by the same expression as P_k^*, except for the state vector ξ_k^o and $\dot{\xi}_k^o$, which are now ξ_k and $\dot{\xi}_k$. Let us denote P_k^o the driving torque vector when the states belong to the nominal trajectory and when $\Delta\theta_p = 0$. Thus the deviation $\Delta P_k = P_k - P_k^o$ becomes

$$\Delta P_k = (S_H(\xi_k)\dot{\xi}_k + S_h(\xi_k))\Delta\theta_p + \Delta P^N(\xi_k, \xi_k^o, \dot{\xi}_k, \dot{\xi}_k^o) \tag{3.3.32}$$

where ΔP^N represents the nonlinear component which is the consequence of the difference of the matrices H and h at the instantaneous states and the corresponding states on nominal trajectory. When the system states under the feedback reaction reach the desired trajectory, i.e. $\xi_k \to \xi_k^o$, then $\Delta P^N \to 0$. The expression enables the derivation of the estimator for the parameter $\Delta\theta_p$. Let us denote by $S_p(\xi_k)$ the matrix

$$S_p(\xi_k) = S_H(\xi_k)\dot{\xi}_k^o + S_h(\xi_k) \tag{3.3.33}$$

Following the quasi-gradient procedure [38, 39], the estimates can be obtained from the recursive algorithm

$$\Delta\theta_p(t_{k+1}) = [I - a_p(t_k)S_p^T(\xi_k)S_p(\xi_k)]\Delta\theta_p(t_k) + p(\xi_k) \tag{3.3.34}$$

where $a_p(t)$ represents a positive real function which determines the transient behaviour during the estimation process, and $p(\xi_k)$ represents the dynamic correction due to the influence of the input variations:

$$p(\xi_k) = a_p(t_k)S_p^T(\xi_k)\Delta P_k \tag{3.3.35}$$

or

$$p(\xi_k) = a_p(t_k)S_p^T(\xi_k)D_a\Delta u_k \tag{3.3.36}$$

with Δu_k being the input variation, and D_a the matrix which depends on the actuator parameters.

A general block diagram of the algorithm is given in Fig. 3.12. The efficiency of the dynamic estimator and the complete adaptive algorithm may be best recognized through numerical examples. We shall present two examples. The first refers to a hydraulic 6-degree-of-freedom manipulator, presented in Fig. 3.13. Manipulator parameters (link lengths, masses and moments of inertia) are given in Table 3.1.

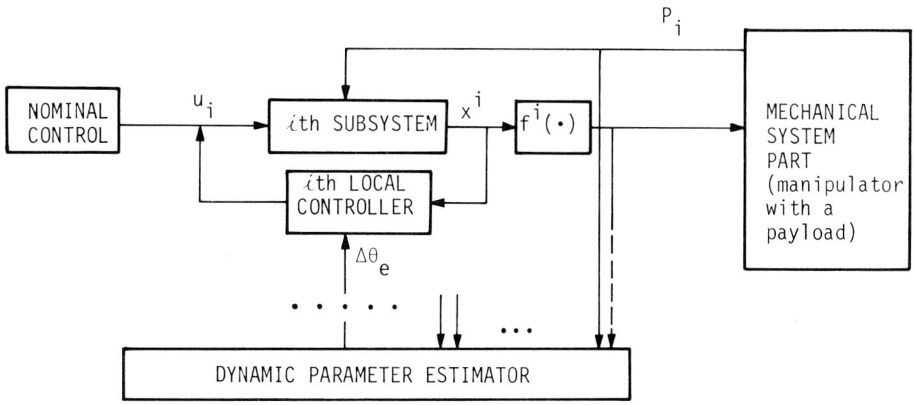

Fig. 3.12. General block-scheme

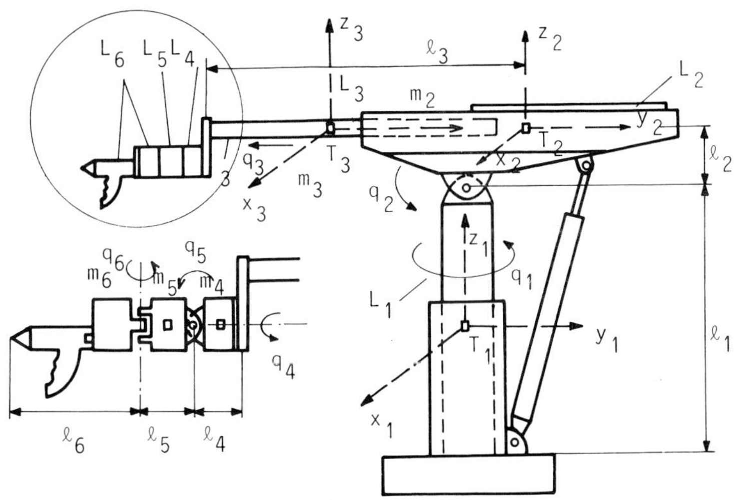

L_i - link i

ℓ_i - length of link i

q_i - joint coordinate i

x_i, y_i, z_i - coordinate frame of link i

Fig. 3.13. Manipulator UMS 3B

LINK	1	2	3	4, 5	6
MASS m_i (kg)	-	27.8	22.34	2.33	0.33
LENGTH ℓ_i (m)	1.2	0.142	1.142	0.14	0.26
J_{ix} (kgm^2)	-	2.98	2.98	0.004	.0007
J_{iy} (kgm^2)	-	-	-	0.004	.0009
J_{iz} (kgm^2)	0.322	3.701	1.21	0.004	.0009

Table 3.1. Parameters of the mechanism in Fig. 3.13.

Hydraulic cylinders, controlled by hydraulic servovalves, are modeled by models of order n_i=3. The state vector of the hydraulic actuator is $x^i = [\ell^i, \dot{\ell}^i, p^i]^T$, where ℓ^i is the piston travel, $\dot{\ell}^i$ is the piston velocity, p^i is the oil pressure in the cylinder. The first three manipulator degrees of freedom are driven by means of hydraulic cylinders while the remaining joints are powered by revolute vane actuators. In the second case, the state vector of subsystems S^i (i=4,5 and 6) is taken as $x^i = [q_i, \dot{q}_i, p^i]^T$, where q_i, $\dot{q}_i$ are the corresponding angle and angular velocity. The actuator inputs are servovalve coil currents which are constrained in amplitude. As the first two d.o.f. are revolute, but driven by linear actuators, the connection between the actuator model state coordinates x^i and the state coordinates of the mechanical system part model (i.e. the corresponding angles and angular velocities) is not linear. These nonlinear relations are shown in Fig. 3.12 by vectorial function f(•).

The nominal trajectory for the particular manipulation task is given in the following way: the tip of the manipulator should move along a straight line from the point A (whose coordinates with respect to the reference coordinate system are $[0, 1.14, 1.342]^T$[m]) to the point B (with coordinates $[-0.31, 1.14, 1.65]^T$[m]) while the angles of joints 4, 5 and 6 should be kept constant ($q^4 = 0.$, $q^5 = -1.57$[rad], $q^6 = 0.$). The triangular velocity profile of the manipulator tip is accepted and the time duration of the movement is $\tau = 0.8$ s. The only unknown parameter is the payload mass $\theta = m_p$ which can take values belonging to the region $\Theta = \{m_p: m_p\ (0, 10\ \text{kg})\}$.

First we shall try to synthesize non-adaptive control which could withstand all variations of the payload in the assumed range. We have synthesized local controllers for the maximal allowed mass of the payload, θ_d = 10 kg. The local gains are given in Table 3.2. We have simulated the dynamics of the manipulation system (from Fig. 3.13) while tracking nominal trajectory with the synthesized local controllers. If the payload is m_p = 10 kg, the tracking is satisfactory, but when m_p = 0, the tracking is poor, since local gains are too high for the manipulation system without payload. The results of simulation are shown in Fig. 3.14 where $\Delta q_i(t)$ denotes deviations of the joint coordinate i from its nominal value. We have tried to stabilize the system with local contollers synthesized for the mass of the payload θ_d = 5 kg. The tracking of nominal trajectory (see Fig. 3.14) with the local gains given in Table 3.2. is satisfactory for m_p = 5 kg and m_p = 0 kg. The tracking is poor for m_p = 10 kg due to the low gains. This means that we cannot find unique local controllers which are able to withstand all assumed parameter variations.

	SUBSYSTEM	1	2	3	4	5	6
θ_d=0	position gain velocity gain	50.81 10.86	161.77 25.40	154.35 2.30	35.11 0.54	35.09 0.52	35.00 0.35
θ_d=5kg	position gain velocity gain	89.41 19.17	212.89 44.03	154.89 2.65	36.10 2.21	36.10 2.18	35.28 0.87
θ_d=10kg	position gain velocity gain	130.89 28.08	349.16 74.40	154.42 3.01	37.86 4.03	37.83 4.00	35.57 1.37

Table 3.2. Local gains

In order to stabilize manipulation system for all possible payloads we have synthesized adaptive control algorithm as it has been presented before. The local gains have been synthesized for various payload masses (Table 3.2). Fig. 3.15 shows the simulation result of parameter estimation technique. The difference between the real mass step disturbance and the estimated value is quite small. This difference, obviously, depends on constant $a_p(t)$, and the nominal trajectory (velocities and accelerations) of mechanical segments. In this example $a_p(t)$ is substituted by a constant (a_p = 1). The influence of the nominal trajectory itself on the estimated value is rather low. For example, at t ≃ 0.4 s the end-effector acceleration changes from 3.8 m/s^2 influencing the estimated value only about 10 %. Fig. 3.14 shows the tracking of the desired nominal trajectory with adaptive (as well as fixed) control. It can be seen that we now achieve satisfactory tracking of nominal trajectory for arbitrary payload mass in a given range. However, such control is more complex for implementation than the simple robust nonadaptive control law presented above.

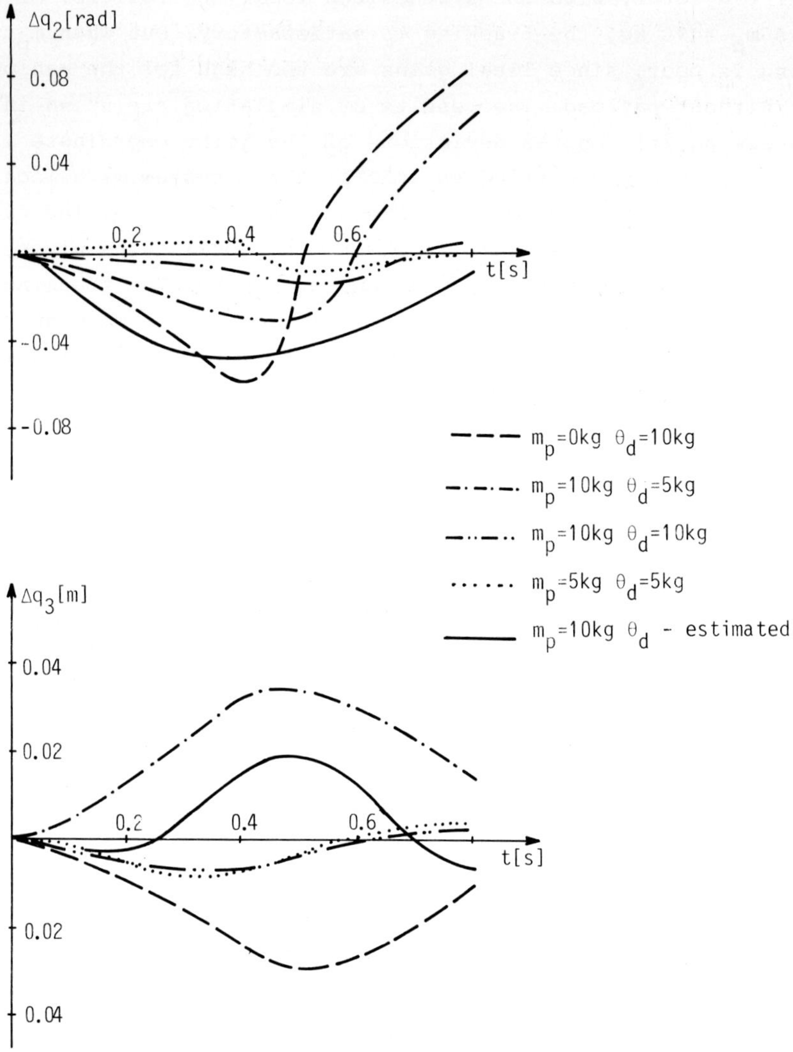

Fig. 3.14. Simulation with fixed and adaptive local regulators

The number of floating-point additions and multiplications that have to be performed in each sampling period for nonadaptive control in this particular example is

$$n_{ADD} = 114, \qquad n_{MUL} = 96$$

and for the identification of payload mass

$$n^e_{ADD} = 28, \qquad n^e_{MUL} = 64.$$

Using the INTEL 8086 microprocessor with 8087 coprocessor, we can achieve sampling period of about 10 ms for nonadaptive and 16 ms for adaptive control.

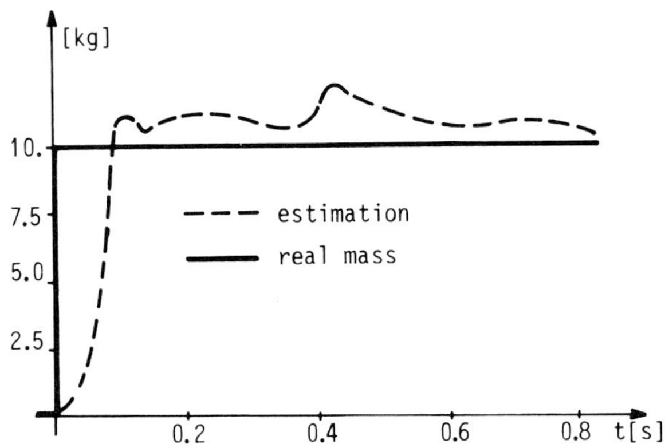

Fig. 3.15. Payload mass estimation

Let us now present the second example. We consider a 6-degree-of-freedom manipulator driven by permanent-magnet DC motors, Fig. 3.16.

Let us consider a straight line trajectory from the initial point T = T_o(0., 0.45, 0.40) [m] to the point T = T_1(-0.283, 0.45, 0.683) [m] in the time interval $t \in [0, \tau]$ with τ = 2s. T denotes a point in the basis of the end-effector (see Fig. 3.16). The velocity profile of the point T along the trajectory is accepted to be triangular. The other three degrees of freedom, which determine the orientation of the workpiece, change from the angular position (q_4^o, q_5^o, q_6^o) = (0., 1.57, 0.) [rad] to (q_4^1, q_5^1, q_6^1) = (0.4, 0.8, 0.6) [rad] (Fig. 3.17). The time-history of driving torques (forces) along this nominal trajectory, is shown in Fig. 3.18. Nominal input signals are shown in Fig. 3.19.

Let us further illustrate the parameter estimation according to dynamic relation (3.3.34). The feedback gains are accepted as G_{pi} = 10 and G_{vi} = 2 for all subsystems ensuring the practical stability of the manipulation system (with a 1kg workpiece) according to the following conditions: $|q_i(t)| \leq 0.02$ rad (i=1,4,5,6) and $|q_i(t)| \leq 0.002$ m (i = 2,3) for $t \geq 0.6$ s. The simulated manipulation task involves a) gripping an object with the mass equal to 3 kg at time instant t = 0, b) travelling along the prescribed trajectory (Fig. 3.17), c) putting the object off after ≃0.3 s, and d) travelling along the same trajectory to the end of the path. The results of simulation are presented in Fig. 3.20. Obviously, the time of adaptation is approximately 150 ms. It should be noted that the dynamic estimator has "worked" off-line, i.e. it has not affected the control of subsystems.

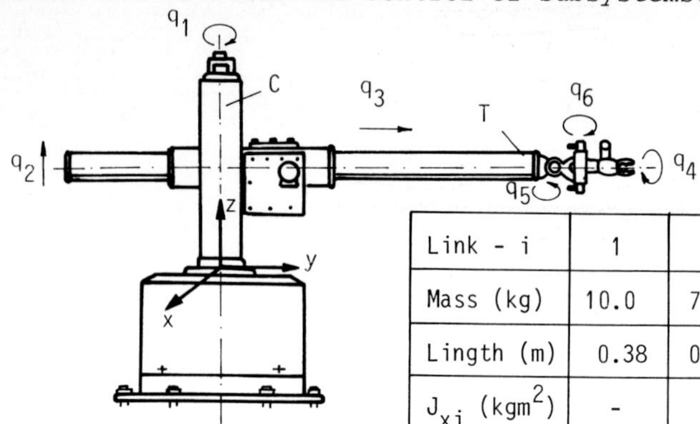

Link - i	1	2	3	4, 5	6
Mass (kg)	10.0	7.0	4.15	0.5	2.
Lingth (m)	0.38	0.02	0.45	0.05	0.2
J_{xi} (kgm^2)	-	-	-	0.01	.0069
J_{yi} (kgm^2)	-	-	-	0.001	.0004
J_{zi} (kgm^2)	0.029	0.055	0.318	0.0015	.0069

Fig. 3.16. Industrial manipulator

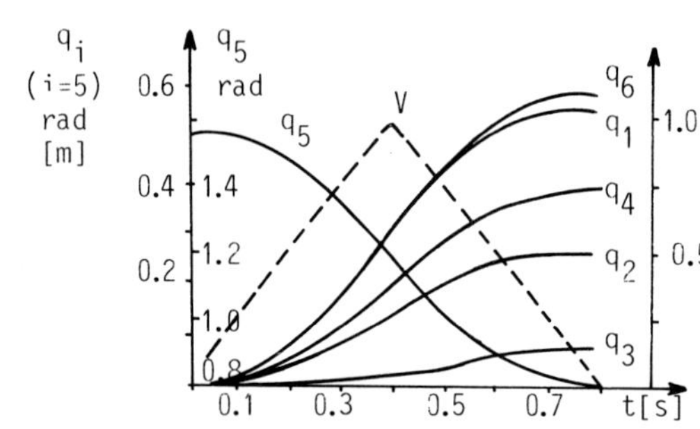

Fig. 3.17. Tip velocity distribution and corresponding joint coordinates (angular for i=1,4,5,6 and linear for i=2,3)

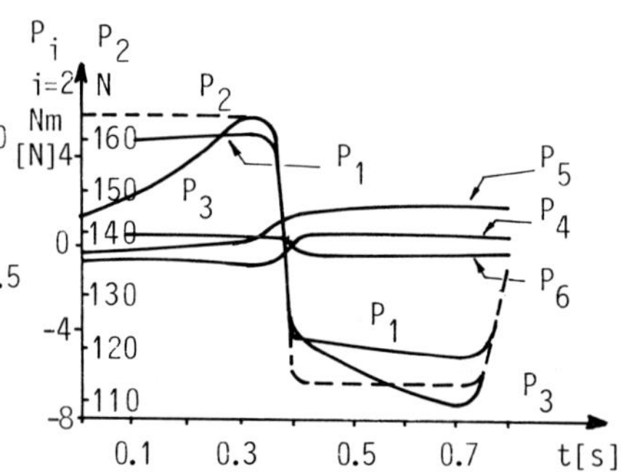

Fig. 3.18. Driving torques (i=1,4,5,6) and forces (i=2,3)

Further, the complete system including the adaptation of all feedback gains proportional to the (on-line) mass estimation is simulated. Gain boundaries are accepted to be $G_{pi}^{min} = 5.$, $G_{pi}^{max} = 20.$, $G_{vi}^{min} = 1.$, $G_{vi}^{max} = 4$. These gain boundaries correspond to the extreme cases: manipulation without payload and with the payload of 3 kg, ensuring the proposed practical stability conditions. The on-line action of the dynamic

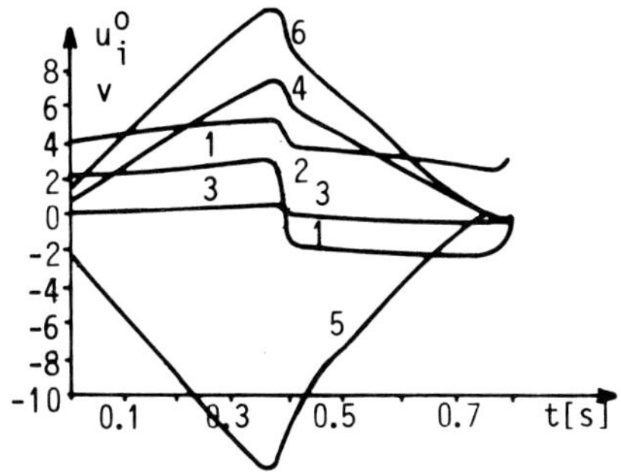

Fig. 3.19. Nominal control signal

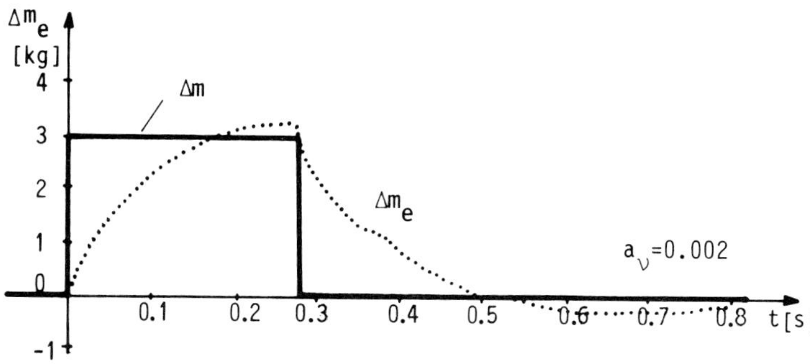

Fig. 3.20. Off-line estimation (Δm - real load, Δm_e - estimated load)

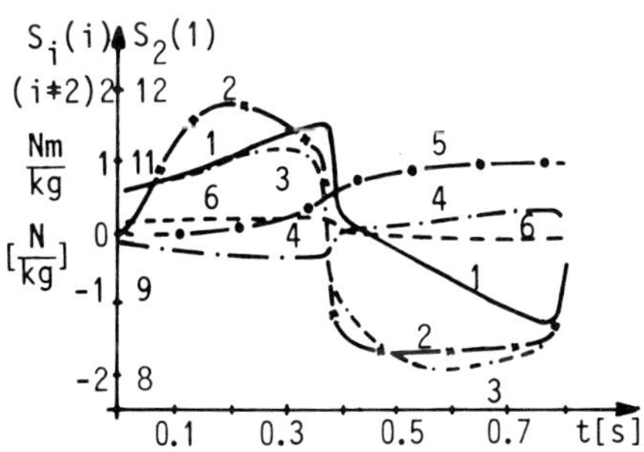

Fig. 3.21. Sensitivity functions on-line generated

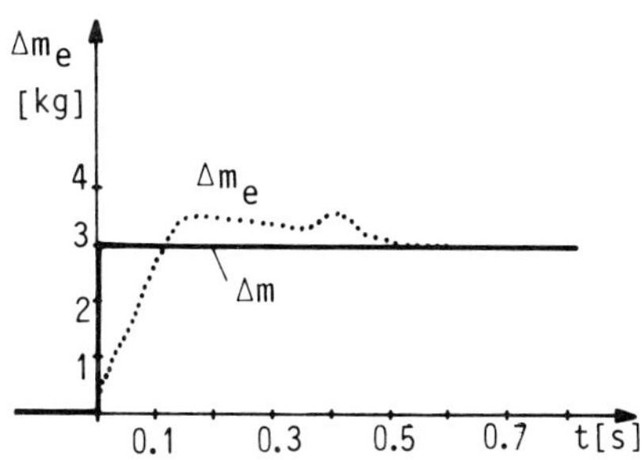

Fig. 3.22. On-line estimation

estimator is represented in Figs. 3.21, 3.22. For that, the gripping of an object (3 kg) occurred at time instant $t=t_o=0$. The sensitivity functions $S_i(t)$, j=1,...,6, formed in real-time by the dynamic estimator are shown in Fig. 3.21. The estimated mass of the manipulated object is shown in Fig. 3.22. Finally, a comparison of adaptive and nonadaptive regulators ($G_{pi} = G_{pi\ min}$, $G_{vi} = G_{vi\ min}$) is shown in Fig. 3.23. From these simulation results it follows that a) the time for on-line adaptation is about 30% shorter than that for non-adaptive control structure (now, it is about 100 ms), and b) the tracking of prescribed trajectory is much better than without adaptation. This is obviously the consequence of increased (adapted) local gains of subsystems.

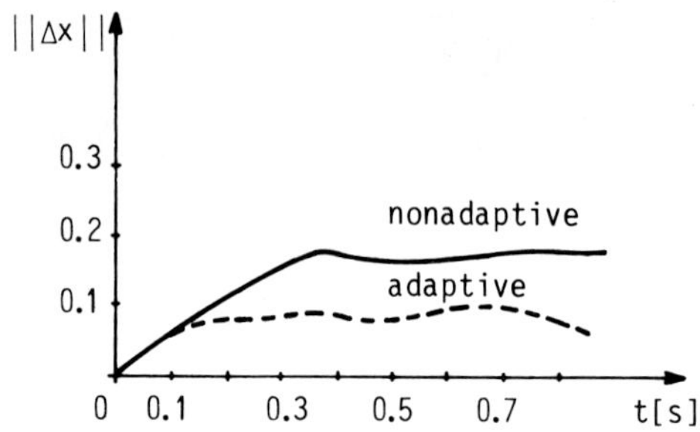

Fig. 3.23. Comparison of adaptive and nonadaptive control

References

[1] Astrom K.J., Borisson U., Ljung L. and Wittenmark B., "Theory and Applications of Self-tuning Regulators", Automatica - The Journal of IFAC, Vol. 13, No 5, 1977.

[2] Astrom K.J., Eykoff P., "System Identification - a survey", Automatica - The Journal of IFAC, Vol. 7, 1971.

[3] Astrom K.J., Introduction to Stochastic Control Theory, Academic Press, New York, 1970.

[4] Astrom K.J., Wittenmark B., "On Self-tuning Regulators", Automatica - The Journal of IFAC, Vol. 9, 1973.

[5] Ljung L., Soderstrom T. and Gustvsson I., "Counter-examples of General Convergence of a Commonly used Identification Method", IEEE Trans. on Automatic Control, Vol. AC-20, 1975.

[6] Ljung L., "On Positive Real Transfer Functions and the Convergence of Some Recursions", IEEE Trans. on Automatic Control, Vol. AC-22, 1977.

[7] Nahorski Z. and Vidal R., "Stabilization of Linear Discrete-time System Using Simultaneous Identification and Control", Int. J. Control, Vol. 19, 1974.

[8] Ljung L., Gustavsson I., Soderstrom T., "Identification of Linear Multivariable systems, Operating Under Linear Feedback Control",

[9] Young P.C., "An Instrumental Variable Method for Real Time Identivication of a noisy process", Automatica - The Journal of IFAC, Vol. 6, 1970.

[10] Koivo A.J., Guo T.H., "Adaptive Linear Controller for Robotic Manipulators", IEEE Trans. on Automatic Control, Vol. 28, No. 2, 1983.

[11] Leininger G., Wang S., "Pole Placement Self-Tuning Control of Manipulators", IFAC Symp. on C.A.D. of Multivariable Technological Systems, W. Lafayette, 1982.

[12] Leininger G., "Self-tuning Control of Manipulators", International Symp. on Advanced Software in Robotics, Liege, Belgium, 1983.

[13] Landau Y.D., Adaptive Control, M. Dekker, New York, 1979.

[14] Landau Y.D., "A Survay of Model Reference Adaptive Techniques. Theory and Applications", Automatica - The Journal of IFAC, Vol. 10, No. 3, 1974.

[15] Courtiol B., Landau Y.D., "High Speed Adaptation System for Controlled Electrical Drives", Automatica - The Journal of IFAC, Vol. 11, 1975.

[16] Tsypkin Ya.Z., Adaptation and Learning in Automatic Systems, Academic Press, New York, 1971.

[17] Popov V.M., Hyperstability of Control Systems, Springer-Verlag, New York, 1973.

[18] Landau Y.D., "A Hyperstability Criterion for Model Reference Adaptive Control Systems", IEEE Trans. Autom. Control, AC-17, No. 2, 1972.

[19] Landau Y.D., "A Generalization of the Hyperstability Conditions for Model Reference Adaptive Systems, IEEE Trans. Autom. Control, AC-17, No. 2, 1972.

[20] Dubowsky S., DesForges D., "The Application of Model Referenced Adaptive Control to Robotic Manipulators", Journal of Dynamic Systems, Measurement, and Control, Vol. 101, 1979.

[21] Dubowsky S., "On the Adaptive Control of Robotic Manipulators: The Discrete Time Case, Proc. 1981 Joint Auto. Cont. Conf., Charlottesville, VA, 1981.

[22] Dubowsky S., "On the Development of High Performance Adaptive Control Algorithms for Robotic Manipulators", 2nd Inter. Symp. on Robotics Research, Kyoto, 1984.

[23] Takegaki M., Arimoto S., "An Adaptive Trajectory Control of Manipulators", Int. J. Control, Vol. 34, No. 2, 1981.

[24] Arimoto S., Kawamura S. and Miyazaki F., "Can Mechanical Robots Learn by Themselves", 2nd Inter. Symp. on Robotics Research, Kyoto, 1984.

[25] Horowitz R., Tomizuka M., "An Adaptive Control Scheme for Mechanical Manipulators - Compensation of Nonlinearity and Decoupling Control", ASME, Winter Annual Meeting, 1980.

[26] Belestrino A., De Maria G. and Sciavicco L., "An Adaptive Model Following Control for Robotic Manipulators", Journal of Dynamic Systems, Measurement, and Control, Vol. 105, 1983.

[27] Kirćanski M.N., "Contribution to Dynamics and Control of Manipulation Robots", Ph. D. Thesis, Beograd, 1984.

[28] Vukobratović K.M., Kirćanski M.N., "An Approach to Direct and Indirect Adaptive Control of Manipulators", Proc. 7th IFAC/IFORC Symp. on Identification and System Parameter Estimation, 1985.

[29] Lee C.S.G., Lee B.H., "Resoved Motion Adaptive Control for Mechanical Manipulators", Journal of Dynamic Systems, Measurement, and Control, Vol. 106, 1984.

[30] Lee C.S.G., "Robot Arm Kinematics, Dynamics, and Control, IEEE Trans. on Computers, Vol. 15, No. 12, 1982.

[31] Anex Jr.R.P., Hubbard M., "Modelling and Adaptive Control of a Mechanical Manipulator", Journal of Dynamic Systems, Measurement, and Control, Vol. 106, 1984.

[32] Vukobratović M., Kirćanski N., "Computer-oriented Method for Linearization of Dynamic Models of Active Spatial Mechanisms", Journal of Mechanisms and Machine Theory, Vol. 16, No. 2, 1981.

[33] Timofeev A.V., Ekalo Yu.V., "Stability and Stabilization of Programmed Robot Movements", (in Russian), Avtomatika i Telemehanika, No. 20, 1976.

[34] Timofeev A.V., "Finite-convergent Local-optimal Algorithms for Solving Inequalities Arrising in Adaptive System Synthesis", (in Russian), Techniceskaya Kibernetica, No. 4, pp. 9-21, 1975.

[35] Liobacevsky B.D., "A Recursive Algorithm of Adaptive Control of Linear Discrete Dynamic Systems", Avtomatika i Telemehanika, No. 3, pp. 83-94, 1974.

[36] Vukobratović K.M., Kirćanski M.N., "An Engineering Concept of Adaptive Control for Manipulation Robots via Parameter Sensitivity Analysis", Bulletin T. LXXXI de L'Academie Serbe des Sciences et des Arts, No. 20, Beograd, 1982.

[37] Vukobratović K.M., Kirćanski M.N., "An Engineering Approach to Adaptive Control Synthesis for Multidegree-of-freedom Manipulation Robots", Proc. of the 3rd IFAC/IFORS Symp. on Large Scale Systems: Theory and Applications, Warsaw, 1983.

[38] Vukobratović K.M., Kirćanski M.N., "An Approach to Adaptive Control of Robotic Manipulators", IFAC Journal - Automatica, Dec. 1985.

[39] Vukobratović K.M., Stokić M.D. and Kirćanski M.N., "Towards Nonadaptive and Adaptive Control of Manipulation Robots", IEEE Trans. on Automatic Control, Vol. AC-29, No. 9, 1984.

[40] Vukobratović K.M., Kirćanski M.N., "Computer Assisted Sensitivity Model Generation in Manipulation Robots Dynamics", Journal of

Mechanisms and Machine Theory, Vol. 19, No. 2, 1984.

[41] Billings S.A., Voon W.S.F., "Least Squares Paramter Estimation Algorithms for Non-linear Systems", Int. J. Systems Sci., Vol. 15, No. 6, 1984.

[42] Volterra V., Theory of Functionals, Glasgow: Blakie, 1930.

[43] Leontaritis I.J., Billings S.A., "Input-output Parametric Models for Non-linear Systems, Part II: Stochastic Non-linear Systems", Int. J. Control, Vol. 41, No. 2, 1985.

[44] Clarke D.W. and Gawthrop P.J., "Self-tuning Controller", Proc. IEE, Vol. 122, No. 9, 1975.

[45] Clarke D.W. and Gawthrop P.J., "Self-tuning Control", Proc. IEE, 126, No. 6, 1979.

[46] Wellstead P.E., Edmunds J.M., Prager D., and Zanker P., "Self--tuning Pole/Zero Assignment Regulators", Int. J. Control, Vol. 80, No. 1, 1979.

[47] Lindorff D.P. and Caroll R.L., "Survey of Adaptive Control Using Lyapunov Design", Int. J. Control, Vol. 18, 1973.

[48] Narendra K.S. and Kudva P., "Stable Adaptive Schemes for System identification and Control", Parts I and II, IEEE Trans. Syst. Man and Cybernetics, Vol. 4, 1974.

[49] Stoten D.P., "The Adaptive Control of Manipulator Arms", Mechanism and Machine Theory, Vol. 18, No. 4, 1983.

[50] Balestrino A., De Maria G. and Sciavicco L., "Hyperstable Adaptive Model Following Control of Nonlinear Plants", Systems & Control Letters, Vol. 1, No. 4, 1982.

[51] Vukobratović M., Kirćanski N., "Computer - Assisted Generation of Robot Dynamic Models in an Analytical Form", Journal of Appl. Mathematics and Mathematical Applications - Acta Applicandae Mathematicae, Vol. 3, 1985.

[52] Kirćanski N., "Computer-Aided Procedure of Forming Robot Motion Equations in Analytical Forms", VI IFToMM Congr. on Theory of Machines and Mechanisms, New Delhi, 1983.

[53] Vukobratović K.M., Kirćanski M.N., Real-time Dynamics of Manipulation Robots, Springer-Verlag, 1985.

[54] Auslander D.M. et al., "Direct Digital Process Control Practice and Algorithms for PM Applications", Proc. IEEE, Vol. 66, 1978.

[55] Ortega R. and Kelly R., "PID Self-Tuners: Some Theoretical and Practical Aspects", IEEE Trans. on Ind. Elect., Vol. IE-31, No. 4, 1984.

[56] Chiu K.Ch. et al., "Digital Control Algorithms: Parts 1, 2 and 3", Instrum. Contr. Syst., Oct., Nov. and Dec. 1973.

Chapter 4
Computer-Aided Control Synthesis

4.1 Introduction

This chapter is dedicated to computer-aided synthesis of non-adaptive and adaptive control for manipulation robots. In Chapters 2 and 3 we have estabilished a theoretical background for the synthesis of control for robotic systems. We have developed algorithms for the synthesis of control at executive control level for an arbitrary type of robotic system and we have underlined, that these algorithms are very suitable for implementation on a digital computer. Thus, the algorithms which have been elaborated in previous chapters are included in the software package for computer-aided synthesis of control for manipulation robots. This software package is developed in the Robotics Department of Mihailo Pupin Institute [1-3].

As we have emphasized in previous chapters, the algorithms for the synthesis of control for manipulation robots can be made considerably faster and more effective if we allow the user to utilize his experience in control design and his knowledge on robotic systems. Thus, such a package has been developed as allows permanent user's parcitipation during the control synthesis. In each step of control synthesis the user can interrupt the algorithm and make a decision on how to continue the synthesis.

In this chapter we shall briefly describe the complete software package for computer-aided synthesis of control for manipulation robots. Since the theoretical background of our approach to control synthesis for robotic systems has been presented in detail in the previous chapters, we shall not explain again the procedures employed to compute control parameters.

We shall especially emphasize the interaction between the package and the user in each step of the algorithm. Actually, the user is enabled to arrange the control synthesis in accordance with his own engineering judgment and his experience about system. The main advantage of our approach to control synthesis over the other approaches (which have been surveyed in Paragraphs 2.3 and 3.1 - 3.3) is the fact that our approach enables the control synthesis to be done step by step (or better to say, "piece-by-piece"): we first synthesize local controllers for decoupled subsystems; then, we analyze the stability of the global robotic system, and then, if necessary, we may introduce global feedback loops. It is obvious that this control synthesis might be performed completely automatically without the participation of the user (save that he has to set requirements i.e. define control task). However, it was recognized a relatively long time ago that the attempts to synthesize control for large-scale systems fully automatically suffer from many drawbacks [4, 5]. Actually, it has been shown that it is very difficult to specify all requirements and all conditions that play certain roles in a specific control task. On the other hand, the user can sometimes find a solution much faster by using his experience. Thus, it is recommendable to avoid a completely automatic synthesis of control and include the user in the control synthesis, leaving all crucial (final) decisions to him. These facts have been taken into account in developing our software package for computer-aided synthesis of control for manipulation robots. Actually, this package can be viewed as an assistive device which can help the user to synthesize the control for a specific manipulator.

In Paragraph 2 we shall describe the computer-aided synthesis of non-adaptive and adaptive control for manipulation robots; we shall specify the data that have to be imposed by the user of the algorithm and describe the interaction between the user and the package. In Paragraph 4.3 we shall present an example of the control synthesis done by the package. We shall describe how we can synthesize control for the specific manipulation system. Finally, in Paragraph 4.4 we shall discuss the possibilities offered by this package to the user.

4.2 Software Package for Computer Aided Synthesis

In this paragraph we shall describe the software package for computer-aided synthesis of control at the tactical and executive control levels for manipulation robots. It is based upon our approach to control synthesis which has been explained in previous chapters. Actually, in this book we have presented control synthesis at the executive control level only, while the synthesis of tactical control level has been discussed in Vol. 3 of this series [6]. However, the synthesis of the executive control level requires a desired movement (trajectories) to be set, so we must include both control levels in this package. We shall briefly explain the synthesis of nominal trajectories at the tactical control level and pay attention to describing the part of the package dedicated to the synthesis of the executive control level.

The software package allows the control synthesis for an arbitrary type of non-redundant manipulation robots. This means that manipulation robots with up to six degrees of freedom are allowed. Actually, this limit is concerned with tactical control level since the solution of the redundancy problem is not included in the package. However, our approach to the synthesis of the executive control level can be applied to redundant manipulation robots, since procedures for the synthesis of local controllers, stability analysis and global control synthesis are not limited with respect to the number of degrees of freedom of the robot.

As we have explained in previous chapters, our approach to control synthesis is based on employing the model of robot dynamics. In order to enable control synthesis for an arbitrary type of manipulation robot, the package includes the algorithm for automatically setting the mathematical model of robot dynamics. As described in Chapter 1, there are various methods for automatic setting of the mathematical models of robot dynamics on a digital computer (see, also books [7, 8]). However, we can divide these methods into two general groups: numerical and analytical. In [7] there were described several numerical methods for automatically setting robot dynamic models. In [8] and in Chapter 1 of this book we have presented our approach to automatically setting the analytical model of robot dynamics. Both approaches (numerical and analytical) have their advantages and disadvantages. The main advantage of the analytical model is a significant reduction of the computer processing time necessary for computing inertia matrix $H(q)$ and vector of gravity, centrifugal and Coriolis forces $h(q, \dot{q})$, once the analytical model has been obtained. Thus, when we obtain the analytical model, control synthesis is much faster since the computer time is highly reduced with respect to numerical models which have to go through all recursive relations at each time-they have to compute H and h. On the other hand, the disadvantage of the method for setting the analytical model of robot dynamics lies in the fact that a long processing time is required to obtain the analytical model for a particular robot. Thus, if we use the analytical approach, it is rather difficult to explore various structures of manipulators and to change parameters or structure of the robot during the control synthesis. The numerical methods for setting dynamical model of the robot do not suffer from this disadvantage: since the mathematical model is assembled each time we have to compute matrices H and h (for given q, $\dot{q}$) quite generally, these methods do not require any pre-computation and preparing when the parameters or structure of the robot change. Thus, these methods are much more convenient if we want to explore various robot structures and parameters. Obviously, these methods take a much longer computation time to synthesize control in comparison with analytical methods.

Because of the above-explained advantage and disadvantage of both approaches, we have decided to include two procedures for setting the mathematical models of robot dynamics in our software package: a numerical approach via D'Alambert principle which has been developed by Vukobratović and Stepanenko [9] and an analytical approach which has been developed by Vukobratović and Kirćanski N. [9]. The user may decide which of these procedures will be employed in particular run. If the parameters and structure of the robot are well-known and defined it is recommendable to employ the analytical method: it will take some period of processing time at the start of the trial to compute the analytical model of robot dynamics, but atterwards the control synthesis will be much faster since this model requires a much shorter processing time. However, if the parameters and structure of the robot are not precisely defined and the user wants to explore various possible structures and parameters of his robot which may still be under design (or, may be redesigned if it suffers from some drawbacks), it is better to employ numerical method for automatic setting of dynamic model of robot. This procedure will take a longer processing time for computing the model matrices, but the variation of robot parameters and structure will not cause any additional time delay (as it would be the case with the analytical method). Obviously, the user can easily combine both methods: he may use the numerical method as long as he wants to vary robot parameters and structure; once he has "fixed" the robot structure and parameters he may pass to the analytical method and obtain the analytical model of his robot. Thus, he can considerably reduce the computation time.

It should be noticed that the above consideration holds for off-line control synthesis (which is actually performed in our package). If the dynamic model is to be used for on-line computation of the control for manipulation robot, then there is no doubt the analytical model will take a much shorter computation time than the numerical, so the analytical approach should be used for real-time control, as it has been explained in Chapter 1. Of course, our package can be employed to obtain the analytical model of a particular robot which will be used in the synthesized controller (assuming that the controller synthesized for a particular control task requires the dynamic model of the robot).

Since we may regard the kinematic model of the robot as a part of the dynamic model, all above considerations hold for kinematic models. In our package we include both the numerical approach to automatic setting of kinematic model and the procedure for automatic setting of analytical kinematic model on a digital computer (see [6]).

A global flow-chart of the package for control synthesis is presented in Fig. 4.1. The order of programming blocks is conditional. The package is organized in such a way that the user can call and access each of the blocks according to his own convenience. In the text to follow we shall briefly present the interaction between the user and each of software blocks.

The algorithm enables the control synthesis for an arbitrary manipulation configuration and arbitrary manipulator types. Thus, the user can synthesize control either for: a) an already built manipulator (evaluation of manipulator), or b) for a manipulator which is under design. In case a) the algorithm can help to improve control for the existing manipulator and determine the equipment necessary for this improvement. In case b) the user can investigate the influence of the choice of various manipulator parameters (link masses distributions, actuator parameters and so on) upon the control synthesis and the performance of the manipulator, in order to find out the best solutions. In both cases the user can determine the simplest control law which satisfies the requirements imposed in a particular manipulation task. The control synthesis can be performed either for a particular pre-determined task (which has to be realized by the manipulator), or, the user can impose some test-task which ensures that the control synthesized is satisfactory for a wide class of tasks.

The interactions between the user and each of software block are basically very similar to one another. Each has two basic modes: the first is "input" mode, and the second is "changing" mode. When the user starts the control synthesis for the first time for his particular manipulation system, he enters the "input" mode. In this mode he has to

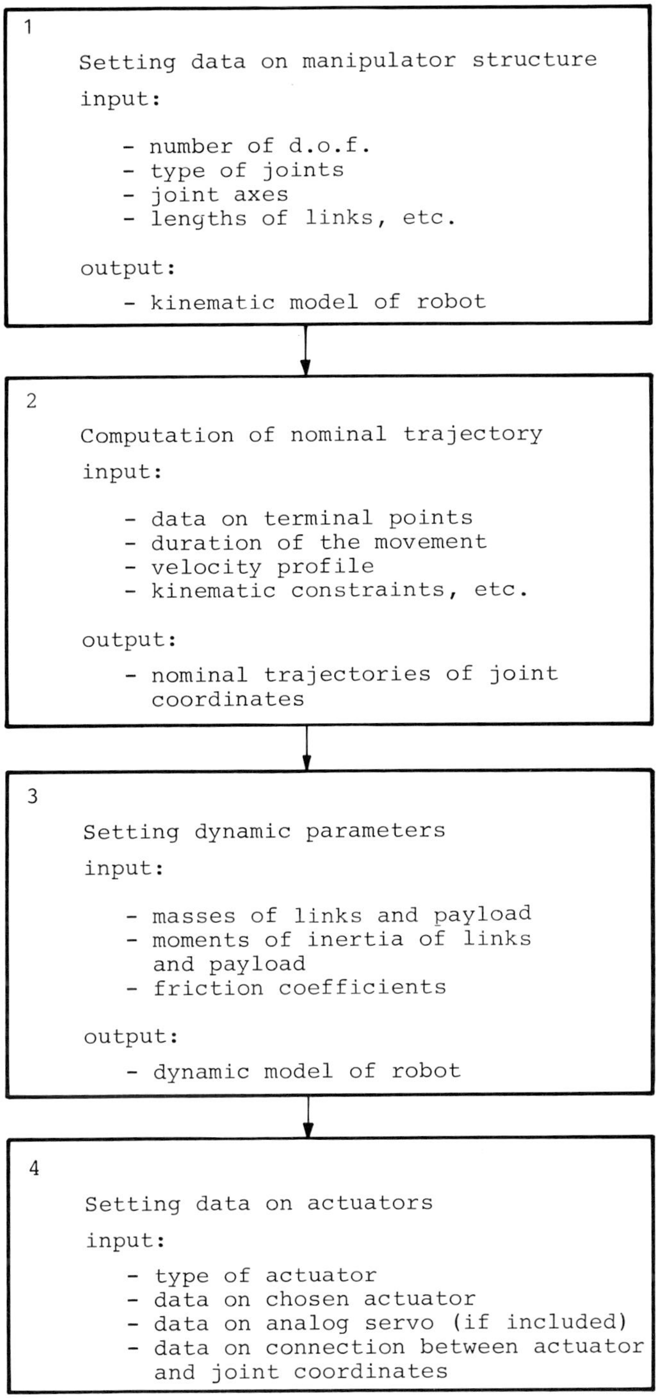

Fig. 4.1. Flow-chart of the software package for computer-aided control synthesis

4

Cont.

- data on connections between actuator load and driving torques around joint

output:

- complete model of the robot (actuators + mechanical part of the robot)

↓

5

Synthesis of nominal dynamics

input:

- centralized or decentralized nominal control
- estimates of inertia and gravity terms

output:

- nominal driving torques
- nominal trajectory of state vector
- nominal programmed control
- power and energy consumption

↓

6

Control task setting

input:

- desired stability degree, or
- tolerances in the terminal points of nominal trajectory (in external coordinates)

output:

- tolerances in joint (internal) coordinates and desired stability degree

↓

(7)

Fig. 4.1. Cont. (1)

(7)

7

Local control synthesis

input:

- local information control structure
- parameters of actuators θ^o
- procedure for local control synthesis
- initial values of the parameters of local controller (weighting matrices, or poles of closed-loop subsystem)
- automatic or interactive synthesis

Synthesis of local controller acc. to algorithm of Fig. 2.20.

---

output:

- local feedback gains

8

Stability analysis of decentralized control

input:

- mode of control synthesis automatic or interactive
- initial value of θ^o and variation $\Delta\theta$
- initial value of d^o and variation Δd

Synthesis of decentralized controller based on the stability analysis of linearized and nonlinear model - Fig. 2.21

If the stability is not guaranteed by decentralized controller:

- go back to (7) and re-select local control (change parameters or procedures or requirements), or
- introduce global control
- change nominal trajectory
- change actuators

(9)

Fig. 4.1. Cont. (2)

(9)

9

Global control synthesis

input:

- automatic or interactive mode of control synthesis
- form of the global control
- implementation of global control (bang-bang global control, force feedback or on-line computation)
- approximate model for on-line computation
- parameters d* for computation
- global gain

Synthesis of global control on the basis of nonlinear model stability analysis - Fig. 2.22

If unique global control cannot be found:

- reselect actuators
- change nominal trajectory and/or tolerances
- introduce adaptive control

output:

- global control law and feedback gains

10

Adaptive control synthesis

input:

- determine the values of parameters for local control synthesis

---

Call blocks (7), (8) for decentralized control synthesis for given parameters

---

- determine transient behaviour during estimation process

Call block (12) for simulation with adaptive control law

---

output:

- variable local gains for adaptive control

Fig. 4.1. Cont. (3)

(11)

11

11

Discrete-time control synthesis

input:

- sampling period

Synthesis of feedback gains in discrete-time domain

- type of microprocessor

output:

- feedback gains in discrete-time domain
- number of additions and multiplications to compute control law
- time required for control computation
- number of μP for control implementation

12

Simulation

input:

- initial conditions
- type of nominal control
- form of local controllers
- form of global control
- adaptive or non-adaptive control
- continuous time or discrete-time control

output:

- deviation of state coordinates, from nominal trajectory
- driving torques developed during tracking
- input signals developed during tracking
- power and energy consumption

Final decision on control law and equipment necessary for its implementation

Fig. 4.1. Cont. (4)

impose all data which are required by the algorithm in the particular step of control synthesis (for example, data on the structure of the chosen manipulator or data on actuators, etc - see the text to follow). All data imposed by the user are memorized in the appropriate files on the corresponding memory unit. Afterwards, if the user wants to change these data he has to enter "changing" mode and he can easily change the previously imposed data for this manipulation system. The conversation between the user and computer is very simple to handle. The algorithm checks all data imposed by the user and detects possible errors. This enables easy synthesis of control for manipulation robots of various types and enables the user to investigate the influence of various manipulator parameters upon the system behaviour [10].

The user can arrange the control synthesis according to his own convenience, as has already been stated. However, some order has to be taken into account; for example, nominal kinematics cannot be synthesized before the user has imposed data on manipulator configuration. The algorithm takes care about this fact, so the user is released of these problems. The user can easily go through the algorithm and iteratively improve the control.

4.2.1. Block for setting data on manipulator structure

The algorithm enables the control synthesis for nonredundant manipulation robots of arbitrary structure. In this block the user sets data on the configuration and geometry of a specific manipulator for which he wants to synthesize the control. The user has to set the following data:

1. The number of degrees of freedom of the manipulator (minimum 3, maximum 6).

2. Type of each joint (revolute or linear).

3. Projections of the joints axes in the internal coordinate systems (each internal coordinate system is connected to one manipulator link, so it is easy to set these data on joint axes' projections).

4. Projections of the links axes in the internal coordinate systems.

5. Lengths of the links (distances between the joint axes).

6. Distances between the joint axis and the corresponding link's center of gravity.

Using these data the algorithm automatically sets the kinematic model of the manipulator. The algorithm also forms a file with the data imposed by the user.

4.2.2. Block for calculating desired kinematic trajectory

The algorithm calculates nominal kinematic trajectories, according to user's desires. The user has to impose the following data:

1. Data on terminal points of each trajectory part. The coordinates of the terminal points can be imposed either in internal coordinates $q \in R^n$ (angles or displacements of each joint of the manipulator) or in external coordinates $s \in R^n$. The following options on external coordinates are available:

   (a) the coordinates of the manipulator tip in the absolute coordinate system (connected to the first joint) and the internal coordinates (angles) of the manipulator gripper,

   (b) the coordinates of the manipulator tip in the absolute coordinate system and the Euler's angles of the last member with respect to the axes of the absolute coordinate system,

   (c) the coordinates of the manipulator tip in the absolute coordinate system and the projections of a unit vector connected to the last member of the manipulator (upon the axes of the absolute coordinate system).

The user has to impose data on the coordinates of the terminal points of each part of desired trajectory. The user may set up to 30 trajectory parts, i.e. 30 points through which the manipulator has to pass.

2. Duration of the movement on each trajectory part (period τ).

3. The profile of the velocity of coordinates between the imposed terminal points. The velocity profile can be triangular or trapezoid. In the latter case the user should impose duration of acceleration

and deceleration.

4. The user can choose whether the manipulator should stop in the imposed terminal point or pass through it at a constant velocity.

5. The user can impose various kinematic constraints upon the internal coordinates of the manipulator, and various obstacles and constraints upon the working space.

The computation of the trajectories of internal coordinates, if external trajectory is defined by terminal points, is performed as follows.

The relation between the external coordinates s and the internal coordinates q is given by:

$$s = f_s(q) \tag{4.2.1}$$

where f_s: $R^n \rightarrow R^n$. If the user sets the trajectory in the external coordinates $s^o(t)$, $\forall t \in T$, the algorithm calculates the trajectory in the internal coordinates $q^o(t)$. Namely, using (4.2.1) it is obtained:

$$\dot{q}(t) = J^{-1}\dot{s}(t) \tag{4.2.2}$$

where $J \in R^{n \times n}$, $J = [\frac{\partial f_s}{\partial q}]$ is Jacobian matrix. In (4.2.2) it is assumed that J is a nonsingular matrix. Using (4.2.2) and set $s^o(t)$, $\dot{q}^o(t)$ is calculated and $q^o(t)$ is obtained by numerical integration of $\dot{q}^o(t)$. The Jacobian matrix for each point $q^o(t)$ is obtained from the kinematic model of the robot which is automatically set on the computer using the parameters on the mechanism structure, as explained in the previous block.

The algorithm calculates nominal kinematics so as to ensure the motion of the manipulator tip along a straight line between the terminal points, while the velocity of the tip changes according to the chosen velocity profile. According to our experience, this way of imposing nominal kinematics is quite satisfactory to enable user to synthesize nominal trajectories for most manipulation tasks which can be encountered in industrial practice.

For more details on the synthesis of nominal trajectories using the kinematic model of the robot see book 3 of this series [6]. Obviously, nominal trajectories might be synthesized using the dynamic model of the robot. For example, distribution of the tip velocity might be computed by minimizing the energy required for the movement [6, 11]. Obviously, such an approach requires a much longer processing time than the synthesis based on the kinematic model.

The algorithm forms files with data imposed by the user, and with data on the calculated nominal trajectory. During calculation of the nominal trajectory the algorithm takes care about the imposed constraints and the manipulator kinematic capabilities. A graphical display of the calculated trajectory (internal angles and velocities) is also available.

4.2.3. Block for setting dynamic parameters

In order to set the mathematical model of the manipulator dynamics, the algorithm needs data on the manipulator dynamics, such as:

1. Masses of all links and payload.
2. Moments of inertia of all members (with respect to the main axes of inertia) and of the payload.
3. Data on friction in each manipulator joint.

The user might define allowable variations of all these parameters D, i.e. for each parameter the user might define the minimum and maximum allowable values.

The algorithm automatically sets the dynamic model of the mechanism and forms files with the imposed data the dynamic model is set in form (2.2.1).

Actually, the algorithm computes the matrix $H(q,d)$ and vector $h(q, \dot{q}, d)$ for given $q(t)$, $\dot{q}(t)$ and given parameters d (geometry parameters, masses, moments of inertia, etc.). Each time during the control synthesis when we need the model of dynamics of the robot we have to set the values of $q(t)$, $\dot{q}(t)$ and the algorithm will compute the corresponding matrices H and h. As we have explained above, the user can choose the procedure which will be applied for setting the model: either the numerical or the analytical procedure. In the latter case the package will take some time to prepare the analytical model of the robot at the very start of the control synthesis for a particular robot.

4.2.4. Block for setting data on actuators

The user has to impose data on the actuators for each joint of the mechanism. He can choose the type of each actuator: D.C. electro-motor or hydraulic linear or vane actuator. If he chooses D.C. motor he has to impose the following data (all these data can be picked from the data sheets for the chosen actuator):

1. Viscous friction coefficient (B_C^i),
2. Electro-motor coefficient of the motor (C_E^i),
3. Moment coefficient of the motor (C_M^i),
4. Moment of inertia of rotor (J_r^i),
5. Gear speed and moment ratio (N_r^i),
6. Resistance of the rotor (R_r^i),
7. Inductance of the rotor (L_r^i),
8. Upper and lower amplitude constraint upon the actuator input (u_m^i).

Using these data the package computes the matrices of the time-invariant linear models of actuators S^i. The matrices A^i and vectors b^i, f^i are computed according to Eq. (2.A.2), and the package assumes the models of actuator to be in linear form (2.2.2).

When the user chooses hydraulic actuators, he has to type the following data:

1. Viscous friction coefficient in the cylinder (B_C^i),
2. Piston area (A_C^i),
3. Cylinder (working) volume (V_u^i),
4. Mass of the piston (including the rod mass) (m_t^i),
5. The oil compressibility coefficient (β^i),

6. Gradient of the servovalve flow-pressure characteristic (k_c^i),

7. Proportionality coefficient of the oil flow and servovalve coil current (K_q^i/C_1^i),

8. Upper and lower amplitude constraint upon the actuator input (u_m^i).

Using these data the algorithm computes matrices A^i and vectors b^i, f^i according to Eq. (2.A.6) and assumes the linear time-invariant models of actuators S^i (2.2.2).

In both cases the user might impose the range of all parameters θ (if their variation is expected), i.e. for each parameter the user might define minimum and maximum allowable values.

In both cases the user can choose the order n_i of the model of each actuator. As we have explained in Appendix 2.A., the model of D.C. motor might be taken to be 2 or 3, and the user might select the order he prefers. Obviously, if the user selects the order of the model of D.C. motor to be 2, then the matrices of the model are defined by (2.A.3), and the inductance of the rotor coil is neglected. (Thus, the inductance need not be set). The situation is similar with the hydraulic actuator.

It should also be noted that if the user desires to apply the hydraulic actuator, he might choose between linear and vane actuators. The data required in both cases are as listed above, just the units are different depending on the type of the hydraulic actuator selected.

The package allows a special option to include already implemented analog servosystems.

If the implemented actuators already have analog servosystems*) the user has to set data on feedback gains. The assumed structure of the analog servosystem is presented in Fig. 4.2. The user has to set the following data:

---

*) This situation appears if we have already built the manipulator and implemented local analog servosystems, and now we want to introduce additional control through a microprocessor unit in order to improve robot's performance. Thus, the analog servosystem gains are assumed to have been synthesized previously and the user can type their values.

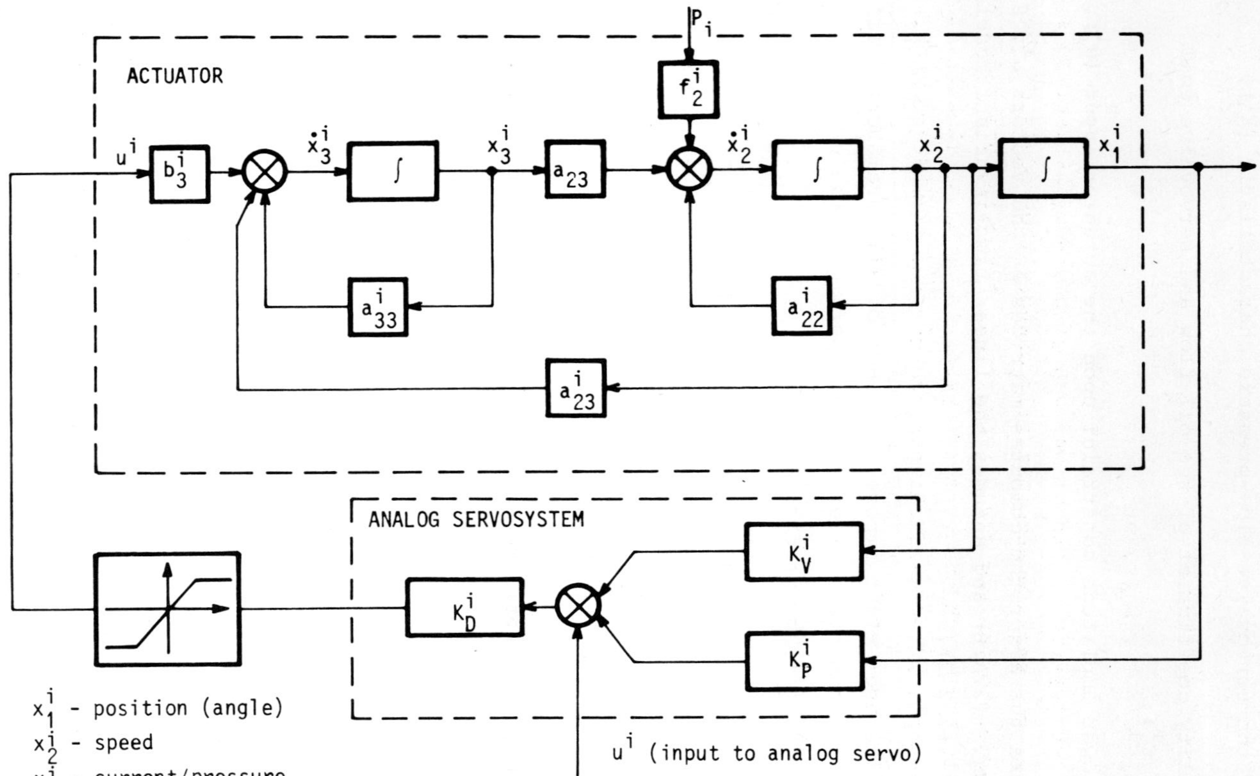

x_1^i - position (angle)
x_2^i - speed
x_3^i - current/pressure
K_P^i - position analog feedback gain
K_V^i - speed analog feedback gain
K_D^i - gain in direct branch

Fig. 4.2. Scheme of the actuator with analog servosystem

1. gain K_D^i in direct branch,
2. gain K_P^i in feedback loop by position,
3. gain K_V^i in feedback loop by speed.

If the analog servosystem exists in the ith joint, and the user sets the above gains, the package automatically changes the models of the actuators in order to include the servosystems. Actually, the model of the actuator together with analog servosystem is given by (2.2.2), too, but the matrices A^i, b^i, f^i are changed and get the form (for the third order models of actuators):

- for D.C. motors:

$$A^i = \begin{bmatrix} 0 & 1 & 0 \\ 0 & -B_C/J_r & C_M/J_r \\ -K_DK_P/L_r & -(C_E+K_DK_V)/L_r & -R_r/L_r \end{bmatrix} \quad b^i = \begin{bmatrix} 0 \\ 0 \\ \frac{K_D}{L_r} \end{bmatrix} \quad f^i = \begin{bmatrix} 0 \\ -1/J_r \\ 0 \end{bmatrix} \tag{4.2.3}$$

- for hydraulic actuators

$$A^i = \begin{bmatrix} 0 & 1 & 0 \\ 0 & -B_C/m_t & A_C/m_t \\ -K_DK_Pb_3' & -4\beta A_C/V_u-K_DK_Vb_3' & -4\beta(k_C+C_{tc})/V_u \end{bmatrix} \quad b^i = \begin{bmatrix} 0 \\ 0 \\ b_3'K_D \end{bmatrix}$$

$$f^i = (0, \ -1/m_t, \ 0)^T, \quad b_3' = 4\beta K_q/(V_uC_1) \tag{4.2.4}$$

Obviously if the analog servosystems have not yet been synthesized, but still have to be selected by the package, the user cannot impose the above data and the matrices of actuators models remain unchanged.

The user also has to impose the following data:

1. He has to choose the type of connection between the manipulator joint coordinate (angle or displacement) and the coordinate (displacement) of the corresponding actuator. When D.C. motor or vane hydraulic actuators is implemented in rotational joint, or when linear hydraulic actuator is applied in the linear degree of freedom

of the manipulator, this connection is trivial. However, in the opposite case when rotational actuators are applied for linear degrees of freedom or vice versa, this connection may be complex, so the user has to impose the type of connection and the parameters of the chosen connection. These connections are given by relation (2.2.4). There are several types of connections which are included in the package (such as cosines formula, tangens relation etc.) and the user has to depict the one which corresponds to his robot (for each joint he has to select the right type of the connection).

2. The user also has to choose the type of connection between the driving torque in the manipulator joint and the load acting upon the corresponding actuator. The relation between the driving torque and actuator load is defined by (2.2.7). The user has to select among the preassumed possible types of connection which correspond to the connection in each joint of his manipulator.

Obviously, these relations between the joint coordinates and actuator coordinates and the relations between the joint driving torque and load in actuator reflect the mathematical model of the robotic system, as it has been explained in Paragraph 2.2.2. Thus, the load distribution vectors f^i are changed to include relations (2.2.7), so these vector are not constant but become functions of q_i.

The package forms files with all these data imposed by the user. In this way the user has set all the data required to specify the model of the particular robot, and the package can operate with the complete model of the robot when it is needed during the control synthesis.

4.2.5. Block for synthesis of nominal dynamics

In this block nominal dynamics is synthesized in the following manner:

1. Using the calculated nominal kinematic trajectories $q^o(t)$, $\dot{q}^o(t)$ $\forall t \in T$, the package computes nominal driving torques on the basis of the model of the robot mechanism dynamics:

$$P^o(t) = H(q^o(t), d^o)\ddot{q}^o(t) + h(q^o(t), \dot{q}^o(t), d^o) \tag{4.2.5}$$

where accelerations $\ddot{q}^o(t)$ are obtained by numerical differentiation of the nominal velocity $\dot{q}^o(t)$. The user has to define for which values of the parameters in a given range $d^o \in D$ the nominal driving torques will be computed.

2. If the user has selected that the order of the actuator model is 3, this means that each actuator model has three state coordinates ($(\ell^i, \dot{\ell}^i, i_r^i)^T$ - in the case of the D.C. motors, and $(\ell^i, \dot{\ell}^i, p^i)^T$ - in the case of hydraulic actuators, where i_r^i denotes rotor current and p^i denotes pressure difference in the cylinder). Since $q^o(t)$, $\dot{q}^o(t)$ define $\ell^o(t)$, $\dot{\ell}^o(t)$ through set relations (2.2.4), the package computes the nominal trajectory of the third state coordinate of each actuator (let us denote the state coordinates by $x^i = (x_1^i, x_2^i, x_3^i)^T$). Since the matrices A^i are of equal form regardless of whether D.C. motors of hydraulic actuators are selected, the nominal trajectory of the third state coordinate $x_3^{io}(t)$ in both cases is computed as:

$$x_3^{oi}(t) = (\dot{x}_2^{oi}(t) - a_{22}x_2^{oi}(t) - f_2^i(q^{oi}(t))\cdot P_i^o(t))/a_{23}^i \qquad (4.2.6)$$

where a_{kj}^i and f_2^i denote the corresponding elements of the matrices A^i and f^i, respectively.

Obviously, if the user has selected the order of the ith actuator's model to be 2, there is no need to compute nominal trajectory of the third state coordinate.

3. The package computes nominal programmed control, i.e. nominal imputs to each actuator which has to be supplied to actuators in order to develop nominal driving torques $P_i^o(t)$ and nominal motion of the robot (i.e. to drive the robot along nominal trajectory $x^o(t)$, $\forall t \in T$). The programmed nominal control is obtained in the form:

- if the order of the actuator model is $n_i=3$

$$u^{oi}(t) = (\dot{x}_3^{oi}(t) - a_{32}^i x_2^{oi}(t) - a_{33}^i x_3^{oi}(t))/b_3^i \qquad (4.2.7)$$

- if the order of the actuator model is $n_i=2$

$$u^{oi}(t) = (\dot{x}_2^{oi}(t) - a_{22}^i x_2^{oi}(t) - f_2^i(q^{oi}(t))\cdot P_i^o(t))/b_2^i \qquad (4.2.8)$$

In Eq. (4.2.7) and (4.2.8) we have taken into account the forms of the actuator model matrices A^i, b^i, f^i.

The nominal programmed control $u^{oi}(t)$ computed as presented above would realize $x^o(t)$ in ideal conditions, i.e. if no perturbation is acting upon the robotic system and assuming that the mathematical model of the robotic system is ideally identified. However, computation of $u^o(t)$ requires to compute dynamic model of the robot, which might be too complex to be implemented on-line as we have explained in previous chapters and which requires perfect knowledge of all parameters of the robot mechanism. Thus, instead of nominal control synthesized using the centralized model of the robotic system, the package might synthesize nominal control using approximative decentralized model (see Paragraph 2.4) in which coupling between the joints is neglected (i.e. in which the dynamics of the mechanism is neglected). Thus, the package might compute the so-called local nominal control $\bar{u}^{oi}(t)$ according to the following relations:

- if the order of the actuator model is $n_i=3$

$$x_3^{oi}(t) = (\dot{x}_2^{oi}(t) - \tilde{a}_{22}^i x_2^{oi}(t))/\tilde{a}_{23}^i$$
$$u^{oi}(t) = (\dot{x}_3^{oi}(t) - \tilde{a}_{32}^i x_2^{oi}(t) - \tilde{a}_{33}^i x^{oi}(t))/\tilde{b}_3^i \qquad (4.2.9)$$

- if the order of the actuator model is $n_i=2$,

$$u^{oi}(t) = (\dot{x}_2^{oi}(t) - \tilde{a}_{22}^i x_2^{oi}(t))/\tilde{b}_2^i \qquad (4.2.10)$$

where $\tilde{a}_{kj}^i$, $\tilde{b}_j$ denote the corresponding members of the matrices $\tilde{A}^i$, $\tilde{b}^i$ which are defined by (2.4. 7). Actually, $\tilde{A}^i$, $\tilde{b}^i$ are the matrices of the actuator models which include estimation of inertia term $\bar{H}_{ii}\ddot{q}_i$ and estimation of gravity term $\bar{h}_i q_i$, as explained in Paragraph 2.4.

Thus, the user has to make two choices:

a) To choose whether he wants to synthesize nominal control using the centralized model (Eqs. (4.2.7), (4.2.8)), or the decentralized model (Eqs. (4.2.9), (4.2.10)).

b) To choose the estimates of the inertia term $\bar{H}_{ii}\ddot{q}_i$ and gravity terms $\bar{h}_i q_i$ which should be associated to the subsystem model (for each joint).

Actually, the package computes $\min_{t\in T} H_{ii}(q^o(t))$ and $\min_{t\in T} |\partial(g_i(q^o(t))/\partial q_i|$

and the user has to select the part of this value that will be taken in the subsystem model; the user types the part of the inertia estimates that he wants to associate to subsystem model K_I and the package takes $\bar{H}_{ii}$ to be $K_I \min_{t \in T} H_{ii}(q^o(t))$ a similar situation arises with the estimation of the gravity term.

4. The package checks whether the chosen actuators can perform the imposed motion. Namely, the package checks whether the computed nominal control $u^{oi}(t)$ satisfies the amplitude saturation constraint upon the actuator input (2.2.3) for all subsystems and over the desired time interval T. If the computed nominal control for the *i*th actuator $u^{oi}(t)$ at some instant on the trajectory exceed the amplitude bound u_m^i, this means that the chosen *i*th actuator cannot realize the desired nominal trajectory (even in an ideal case if no perturbation is acting upon the system). In that case, the user might decide whether:

   a) to go back to block for input data on actuator parameter and change the *i*th actuator, or

   b) to go back to block for calculation of nominal kinematic trajectory and change the trajectory (often he has to decrease the desired speed of motion by increasing the duration of manipulator motion).

5. The package computes the power and energy demanded to implement the desired nominal trajectory.

The algorithm forms a file with all calculated data on nominal dynamics (nominal angles or displacements, nominal velocities, driving torques, inputs, power and energy consumptions). A graphic display of these data is also available.

4.2.6. Block for setting control task

The user has to impose requirements concerning the quality of tracking the nominal trajectory, i.e. he has to impose the requirements upon the control task. Two ways of imposing the control task are available:

1. The user can directly set the desired degree of stability α of the

manipulation system. In this case the algorithm has to ensure that the poles of the system are to the left from the imposed stability degree α in a complex plane. This means that the slowest response of the system must be faster than exponent of the prescribed stability degree α (i.e. it must be satisfied $||\Delta x(t)|| < ||\Delta x(0)||\exp(-\alpha t)$).

2. The second way is to impose tolerances in the terminal points of the trajectory parts which have been set in block for computation of nominal trajectory. The tolerances have to be imposed in the same type of external coordinates (or internal coordinates) in which the corresponding terminal points have been imposed.

The user sets tolerances in external coordinates so as to impose allowable deviations $\Delta s(t_i)$ from the nominal values for each coordinate where t_i, $\forall i \in I_1$, $I_1 = \{i: i=1,2,\ldots,m\}$ are control instants on the trajectory, $\forall t_i \in T$. In this way we define the regions in manipulator external coordinates space $X_s(t_i)$, $\forall i \in I_1$, which manipulation robot has to satisfy:

$$X_s(t_i)=\{s(t_i): |s^j(t_i)-s^{oj}(t_i)|<\Delta s^j(t_i), \quad \forall j \in I\}, \forall i \in I_1, \tag{4.2.11}$$

where $s = (s^1, s^2,\ldots,s^n)^T$. On the basis of relation (4.2.1) between the external and internal coordinates, the algorithm maps the regions $X_s(t_i)$ from the external coordinates space to the regions $Q(t_i) \subset R^n$ in the internal coordinates space:

$$F_s^{-1}(X_s(t_i)) \to Q(t_i), \qquad i \in I_1 \tag{4.2.12}$$

Similarly, the user may impose allowable deviations by rates and accelerations in the external coordinates, and then the algorithm determines allowable regions of deviations by rates and accelerations in the space of the internal coordinates $Q^v(t_i)$, $Q^a(t_i)$, $\forall i \in I_1$. If the user does not set allowable deviations by rates and accelerations, the algorithm adopts regions $Q^v(t_i)$, $Q^a(t_i)$, $\forall i \in I$, according to the feasibilities of chosen actuators. Regions $Q(t_i)$, $Q^v(t_i)$, $Q^a(t_i)$, $\forall i \in I$ are thus obtained and mapped by the algorithm, using (2.2.12), into regions $X_s^t(t_i) \subset R^N$ in the state space of the system S, which may be estimated by $X^t(t_i) \supseteq X_s^t(t_i)$ $\forall t \in I_1$ where $X^t(t_i) = X^{t(1)}(t_i) \times X^{t(2)}(t_i) \times,\ldots$ $\ldots,\times X^{t(n)}(t_i)$, $\forall i \in I_1$, $X^{t(j)}(t_i) \subset R^{n_j}$, $\forall j \in I$ *). In this way the regions

*) Here, we assume that $n_i=3$, $\forall i \in I$, which is the most common case. If we choose $n_i=2$, it is sufficient to define regions $Q(t_i)$, $Q^v(t_i)$.

in the state space to which the system state must belong during tracking of the set nominal trajectories are defined, i.e., $x(t_i)\in X^t(t_i)$, $\forall i\in I_i$.

However, in order to simplify control synthesis we prefer to get stability regions in the form assumed in Para. 2.2.2,

$$X^{t(i)'}(t) = \{\Delta x^i(t): \quad ||\Delta x^i(t)|| < \bar{X}^{t(i)}\exp(-\alpha_i t)\}, \quad \forall t\in T \quad (4.2.13)$$

It is very easy to compute the exponential degrees α_i of shrinking the regions $X^{t(i)'}$ so that these regions satisfy:

$$X^{t(i)'}(t_j) \subseteq X^{t(i)}(t_j), \quad \forall t_j\in I_1, \quad \forall i\in I \qquad (4.2.14)$$

Thus, the package determines the stability regions in the following way:

a) First, the user has to set allowable deviations $\Delta s(t_j)$, $\forall j\in I_1$ from nominal trajectory in external coordinates and define the allowable regions $X_s(t_j)$, $\forall j\in I_1$ by (4.2.11).

b) The package maps the regions $X_s(t_j)$ from external coordinate space to regions $Q(t_j)$ in the internal coordinate space using (4.2.12).

c) If the user sets allowable deviations by rates and accelerations in the external coordinates, the package maps these regions into regions by rates and accelerations $Q^v(t_j)$, $Q^a(t_j)$, $\forall i\in I_1$ in internal coordinate space, otherwise, the package adopts regions $Q^v(t_j)$, $Q^a(t_j)$ according to the feasibilities of the actuators [10].

d) The regions $Q(t_j)$, $Q^v(t_j)$, $Q^a(t_j)$, $\forall j\in I_1$ define the regions $X_s^t(t_j)$, $\forall j\in I_1$ in the state coordinate system by relations (2.2.11)-(2.2.13).

e) The obtained regions $X_s^t(t_j)$ are "decoupled" by the package; The package defines the regions $X^t(t_j) = X^{t(1)}(t_j)\times X^{t(2)}(t_j)\times \cdots \cdots \times X^{t(n)}(t_j)$, $\forall j\in I$, where $X^{t(j)} \subset R^{n_i}$, $\forall i\in I$ are the regions in the subsystems state space. (Actually, the package defines the largest regions $X^t(t_j)$ which satisfy both $X^t(t_j) \subseteq X_s^t(t_j)$ and $X^t(t_j) = X^{t(1)}(t_j)\times X^{t(2)}(t_j)\times \cdots \times X^{t(n)}(t_j)$, $\forall j\in I_1$).

f) The package determines "the best" estimations of the regions $X^{t(i)}(t_j)$ by the regions $X^{t(i)'}(t)$ defined by (4.2.13); actually, the package has to find out the smallest numbers α_i and largest $\bar{X}^{t(i)}$ so that the region $X^{t(i)'}(t)$ defined by (4.2.13) satisfies (4.2.14).

Thus, the package gets the regions of practical stability $X^t(t)$ defined as in Chapter 2 and, by this, parameters of the control task (2.2.18) are set.

The package forms a file with data on tolerances in external coordinates (as they were imposed by the user) and a file with calculated tolerances in internal coordinates.

4.2.7. Block for local control synthesis

In the previous paragraph we have described that the algorithm automatically "decouples" the stability regions in subsustems state space. In this way the control task (2.2.18) is decouples into local control tasts (2.4.8). Obviously, this decoupling is approximative since the satisfaction of the local control tasks does not guarantee satisfaction of the global control task (2.2.18). Now, the package can synthesize the local controllers so as to satisfy local control tasks (2.4.8), i.e. for each subsystem (actuator) (2.4.7) the package synthesizes the local controller which satisfies (2.4.8). The algorithm for the synthesis of local controller for decoupled subsystem has been described in Paragraph 2.4.4. The package completely implements this algorithm (Fig. 2.20). Naturally, here we shall not repeat description of the algorithm for local control synthesis. We shall just point out the interaction between the user of the package and the algorithm, which is intended to speed-up the procedure, as mentioned in Paragraph 2.4.4.

In this block the user has to make the following decisions:

1) To choose the information control pattern for each local controller, i.e. to choose the local feedback loops that will be introduced for each subsystem (joint); the user has to state for each actuator:*)

   a) Will feedback by speed be introduced, or not?

   b) Will feedback by rotor current (if D.C. motor is applied) or by cylinder pressure (if hydraulic actuator is applied), be introcuced of not?**)

---

*) The package assumes that a positional feedback loop is introduced in each joint which is commonly the case.

**) Obviously, the introduction of the feedback by current/pressure can be done if the user assumes the order of the actuator model to be 3.

c) Will integral feedback (integral of the position error) be introduced or not? (Actually, this question means: shall we apply dynamic or static controller?)

2) To specify parameters of each subsystem-actuator $\theta^{o} \in \Theta$ for which local controller will be synthesized.

3) To choose the procedure for the local control synthesis. The user has to decide whether the package will synthesize the local controller by minimizing local quadratic controller or by pole-placement*): obviously, this choice is connected with the above selection of information control pattern (i.e. the local optimal regulator requires a complete state feedback so that both feedbacks by velocity and current (pressure) have to be introduced).

4) To specify initial values for the parameters of the local controller which have to be set in the selected procedure; if the user has selected to synthesize local optimal controller he has to specify weithting matrices Q_i, r_i and prescribed stability degree β_i in local standard quadratic criterion (2.4.16); if pole-placement has been selected, the user has to specify poles of the closed-loop local subsystem; obviously, if the user does not want to specify these values, the package might adopt them in accordance to local control task specifications.

5) To decide whether the local control synthesis will be performed completely automatically or in the interactive mode; if the user selects the automatic mode, then the package should iteratively re-select local controller until it reaches the specified local control task (2.4.8)**); if the user decides to synthesize local control in interactive mode, this means that the package would ask the user to re-select parameters of the local controller at each itera-

---

*) Our package actually includes several other procedures for local control synthesis (e.g. Bode's method for synthesis in frequency domain, etc.) but, for simplicity, we shall not discuss these well-known methods since the procedure for stability analysis is invariant with respect to the method for local control synthesis.

**) It should be noted that if automatic mode is chosen, even the above listed data (information control pattern, procedure for local control synthesis, parameters of local control etc.) need not be specified by the user, but the algorithm might adopt them starting from the simplest case and iteratively re-choose them. However, this is usually time consuming, so it is much faster if the user can specify even the initial values of these parameters.

tion until the requirements specified in local control task are met.

6. If the automatic mode of local control synthesis has been chosen, the package iteratively searches for the local control which satisfies local control task (2.4.8); the user has to specify:

   a) The maximum number of iterations for searching of local control by increasing the prescribed stability degree β_i (if local optimal regulator is chosen) or by moving the poles of the closed-loop subsystem to the left in the complex plane (if pole-placement method is used); if the specified number of iterations is exceeded without success, the algorithm will pass to change of information pattern.

   b) Feedback loops which might be added in the local controller in order to achieve satisfaction of local control task (e.g. the user should specify whether or not he allows feedback by velocity or by pressure/current).

   c) Sequence of the methods for control synthesis which will be used by the package in searching for an adequate local control (i.e. the user has to specify the method for control synthesis that will be used next, if the preceding one has failed).

7) If interactive mode has been selected, the user is directly included in control synthesis; this means that after each trial the user is asked to:

   a) Re-set the parameters of the controller, i.e. to specify a new value of prescribed stability degree or new weighting matrices, or new values of closed-loop subsystem poles, or

   b) specify new information control pattern if he wants so, or

   c) specify a new method for control synthesis, if the user prefers to try with another procedure for control synthesis.

It should be noted that since we consider linear time-invariant subsystems it is usually rather simple to synthesize local control which satisfies the local control task. Most frequently the user can determine an adequate local controller in one or two iterations (using the package support, of course). However, we have included all above options in order to make the package general and to provide a basis for the next step of control synthesis (linear and nonlinear stability analysis) in which these options are also included (see the text to follow).

4.2.8. Block for stability analysis of decentralized control

In this block of the software package, stability of the overall robotic system is analysed. Actually, this block represents implementation of the algorithm described in Paragraph 2.5 and presented in Fig. 2.21. As with the previous block, we shall not repeat the description of the algorithm. We shall briefly discuss interaction between the user and the package.

As we have explained in Paragraph 2.5, the stability analysis of the overall robotic system is performed in two steps: first, the linearized model of the robot is analyzed and, then, nonlinear analysis of stability is performed. In both steps the user might be directly included in iterative search for a decentralized controller which satisfies the set conditions of stability, or this search might be performed completely automatically. Here we shall list data that are expected by the user in both modes of decentralized control synthesis:

I The user has to choose the mode of control synthesis that will be executed: automatic or interactive.

II If automatic mode has been chosen, the user might impose the following data:

a) Initial values of actuator parameters $\theta_o \in \Theta$ and parameter variations $\Delta\theta$ which will be used to vary parameters of actuators over the imposed region of allowable parameters Θ.

b) Data required for local control synthesis for each decoupled subsystem in automatic mode (see Paragraph 4.2.8).

c) Initial values of parameters of the mechanical part of the robot $d = d_o \in D$ and parameter variations Δd which will be used to vary parameters of the mechanical part of the robot over the imposed region of allowable parameters D in order to examine the stability of the robotic system for $\forall d \in D$.

Obviously, all above data might be automatically chosen by the algorithm, but it seems that their choice is simple and it might considerably speed-up the iteration process of determining the parameters of the local controllers.

III If interactive mode has been chosen then the user is permanently included in searching for the adequate local controllers. In this the user has to impose the following data:

a) All initial data listed above for automatic mode are also included in interactive mode.

b) If the linearized model of the robot does not satisfy the set stability requirements, the user has to specify new parameters for local controllers; the user might determine the local controller that has to be changed in order to stabilize the overall system*); the user might change information control structure in the corresponding local controller in order to improve the performance of global robotic system; the user might change the procedure for local control synthesis if he presumes that it might lead faster to the solution.

c) If the analysis of the practical stability does not guarantee the practical stability of the global system with the synthesized decentralized controller, the user might also specify the parameters of local controllers in the next search; in this case, since the nonlinear stability analysis is performed subsystem by subsystem, it can be easily determined which local controller has to be changed in order to stabilize the nonlinear model of the robot; the local information control structure might also be changed if the user finds that this is worthwhile; the user might also change the method for local control synthesis according to his convenience.

d) If the algorithm cannot find unique local controllers which can accommodate all variations of actuator parameters Θ and parameters of the mechanical part of the robot D, he might try to change parameters of actuators θ^o for which the local control-

---

*) In automatic mode the algorithm must test the sensitivity of the linearized model to each local controller, i.e. it must change all local controllers in order to find out which of them leads to stabilization of the linearized model (see Paragraph 2.5); however, in interactive mode the user might directly "recognize" the local controller which can improve stability of the overall system.

lers have been synthesized. Similarly, if centralized nominal control is included in the control law, the user might change parameters of the mechanism d^o for which the nominal driving torques are computed (4.2.5). These user's interventions might sometimes help to find unique decentralized control (with or without centralized nominal control) which can withstand all parameter variations.

IV In both modes if too many trials have been made without success, the user has to decide whether he wants to continue searching for decentralized control which satisfies the given control task, or to introduce additional global feedback loops. Namely, as it has been explained in Paragraph 2.5, if the influence of the dynamics of the mechanism is too strong, the satisfaction of the given control task might require very high local feedback gains. We have listed in Paragraph 2.5, some of the drawbacks of "high gain" solution of decentralized controller. It is up to a user to decide whether he wants to continue searching for decentralized controller and "risk" to implement "high gain" local controllers, or to introduce global control. The user should estimate whether the local feedback gains are too high so that they might cause appearance of elastic modes of the robot, or the decentralized controller is still acceptable. Obviously, the user might change the control task if he finds out that the requirements are too hard for a particular robotic system. The user might slow-down the desired trajectory $x^o(t)$ or change the finite regions of practical stability X^I and $X^t(t)$, $\forall t \in T$. The another possibility is to change actuators, i.e. to try to satisfy control task requirements with more powerful actuators which might stabilize the robot even with the decentralized controller only.

As the output of this block, local feedback gains have to be computed, either those which satisfy the stated control task without additional control loops, or the local feedback gains which are acceptable according to the user's opinion but which require introduction of global control in order to satisfy the stated control task. The package forms the file with the computed local feedback gains.

4.2.9. Block for global control synthesis

If the user has decided to introduce global control he has to enter this block and synthesize global control which ensures stability of the overall robotic system. This block is program implementation of the algorithm for global control synthesis which has been described in Paragraph 2.6, the flow-chart of which is presented in Fig. 2.22. Here, we shall briefly discuss interaction between the user and the package.

The list of data and the decisions that might be expected from the user are given below.

I The user has to decide whether the global control synthesis will be performed fully automatically or interactive synthesis will be executed. In automatic mode the user need not set any additional data (except those which he might set during synthesis of local controller - see Paragraph 4.2.8). However, the execution of the automatic mode for global control synthesis might be very time consuming, so it is not recommendable to use this mode. In this block the user's experience about the robot dynamics may be fully utilized.

II If interactive mode has been selected the user might take part in each iteration during searching for adequate global control law and parameter. However, in each step the user may make his own decision or leave to the algorithm to search for adequate solution. Here, we shall list all data that might be imposed by the user, although some of them might be computed by the package.

a) The user might select the form of the global control, i.e. he might select among forms (2.6.24), or (2.6.25), or (2.6.4).

b) The user might select implementation of global control among three possibilities: maximum input amplitude of global control (bang-bang global control), force feedback, or on-line computation of driving torques.

c) If the user has decided to implement on-line computation of the coupling acting upon the corresponding joint (for which the global control is synthesized), he has to select the approximative model of robot dynamics which will be used to compute global control. In Paragraph 2.6, we have listed various approximative models which might be used to compute driving torque acting upon the corresponding joint. In making this decision, the user has to utilize his experience. He may recognize the forces that are significant for the corresponding joint, so that these forces have to be computed on-line. In this way the user may considerably speed-up determination of adequate global control.

d) If the user has decided to implement on-line computation of the coupling, he may determine the parameter values $d^* \in D$ for which the driving torques will be computed. This choice may be very important if we want to find out such global control which will (together with the decentralized controller) withstand all expected parameter variations D. This choice is sometimes very difficult, but the user may find the solution using his experience.

e) The user may also choose the global gain for the particular joint for which the global control will be introduced. Obviously, the package could determine the global gain by iteratively searching for it, but this search might require a long processing time. The user can speed up this process since he may consider the numbers ξ^*_{ij} in (2.6.18) and recognize how the global control changes the interconnections among subsystems. By this he can roughly estimate the global gain and the package may perform just "fine" adjustment of the global gain.

All above listed data may be changed by the user in each iteration. Namely, if the package announces that the robotic system is not practically stable with the selected global control (with previously synthesized decentralized controllers), the user may change any of the above listed data according to his opinion, and the package will test whether the overall robotic system is practically stable.

III In both automatic and interactive mode if the package cannot determine unique decentralized and global control which can withstand all expected parameter variations D, after some predicted number of iterations, the user may decide to stop this procedure and introduce the adaptive controller. Namely, two situations may appear:

a) The package cannot find the global control which can stabilize the robotic system even for parameter values d^* for which the coupling has been on-line computed in global control, then it is very probable that the selected actuators are too weak for the specific robotic manipulator and for the given control task. In that case user may either re-select actuators, or change the control task (i.e. select a slower trajectory and/or require larger tolerances $X^t(t)$, $\forall t \in T$).

b) If the package has found the global control which stabilizes the robotic system for some values of parameters but if the unique control cannot be determined which can withstand all expected parameter variations, then the user should give up searching for non-adaptive control and pass to the synthesis of adaptive control [12].

4.2.10. Block for adaptive control synthesis

The decentralized adaptive control algorithm presented in Paragraph 3.3.3, is implemented in this block of the software package. As explained, the purpose of adaptive control is to introduce a kind of "tuning process" in local gains in accordance with the estimated parameters of a payload. Thus, the user has to impose (Paragraph 4.2.3) the limits of possible payload parameters. For example, for a robot intended for manipulation with up to 60 kg workpieces, the lower bound is m = 0 kg and the upper bound is 60 kg. Of course, if the user wishes to test what will happen when the manipulator is "overloaded", the upper bound might be set to 70 kg or more. The limits of payload moments of inertia should be imposed as well. Then, the local gains corresponding to lower--bound and upper-bound parameters will be automatically determined, as described in Paragraphs 4.2.6 and 4.2.7. The instantaneous gains will be formed to be proportional to the payload parameter estimates (see (3.3.38) - (3.3.30)). The second possibility is to impose more points in which the local gains should be calculated. Then, the function which relates the local gains and the payload parameters will be formed (see Fig. 4.3).

The parameter estimation process is then simulated using (3.3.34). The user should impose the parameter a_p which determines the transient behaviour during the estimation process. The estimation could be realized with or without torque/force measurements. The dynamic correction term $p(\xi)$ in (3.3.34) is then computed using either (3.3.35) or (3.3.36).

The user can simulate the entire process by using the software block for simulation. Thus, the user can get an insight into the quality of both estimation process and tracking (see Figs. 3.14 and 3.15).

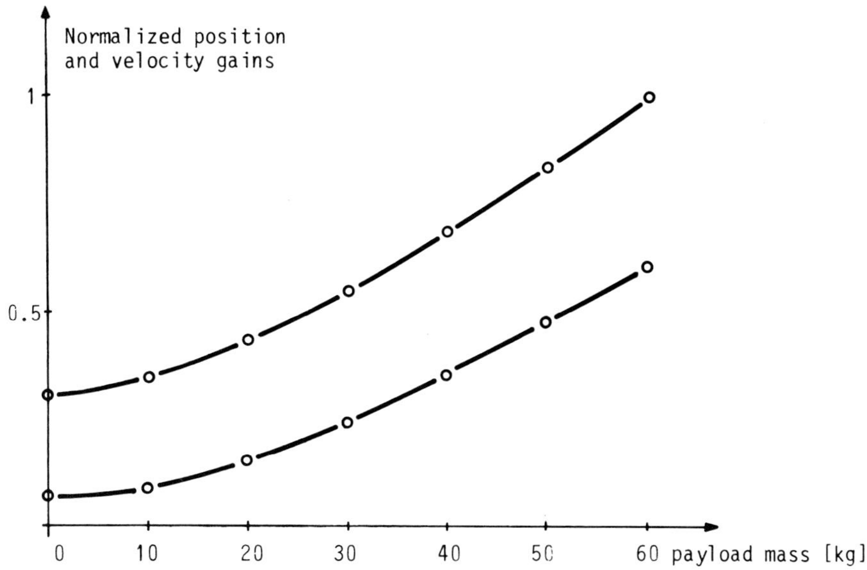

Fig. 4.3. Feedback gains with respect to payload mass

4.2.11. Block for discrete-time control synthesis

The control synthesized in the above described algorithm blocks may be implemented in analog technique or by microprocessors. The latter approach has certain advantages (see Chapter 5). However, microprocessor implementation requires the control synthesis to be considered in discrete-time domain. The control discussed up to now has been synthesized in continuous-time domain. However, the microprocessor implementation introduces delays since the microprocessor requires a period of time to compute the control which will be sent to actuators inputs. Thus, we have to consider the control synthesis in discrete-time domain, i.e. to take into account the sampling period required by the microprocessor to compute the chosen control. This problem has been treated in Vol. 2 of this series [10], and it will also be addressed in Chapter 5.

The algorithm includes a block for the synthesis of control in discrete-time domain. In this block the algorithm synthesizes the parameters of the already chosen control law (i.e. the decentralized and global controller which has been selected in previous blocks) taking into account the sampling period required by the microprocessor. Actually, the user has to select the sampling period and the algorithm computes the parameters (feedback gains) of the previously chosen control law for the discrete-time model of the robot. The algorithm also checks the stability of the system in discrete-time domain, i.e. checks whether the selected sampling period is compatible with the dynamics of the robotic system. In this way the user can verify the performance of the robotic system if the delay in control computation (due to its microprocessor implementation) is taken into account, and select the longest sampling period which is still compatible with the dynamics of the robot [13].

On the other hand, the algorithm can determine the number of floating--point multiplications and additions that have to be performed in each sampling period in order to calculate the selected control law. In this way we get the (approximate) time taken by the selected microprocessor to compute the control in each sampling interval (i.e. if we multiply the number of operations by the time needed for one floating-point multiplication and addition for the particular microprocessor, we obtain the duration of control law computation). Thus, the algorithm can determine the number of microprocessors of the selected type which have to be implemented in order to achieve a desired sampling period for a chosen control law. Obviously, this is an approximate approach, since a considerable processing time is required for data exchange between microprocessors and for some additionals operations. To take this into account, we suggest the user to multiply the above obtained computation time by two. In this way the user can easily select the control equipment necessary to implement various control laws [14]. Evidently, the more complex the control law selected, the more complex end expensive the control equipment required for its implementation. The user is allowed to make a trade-off between control complexity and robot performance. This problem will be discussed in detail in Paragraph 5.3.

4.2.12. Block for simulation

In order to verify the performance of the system with various control laws, the algorithm includes a block for system dynamics simulation. The algorithm enables the user to simulate tracking of the selected nominal trajectory as follows:

1. With specific initial conditions (initial errors) imposed by the user or defined by the algorithm itself (those for which the influence of coupling is the strongest according to stability analysis).

2. With or without nominal control. When the user decides to introduce nominal control he can select:

   a) Nominal control calculated using the complete, exact model of manipulation system [10].

   b) The so-called "decentralized" nominal control - the nominal control using models of actuators only. In this case the calculation of nominal control is much simpler but the dynamics of the mechanism is not taken into account (see Chapter 2).

3. Various forms of local controllers - various information control patterns can be examined.

4. Various forms of global control may be examined by simulation. The user may choose the form of global control which he wants to examine and specify the parameters of the control law. (Obviously, the user should examine the control law which has been synthesized in the above described blocks).

5. The user may simulate the tracking of nominal trajectory by adaptive control. The user has to impose the value of payload parameters and to analyze the estimation process and adjustment of local feedback gains (see Paragraph 4.2.10).

6. The user may examine continuous-time or discrete-time version of the selected control. In the latter case, by simulating the tracking of the nominal trajectory the user can verify the choice of sampling period, i.e. examine whether the chosen sampling period is adequate for a particular robotic system (and selected control law).

As the output of this block the algorithm computes:

1. Deviation of the state coordinates from the nominal trajectory during tracking.

2. The real driving torques developed around the robot joints during tracking the nominal trajectory.

3. The real input signals developed during the tracking of the nominal trajectory by the selected control law.

4. Power and energy consumptions of all actuators for a particular

tracking of the nominal trajectory.

A graphic display of all these values (versus time) is available.

With these results the user can make a final decision on the control law which is the most convenient for his manipulation system and for a particular task. The user can easily analyze the quality of tracking of the imposed nominal trajectory and the energy consumption during this tracking in order to find the best solution.

4.3 Example

In this paragraph we shall present the computer-aided synthesis of control for a specific manipulation robot. The synthesis has been performed using our software package based on the algorithm described in the previous paragraph.

We have selected the manipulation robot GORO-101 with six revolute joints. The robot is shown in Fig. 4.4. To illustrate the computer--aided synthesis of control by our software package, we shall describe the procedure step by step:

1. First, we have to impose data on the geometric parameters of the robot. These data are given in Table 4.1. The geometric structure of the robot is presented in Fig. 4.5.

2. Next, we have to impose data on the dynamic parameters of the robot. These data are given in Table 4.2. We assume that all parameters are fixed and well defined, except for payload parameters which may vary as: $m_p \in (0,\ 5\ kg)$, $J_{px} \in (0.,\ 0.08 kgm^2)$, $J_{py} \in (0.,\ 0.08 kgm^2)$, $J_{pz} \in (0,\ 0.08 kgm^2)$, where m_p is the mass of the payload and J_{px}, J_{py}, J_{pz} are the moments of inertia of the payload with respect to the main axes of inertia. Thus, the set of allowable parameters may be considered as: $D = \{d: m_p \in (0,\ 5\ kg),\ J_{pe} \in (0,\ 0.08 kgm^2),\ e = x, y, z\}$ where $d = (m_p, J_{px}, J_{py}, J_{pz})^T$.

3. The nominal trajectory has been synthesized. We have set the terminal points for the nominal trajectory (through which the robot has to pass) and the duration of movement between each two points. We have chosen to specify the coordinates of the robot tip with respect to the absolute coordinate system and internal (joint) coordinates

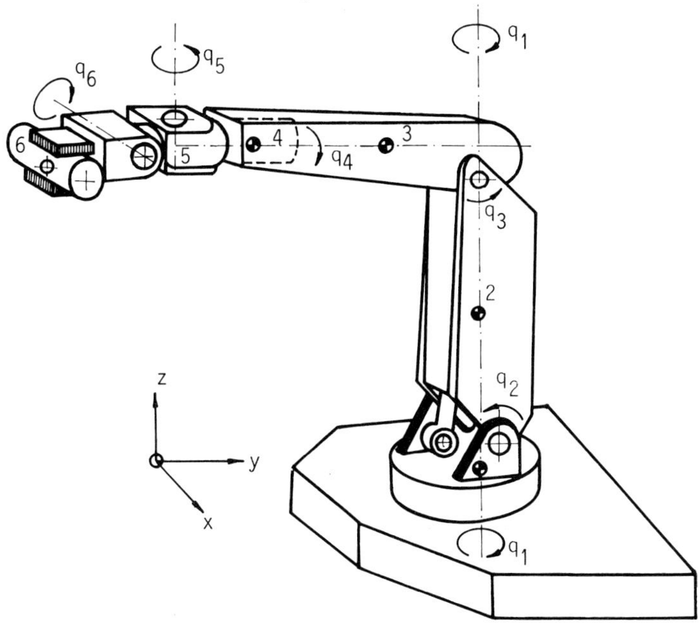

Fig. 4.4. Manipulation system GORO-101

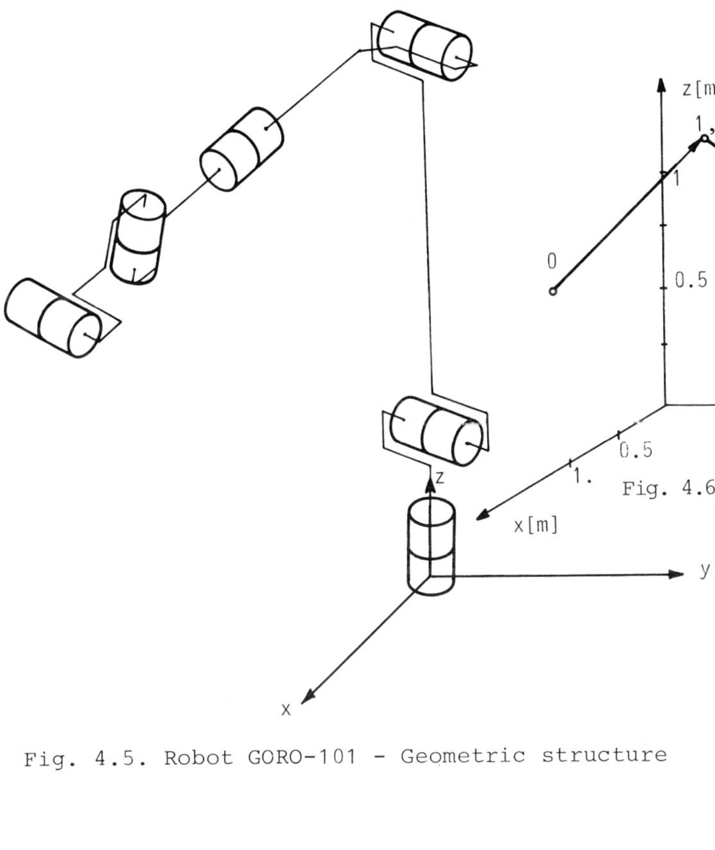

Fig. 4.5. Robot GORO-101 - Geometric structure

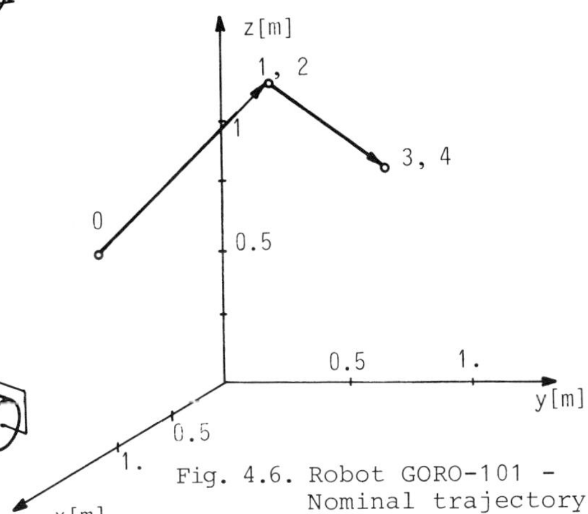

Fig. 4.6. Robot GORO-101 - Nominal trajectory of the robot tip (see Table 4.3)

Joint	Type of joint	Joint unit axis - $\vec{e}_i$	Link vector $\vec{r}_{i,i}$ [m],	Link vector $\vec{r}_{i,i+1}$ [m]
1	R	0., 0., 1.	0., 0., 0.05	0., 0., -0.05
2	R	0., -1., 0.	0., 0., 0.55	0., 0., -0.465
3	R	0., -1., 0.	0., 0., 0.2	0., 0., -0.4
4	R	0., 0., 1.	0., 0., 0.008	0., 0., -0.21
5	R	-1., 0., 0.	0., 0., 0.1	0., 0., -0.121
6	R	0., 1., 0.	0., 0., 0.13	0., 0., -0.004

Table 4.1. Robot GORO-101 - geometry parameters (all vectors are in internal (joint) coordinate systems) (R means rotational joint)

Link	Mass m_i [kg]	Moment of inertia [kgm^2]		
		J_{xi}	J_{yi}	J_{zi}
1	210.	-	-	4.57
2	152.	22.17	21.74	5.738
3	67.	7.2	7.19	1.22
4	18.	1.34	1.335	0.054
5	13.	0.194	0.164	0.066
6	35.	0.203	0.205	0.182

Table 4.2. Robot GORO-101 - dynamic parameters of the mechanism (moments of inertia around main axes of inertia of the link)

Point on trajectory	x [m]	y [m]	z [m]	q_4 [rad]	q_5 [rad]	q_6 [rad]	Duration τ[s]
0	1.173	0.	0.8	1.573	0.	0.	-
1	0.9	0.5	1.4	1.573	0.	0.	0.9
2	0.9	0.5	1.4	1.573	0.	0.	0.3
3	0.6	0.9	1.	0.2	0.	1.573	1.2
4	0.6	0.9	1.	0.2	0.	1.573	0.2

Table 4.3. Robot GORO-101 - terminal points on nominal trajectory (external coordinates - (x, y, z) coordinates of the tip w.r. to absolute coordinate system; trapezoid velocity profile)

of the gripper. Also, we have required a trapezoid profile of the tip velocity along the trajectory (between each two given points), and we have specified the duration of acceleration and decelaration. We have selected the mode in which the robot has to stop at each specified terminal point.

Data on the set points on the nominal trajectory are given in Table 4.3. The positions of the robot along the specified nominal trajectory are presented in Fig. 4.6. The algorithm computes the nominal trajectories in joint coordinates that correspond to a desired path of the robot tip. These joint nominal trajectories are given in Fig. 4.7.

4. Next step is to impose data on actuators in the robot joints. The robot GORO 101 is driven by hydraulic actuators. Data on the selected hydraulic actuators are given in Table 4.4. The matrices of actuators models A^i, b^i, f^i are computed as explained in the previous paragraph, and are presented in Table 4.5.

   The connections between the joint coordinates and actuator coordinate are not trivial for the first three joints (since rotational joints are driven by hydraulic actuators with linear motion of pistons). These connections are specified in Table 4.6. The rotational (vane) hydraulic actuators are used for the gripper joints; for these joints, relations between the mechanism coordinates and actuator coordinates are linear.

   Table 4.7. gives the relations between the driving torques around the mechanism joints and the load acting upon actuators.

5. The algorithm computes the nominal driving torques and the nominal programmed control for the given nominal trajectory. The results of these computations are presented in Figs. 4.8 - 4.11. The nominal driving torques around the robot joints are presented in Fig. 4.8 (for $m_p = 0$, $J_{px} = 0.$, $J_{py} = 0.$, $J_{pz} = 0.$). Fig. 4.9 shows the nominal programmed control synthesized using the complete centralized model. (We have not used this nominal control for our controller, but we have computed it in order to check the capability of the selected actuators to realize the chosen nominal trajectory). Fig. 4.10 gives the nominal power required to drive the robot along the desired nominal trajectory. Nominal energy consumptions are presented in Fig. 4.11.

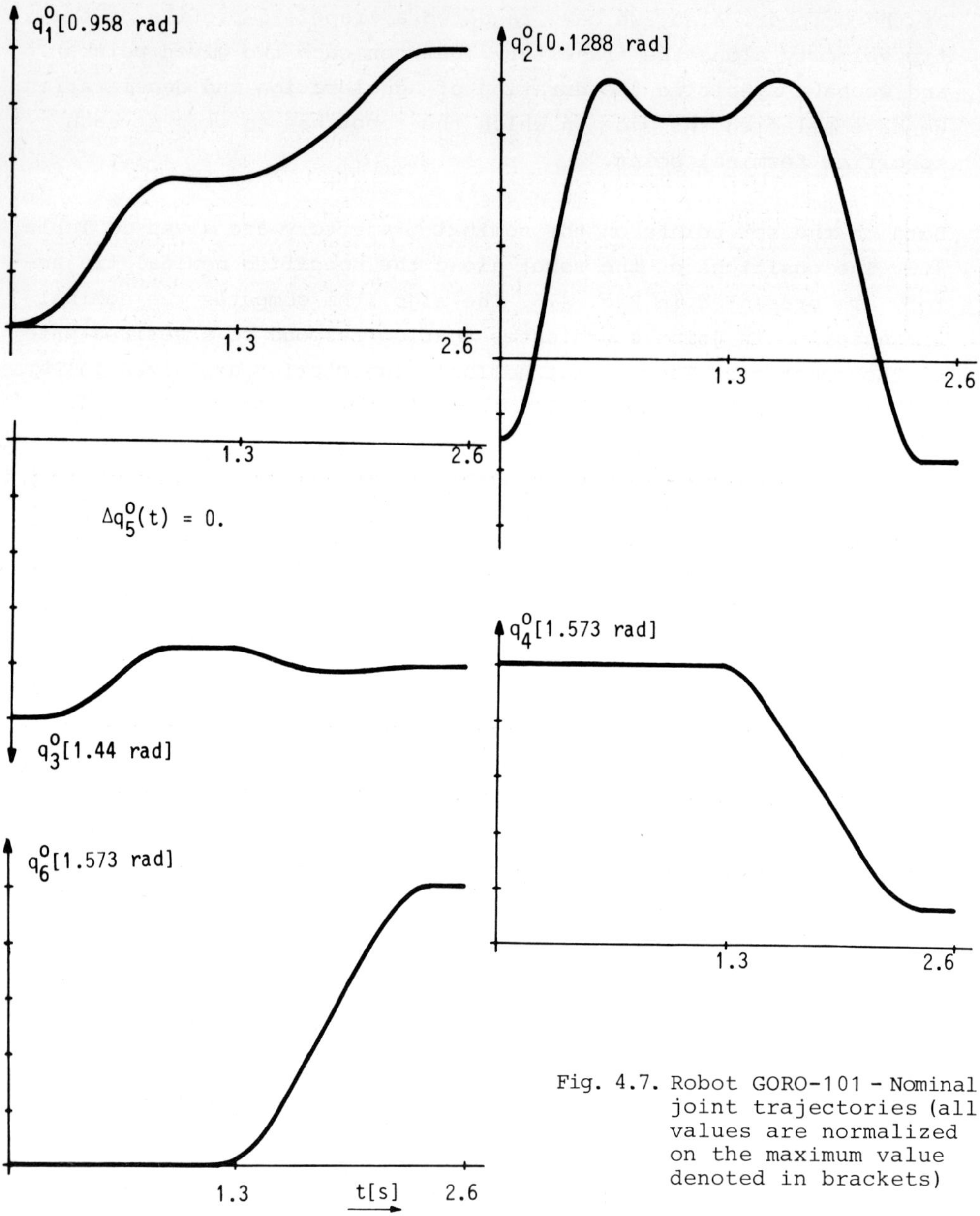

Fig. 4.7. Robot GORO-101 - Nominal joint trajectories (all values are normalized on the maximum value denoted in brackets)

Joint	Piston area [cm^2]	Cylinder volume [cm^3]	Piston mass [kg]	Viscous friction coefficient [N/m/s]	Gradient of servovalve [$cm^3/s/N/m^2$]	Flow-current coeff. [$m^3/s/mA$]	Input ampl. constr. [mA]
1	33.	1690.	16.7	30.	0.000107	0.00026	75.
2	25.	1400.	18.4	30.	0.000107	0.00026	75.
3	12.5	1400.	9.20	10.	0.000107	0.00026	75.
4	7.5	300.	0.051	0.5	0.00027	0.00026	75.
5	7.5	300.	0.051	0.5	0.00027	0.00026	75.
6	3.75	250.	0.026	0.5	0.00027	0.00026	75.

Table 4.4. Robot GORO-101 - actuator parameters

Actuator	a^i_{22}	a^i_{23}	a^i_{32}	a^i_{33}	b^i_3
1	-1.79	19.76	-1327.8	-4.305	104.6
2	-1.63	13.58	-1214.3	-5.197	126.29
3	-1.087	13.587	-607.14	-5.197	126.29
4,5	-9.8	1470.6	-1700.0	-61.200	589.33
6	-19.2	1442.3	-1020.0	-73.440	707.20

Table 4.5. Matrices of the actuators models (robot GORO-101)

Joint	a^i_c	b^i_c	c^i_c	d^i_c
1	0.6913	0.2132	0.62	0.93
2	0.583	-0.29	1.86	0.71
3	0.65	-0.31	-0.17	0.75

Table 4.6. Robot GORO-101 - Connection between actuator (ℓ^i) and mechanism coordinate (q_i) $\ell^i = (a^i_c + b^i_c \cos(q_i + c^i_c))^{1/2} - d^i_c$ joint 4,5,6 the connections are trivial

Joint	a^i_f	b^i_f
1	-0.46	-1.029
2	-0.24	0.
3	-0.55	1.29

Table 4.7. Robot GORO-101 - Connection between actuator load and mechanism driving torque ($M_i = P_i a^i_f / \cos(q_i + b^i_f)$)

Point	Tip [m]	q_4 [rad]	q_5 [rad]	q_6 [rad]	Stability degree
0	0.01	0.05	0.05	0.02	-
1, 2	0.002	0.01	0.01	0.01	2.48
3, 4	0.001	0.01	0.01	0.01	2.97

Table 4.8. Robot GORO-101 - Tolerances around points on nominal trajectory (Table 4.3)

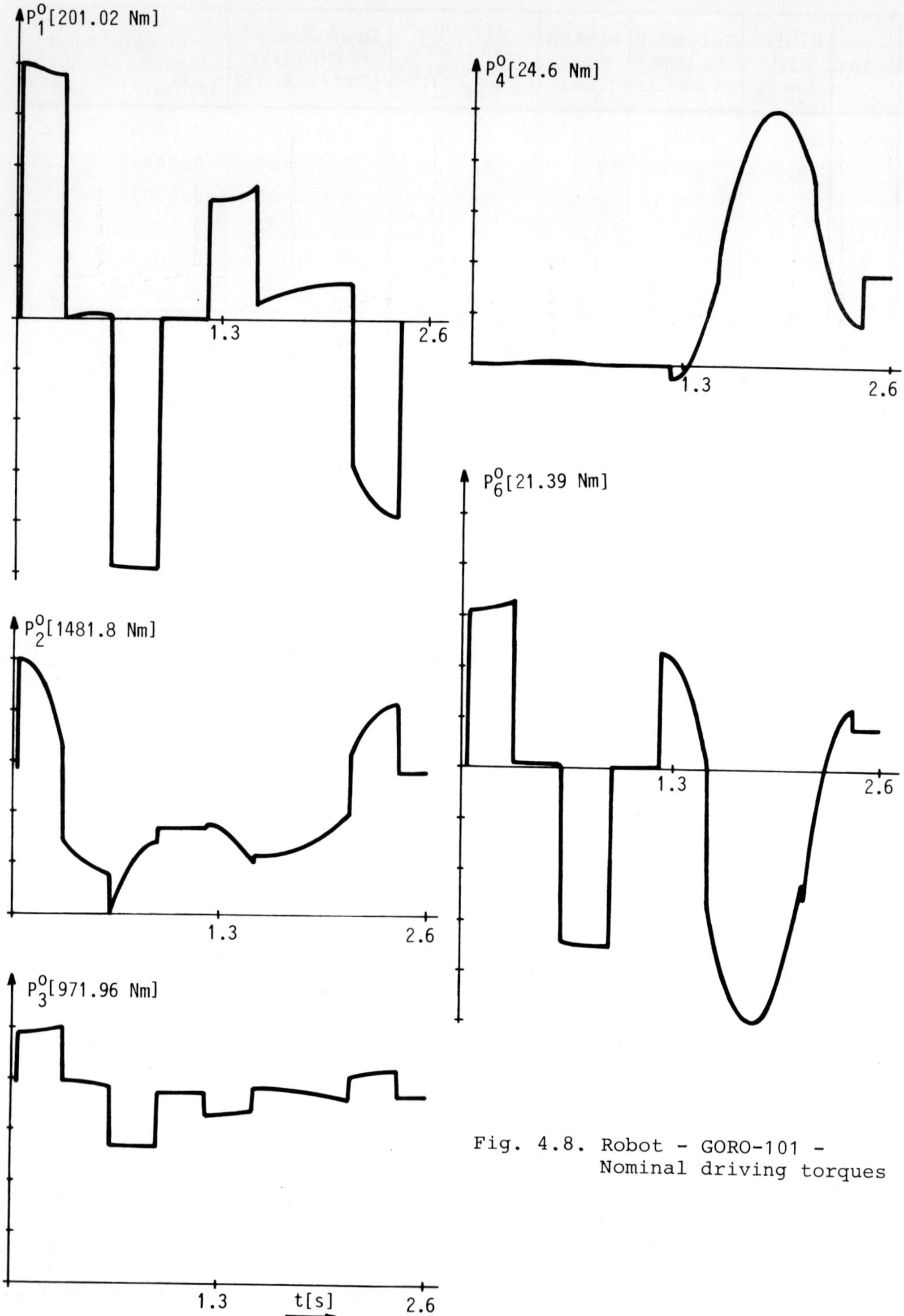

Fig. 4.8. Robot - GORO-101 - Nominal driving torques

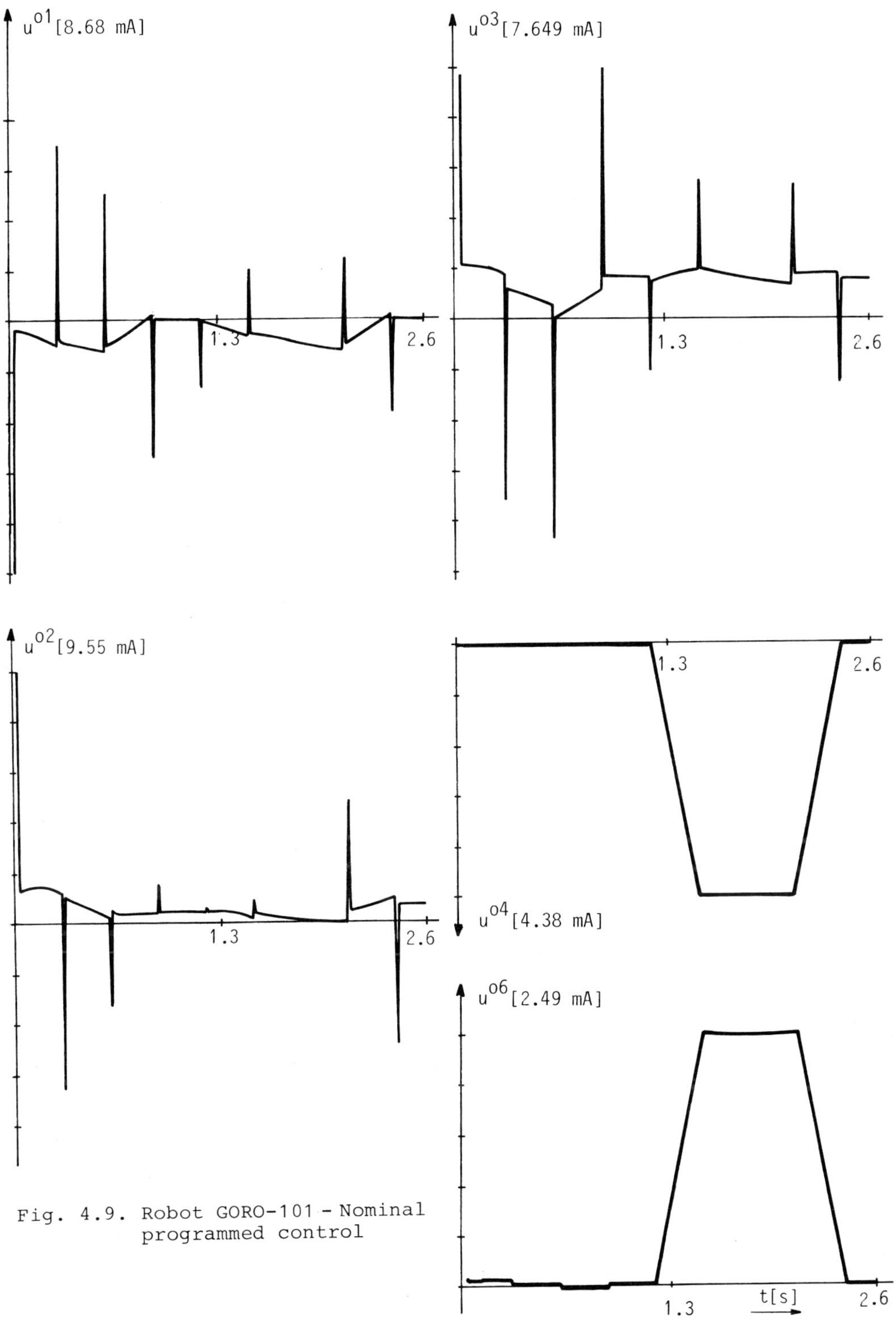

Fig. 4.9. Robot GORO-101 - Nominal programmed control

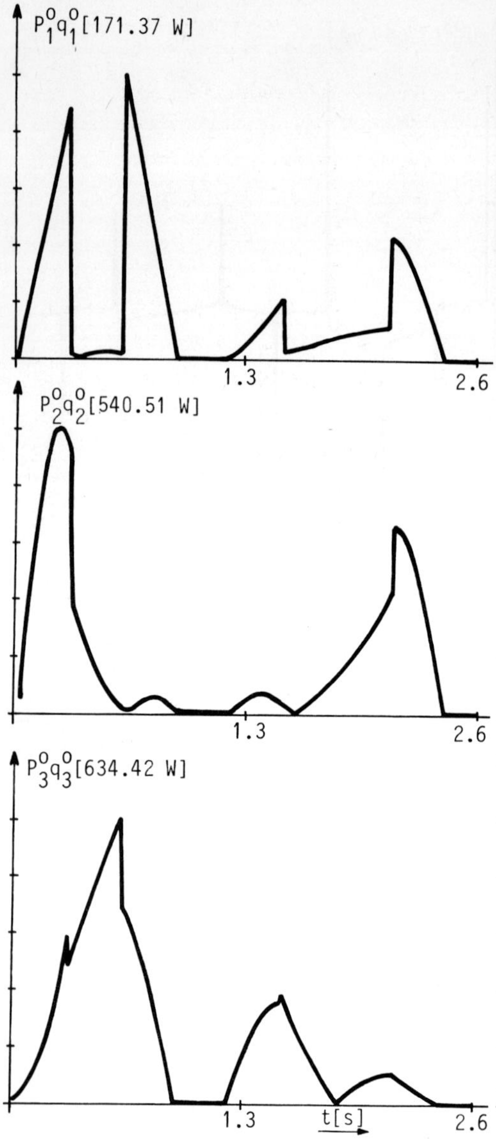

Fig. 4.10. Robot GORO-101 - Nominal power demands

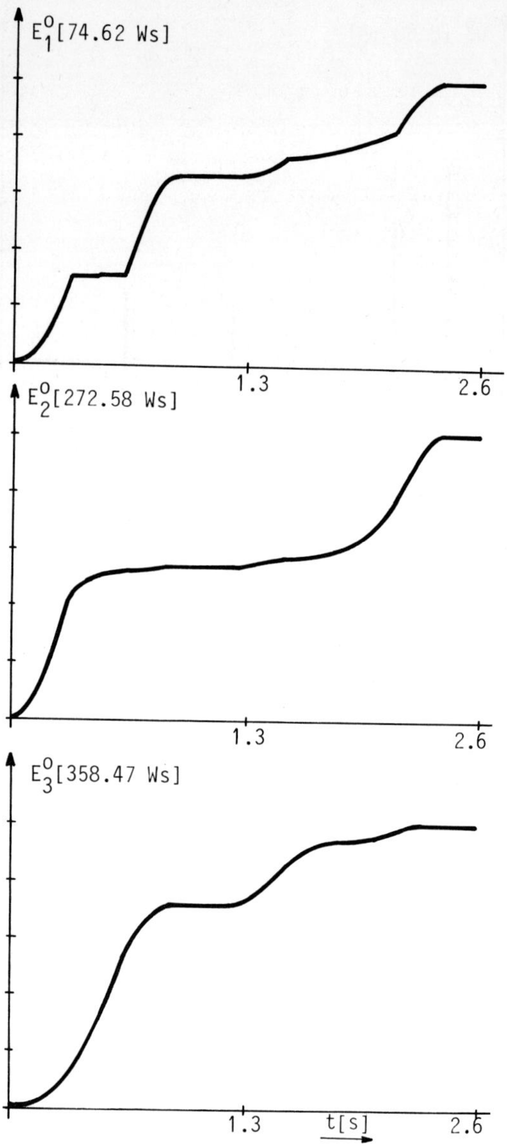

Fig. 4.11. Robot GORO-101 - Nominal energy consumptions

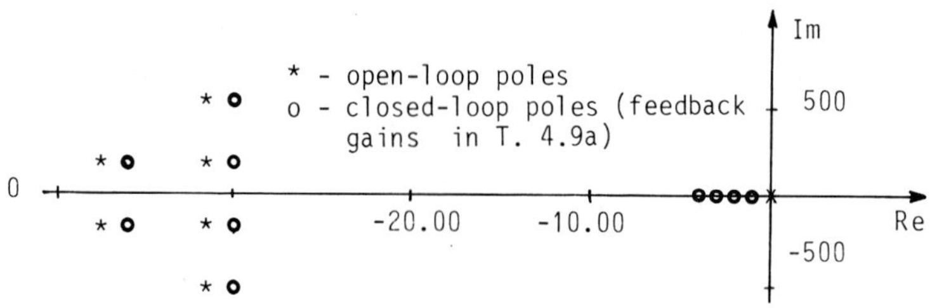

Fig. 4.12. Positions of the poles of the linearized model of robot

6. In the next step we have to specify the control task, i.e. the allowable tolerances around the nominal trajectory. The specified tolerances around the given points on nominal trajectory are given in Table 4.8. The algorithm computes the tolerances in joint coordinates that correspond to the given tolerances around external coordinates and determines the desired stability degree (see Paragraph 4.2.6).

7. Local controllers have to be synthesized next. First we try to synthesize the local controllers by pole-placement. We consider local controllers with various information control structures: (a) with position feedback only, (b) with position and velocity feedback loops, (c) with position, velocity and pressure feedback loop. The results of synthesis are summarized in Table 4.9. The required poles of closed-loop subsystems are given in Table 4.9 for various control structures, together with the obtained feedback gains and poles of the linearized model of the complete robotic system. The positions of the poles of the closed-loop model of the complete robotic manipulators (in the complex plane) are given in Fig. 4.12. The local optimal regulators are also synthesized; the weighting matrices, prescribed stability degrees, obtained local feedback gains and poles of the closed-loop linearized model of the complete robotic system are given in Table 4.10. Stability analysis of the system (next step) shows that we must adopt the local optimal regulators (i.e. feedback by all state coordinates).

8. Next, stability of the nonlinear model has to be analyzed. The package analyzes the practical stability of the robotic system for various values of parameters D. The stability of the system with decentralized controllers synthesized as described in the previous step (and with decentralized nominal control) is analyzed according to the algorithm described in Paragraph 2.5. In Table 4.11 the results of stability analysis are given for decentralized controller with local optimal regulators (Table 4.10). As can be seen from Table 4.11, the stability of subsystems 2 and 5 is not guaranteed with local controllers only. Obviously, the influence of coupling upon these joints is too strong to be withstood by the selected local controllers (although the linearized model is stable). Thus, we may either increase local feedback gains (i.e. increase the prescribed stability degree of local optimal controllers in the corresponding joints) or introduce global control.

Subsystem	Desired poles R_e	I_m	Position feedback gain	Velocity feedback gain	Pressure feedback gain	Poles of linearized closed-loop model R_e	I_m
	1		2	3	4	5	
1	-3. –	0. –	36.63	-	-	-29.8 -35.7	±600. ±281.
2	-3. –	0. –	24.71	-	-	-29.2 -1.35	±277. ±7.8
3	-3. –	0. –	1.22	-	-	-1.03 -0.54	±4.4 ±6.29
4	-3. –	0. –	8.65	-	-	-3.37 -3.14	0 0
5	-3. –	0. –	8.62	-	-	-3.04 -2.98	0 0
6	-3. –	0. –	4.31	-	-	-2.96 -2.35	0 0

a) Positional feedback loop only

Subsystem	1		2	3	4	5	
1	-4. –	0. 15.	113.02	15.18	-	-32.8 -27.	±40. ±14.5
2	-5. –	0. 10.	111.56	11.54	-	-15. -19.4	±31.6 ±16.6
3	-5. –	0. 10.	5.85	0.63	-	-0.7 -3.5	±9.46 ±8.2
4	-12. –	0. 150.	20.3	0.41	-	-20.7 -13.7	
5	-12. –	0. 150.	29.3	0.63	-	-5.12 -5.12	
6	-12. –	0. 150.	14.2	0.65	-	-4.1 -2.3	

b) Position and velocity feedback

Subsystem	1		2	3	4	5	
1	-12. -15.	0. ±7.6	430.	68.	0.36	-157. -119.	±616. ±321.
2	-13. -15.	0. ±20.	871.	155.	0.34	-82. -27.	±328. ±22.
3	-13. -184.	0. 0.	344.	54.8	0.16	-3.3 -6.7	±5.5 ±6.2
4	-13. -157.	0. ±615.	48.	0.37	0.45	-184. -363.	
5	-12. -160.	0. ±722.	45.	0.92	0.17	-32. -13.	
6	-13. -119.	0. ±220.	20.7	0.71	0.25	-12.5 -6.6	

c) Position, velocity and pressure feedback loop

Table 4.9. Robot GORO-101 - Local controllers (pole-placement)

Subsystem	Q_i diag(q^i)	r_i	β_i	K_1^i K_2^i K_3^i	Poles of linearized model of system R_e	I_m
1	1.	5.	9.	1030.	-160.	±615.
	0.1			131.	-122.	±325.
	0.1			0.5	-63.	±300.
2	0.1	0.5	9.5	1750.	-36.	±28.
	1.			231.	-10.	±9.1
	0.1			0.47	-9.5	±9.2
3	1.	5.	9.5	723.	-177.	
	0.1			79.4	-40.	
	0.1			0.175	-18.	
4	0.1	0.5	9.	58.2	-164.	
	10.			0.389	-18.	
	0.1			0.469		
5	0.1	0.5	9.	55.4		
	10.			0.133		
	0.1			0.120		
6	0.1	0.5	9.	39.4		
	10.			0.771		
	0.1			0.270		

Table 4.10. Robot GORO-101 - Synthesis of local optimal regulators (Q_i, r_i - weighting matrices in the local criteria (2.4.16), β_i prescribed stability degrees $K_1^i=(K_1^i,K_2^i,K_3^i)^T$ - feedback gains of local regulators)

Joint	1	2	3	4	5	6	Stable
1	-.448E+0	.132E-2	.328E-1	.292E-1	.634E-1	.156E+0	Y
2	.951E-3	-.113E+1	.144E+2	.152E-4	.638E-1	.163E-2	N
3	.264E-2	.890E+0	-.566E+1	.481E-3	.159E+1	.719E-2	Y
4	.819E-5	.747E-7	.505E-5	-.101E+3	.352E-3	.652E-2	Y
5	.880E-6	.295E-4	.110E-1	.233E-3	.213E+2	.491E-2	N
6	.370E-3	.200E-4	.527E-4	.107E-1	.603E-1	-.235E+1	Y

Table 4.11. Stability analysis when local optimal regulators are applied (see explanation under Table 4.12)

9. If we decide to introduce global control we have to select its form and implementation. First, we try with on-line computation of coupling using the approximate model which includes just gravity term. However, the results are poor. Then, we try with the approximate model which includes gravity and inertia term. The results are better, but still they do not satisfy the imposed requirements. At last we apply force feedback in joints 2, 3 and 5. The results of practical stability analysis are given in Table 4.12. The robotic system is practically stable for all allowable parameter variations with such unique local and global controllers. The applied control law is presented in Table 4.13. The global gains are determined iteratively as explained in Paragraph 2.6.

10. In order to implement the selected control by microprocessors, we consider the synthesis of feedback gains in time-discrete domain. The results of this synthesis (local feedback gains) for various sampling periods are given in Table 4.14. The analysis of the system performance shows that the longest acceptable sampling period is about 10 ms. The number of additions and multiplications required to compute local control is given in Table 4.15. The processor time which is required for the computation of local control by various microprocessors is also listed in Table 4.15. On the basis of these estimates we can determine the number of microprocessors required to implement the local control and to achieve a sampling period of 10 ms. (Since we have chosen force feedback as global control, which requires just a few additional multiplications and additions, we may neglect the time needed for global control implementation). Obviously, this determination of control equipment is approximate.

11. In the last step of control synthesis we simulate the behaviour of the robotic system with various control laws. The results of simulation are given in Figs. 4.13 - 4.14. Fig. 4.13 presents the results of tracking (deviations of actuator coordinates from nominal trajectories), if local controllers are only (Table 4.10) applied. The tracking is poor, which confirms the results of the above analysis (Table 4.11). Fig. 4.14 shows the tracking with the selected local and global control (Table 4.13) for various values of payload parameters. It can be seen that the tracking is satisfactory for various values of parameters, which means that the unique control is robust enough to withstand all allowable parameter variations.

Joint	1	2	3	4	5	6	Stable
1	-.448E+0	.132E-2	.328E-1	.292E-1	.634E-1	.156E+0	Y
2	.956E-7	-.121E+1	.144E-2	.187E-8	.638E-5	.168E-6	Y
3	.264E-2	.890E+0	-.566E+1	.481E-3	.159E+1	.719E-2	Y
4	.819E-5	.747E-7	.505E-5	-.101E+3	.352E-3	.652E-2	Y
5	.196E-7	.882E-6	.243E-3	.108E-4	-.364E+2	.759E-4	Y
6	.370E-3	.200E-4	.527E-4	.107E-1	.603E-1	-.235E+1	Y

Table 4.12. Stability analysis when local regulators and global control (force feedback in 2 nd and 5th joint) are applied stability matrix G defined by $G_{ij}=-\gamma_i\delta_{ij}+\xi^*_{ij}$ (where δ_{ij} is Kronecker's symbol, γ_i is stability degree of *i*th local subsystem, ξ^*_{ij} is estimation of coupling from joint j upon the *i*th joint - see (2.6.18), (-n means 10^{-n}))

Joint	Local nominal	Local	Global
1	$\bar{u}^{o1}(t)$	$-1030.\ \Delta\ell^1$ $-131.\ \Delta\dot{\ell}^1$ $-0.518\ \Delta p^1$	-
2	$\bar{u}^{o2}(t)$	$-1750.\ \Delta\ell^2$ $-231.\ \Delta\dot{\ell}^2$ $-0.47\ \Delta p^2$	$0.97\ (68200\ \Delta\ell^2+7980\ \Delta\dot{\ell}^2$ $+9.\ \Delta p^2)/169\ \Delta\ell^2+9.\ \Delta\dot{\ell}^2\ +$ $+0.1\ \Delta p^2)\cdot P_2$ (measured)
3	$\bar{u}^{o3}(t)$	$-723.\ \Delta\ell^3$ $-79.4\ \Delta\dot{\ell}^3$ $-0.175\ \Delta p^3$	-
4	$\bar{u}^{o4}(t)$	$-58.2\ \Delta q_4$ $-0.389\ \Delta\dot{q}_4$ $-0.469\ \Delta p_4$	-
5	$\bar{u}^{o5}(t)$	$-55.4\ \Delta q_5$ $-0.133\ \Delta\dot{q}_5$ $-0.120\ \Delta p_5$	$0.95\cdot(24\ \Delta q_5+009\ \Delta\dot{q}_5\ +$ $+0.004\ \Delta p^5)/(4.9\ \Delta q_5\ +$ $+0.004\ \Delta\dot{q}_5+0.0017\ \Delta p^5)$
6	$\bar{u}^{o6}(t)$	$-39.4\ \Delta q_6$ $0.771\ \Delta\dot{q}_6$ $-0.270\ \Delta p_6$	-

Table 4.13. Synthesized control laws (local nominal control computed acc. to (4.2.9))

Joint		T=0.03	T=0.02	T=0.01
1	K_1	555.091	684.492	853.196
	K_2	69.318	84.602	104.15
	K_3	0.367	0.413	0.471
2	K_1	876.00	1089.715	1372.214
	K_2	125.59	151.710	185.702
	K_3	0.312	0.352	0.404
3	K_1	105.7	167.934	307.261
	K_2	13.55	20.359	35.314
	K_3	0.043	0.057	0.087
4	K_1	44.33	46.002	47.868
	K_2	-0.05	0.055	0.057
	K_3	0.006	0.001	0.009
5	K_1	44.195	49.09	39.471
	K_2	-0.002	0.246	0.502
	K_3	0.04	0.002	0.062
6	K_1	21.699	19.422	19.221
	K_2	-0.08	0.206	0.257
	K_3	0.024	0.013	0.082

Table 4.14. Local feedback gains synthesized in time-discrete domain for various sampling period

	Number of additions	Number of multiplications
Local nominal control	18	30
Local control	36	18
Total	54	48

μP	Duration [ms]
INTEL-8085	75
INTEL-86/87	2.2
PDP 11/03	12.6
MOTOROLA 6800	∿30.
HP 64000	2.78
LSI-11/2	∿11.56

Table 4.15. Number of additions and multiplications for computation of local control and duration of computation on various microprocessors

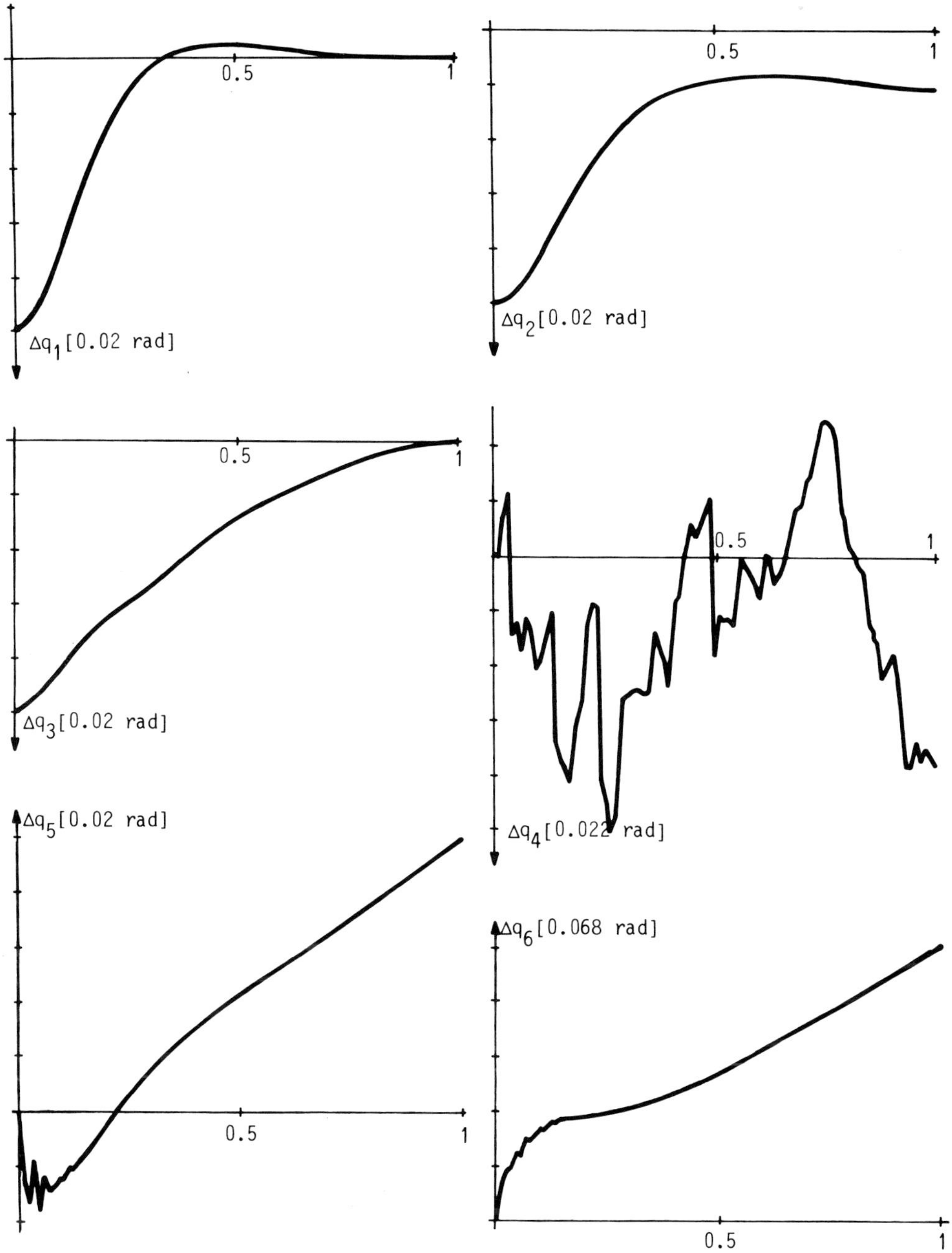

Fig. 4.13. Simulation of tracking of the nominal trajectory with local controllers - deviation of joint angles from nominal trajectory (all values normalized on the maximum value denoted in bracket)

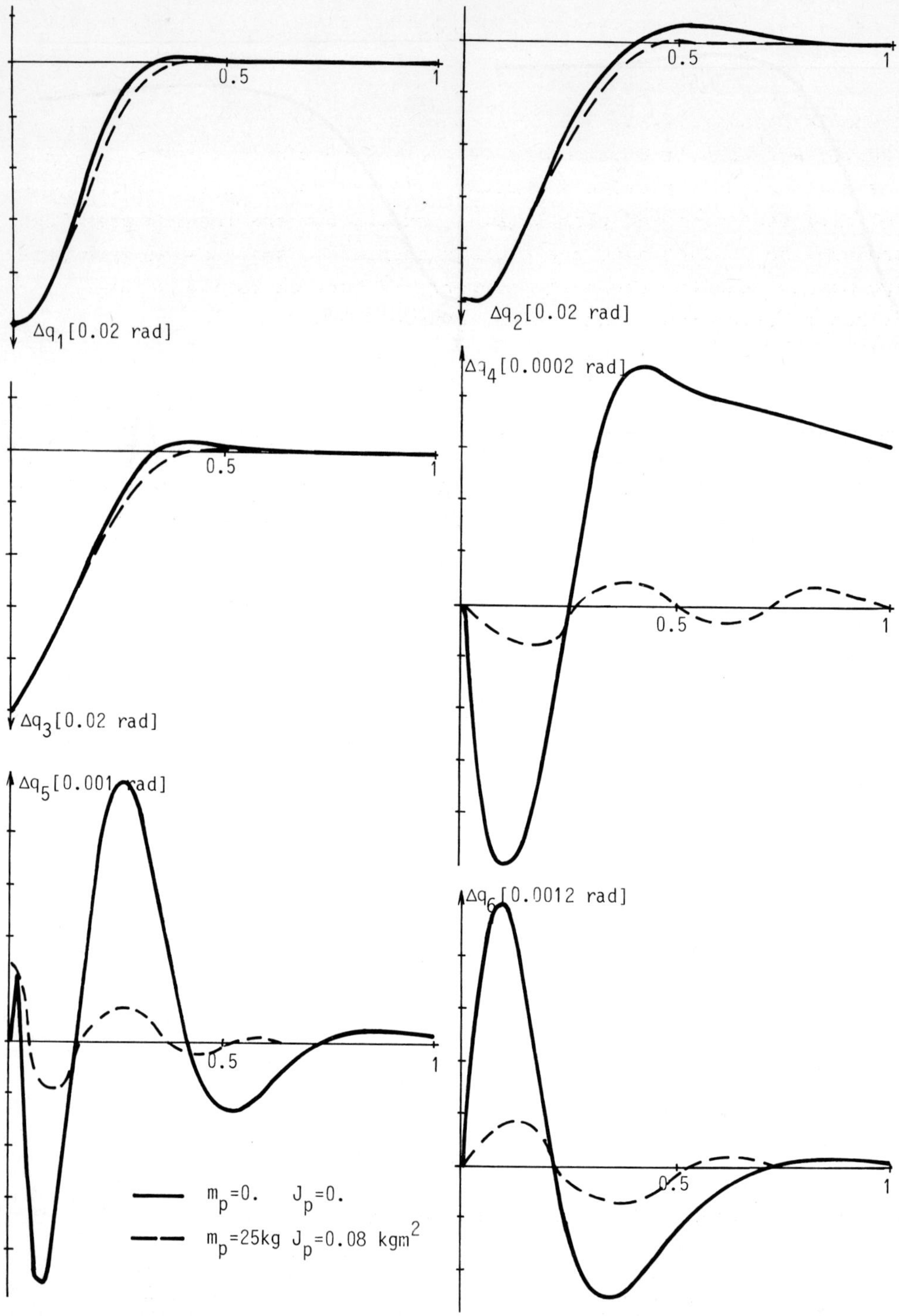

Fig. 4.14. Simulation of tracking of nominal trajectory with local and global control for various payload parameters

Thus, it is not necessary to apply adaptive control in this particular case.

It should be noted that, due to lack of space, we have omitted many intermediate results obtained by the package during the control synthesis (e.g. the iterative selection of local controllers, global control implementation and global gains, etc.), but the results presented here give an insight into the package operation, and show how the package may be used to select the appropriate control law for a particular manipulation robot and a given control task.

Conclusion

In the previous two paragraphs we have briefly described the software package for computer-aided control synthesis of manipulation robots. In Paragraph 4.3 we have described the interaction between the package and the user, and in Paragraph 4.3 we have presented an example of computer-aided control synthesis for a particular robotic system and control task.

In this paragraph we shall briefly discuss the possibilities offered by this package to the user in order to stress the importance and advantages of computer-aided approach to control synthesis.

The user can easily combine all the above mentioned blocks (Paragraph 4.2) and results according to his desires and examine different control laws. In doing this he can easily use his own experience and understanding of the system. For example, if he knows that some joint is weakly coupled with others (or the coupling is zero) he can omit global feedback loops for this actuator. Similarly, if he knows that the influence of gravity forces upon some degree of freedom is weak the on-line calculation of gravity forces for this degree of freedom is not useful. Thus, the use of designer's experience can significantly speed up the control synthesis and be of great help in determination of the best solution. Practically, the synthesis of control is left to the user's decisions. The above package can be regarded as an assistive device which can help the user to verify his solutions.

The algorithm can help the user to make the following decisions and investigations:

1. To verify whether the particular control task can be performed by the chosen manipulator system from kinematic and dynamic points of view.

2. To choose actuators which can ensure implementation of the given control task.

3. To chosen the simplest control structure and law which can satisfy specific requirements.

4. To determine the dynamic parameters of the manipulator system that have to be taken into account in the controller in order to achieve satisfactory system performance.

5. To determine sensors (feedback loops) which have to be implemented (are tacho-generators necessary or not, should force transducers be used or not, etc.).

6. To determine the complexity of the multiprocessor system which has to be implemented (the number of microprocessors and their capabilities).

7. To calculate the parameters of the chosen control law (local and global gains, sampling period etc.).

8. To examine dynamic performance of a particular manipulation system with various controllers and determine how a change in some dynamic parameters of the manipulator can influence the system performance.

9. To examine the robustness of the selected control and decide whether a non-adaptive control law can be applied, or adaptive control is necessary.

10. To synthesize adequate adaptive control if the expected variations of parameters are too large to be accommodated by robust non-adaptive control.

11. To estimate energy consumptions for the realization of various control tasks.

References

[1] Vukobratović K.M., Stokić M.D., Kirćanski V.M., "A Procedure for Interactive Dynamic Control Synthesis of Manipulators", Proc. of IV CISM-IFToMM Symposium on Theory and Practice of Robots and Manipulators, Warsaw, September, 1981.

[2] Vukobratović K.M., Stokić M.D., "A Procedure for Interactive Dynamic Control Synthesis of Manipulators", Transaction on Systems, Man, and Cybernetics, Sept./Oct. issue, 1982.

[3] Vukobratović K.M., Stokić M.D., Kirćanski M.N., Kirćanski V.M., "Software Package for Computer-Aided Synthesis of Control for Manipulation Robots", Preprints of the 9th World IFAC Congress, Budapest, Vol. 6, pp. 62-67, July, 1984.

[4] Chang K-H., Isomura T., Funakubo H., "The Development of a System to Assist in the Design and Control of Robots", Proceedings of the 14th International Symposium on Industrial Robots, pp. 723--733, Gothenburg, 1984.

[5] Queromes J.G., "Computer Aided Design and Robotics: A Full of Promise Cooperation", Proc. of the 12th ISIR, pp. 185-196, Paris, June, 1982.

[6] Vukobratović K.M., Kirćanski V.M., Scientific Fundamentals of Robotics 3, Kinematics and Trajectory Planning, Monograph, Springer -Verlag, Berlin, 1985.

[7] Vukobratović K.M., Potkonjak V., Scientific Fundamentals of Robotics 1, Dynamics of Manipulation Robots, Theory and Application, Monograph, Springer-Verlag, Berlin, 1982.

[8] Vukobratović K.M., Kirćanski M.N., Scientific Fundamentals of Robots 4, Real-Time Dynamics, Springer-Verlag, Berlin, 1985.

[9] Vukobratović K.M., Stepanenko Y., "Mathematical Models of General Anthropomorphic Systems", Mathematical Biosciences, Vol. 17, pp. 191-242, 1973.

[10] Vukobratović K.M., Stokić M.D., Scientific Fundamentals of Robotics 2, Control of Manipulation Robots: Theory and Application, Monograph, Springer-Verlag, Berlin, 1982.

[11] Vukobratović K.M., Kirćanski V.M., "A Method for Optimal Synthesis of Manipulation Robot Trajectories", Trans. of ASME, Journal of Dynamic Systems, Measurement, and Control, Vol. 104, No. 2, pp. 188-193, 1982.

[12] Vukobratović K.M., Stokić M.D., Kirćanski M.N., "Towards Nonadaptive and Adaptive Control of Manipulation Robots", IEEE Trans. on Automatic Control, Vol. AC-29, No. 9, pp. 841-844, 1984.

[13] Vukobratović K.M., Stokić M.D., Kirćanski V.M., "Contribution to Dynamic Control of Industrial Manipulators", The XI Intern. Symp. of Industrial Robots, Tokyo, 1981.

[14] Vukobratović K.M., Stokić M.D., "Approximative Dynamic Models in Hierarchical Control of Robots Systems", The V IFToMM Symp. on Theory and Practice of Robots Manipulators, Udine, 1984.

Chapter 5
Implementation of Control Algorithms

5.1 Introduction

In previous chapters we have presented the synthesis of non-adaptive control (Chapter 2) and adaptive control (Chapter 3) for the executive control level of manipulation robots. We have also presented (Chapter 4) the algorithm (software package) for computer-aided synthesis of control for robotic systems. In this chapter we shall consider the implementation of various control laws.

In principle, we may consider three basic techniques for control implementation: analog technique, microprocessor-based, and hybrid technique. In the last decade (5-10 years) microprocessor implementation of control has become predominant, thanks to tremendous improvements in microprocessors technology. The implementation of control by digital technique has certain advantages over the classical analog technique. Let us mention some of them: a) The implementation of control by microprocessors is much simpler than that by analog technique. This is especially true if complex control laws which require a lot of multiplications and additions, have to be implemented. For example, if we want to compute the dynamic model of the robotic system for global control, this is very complex to implement by analog technique; on the other hand, on-line computation by microprocessors is much simpler (although it may require several microprocessors to operate in parallel - see Paragraph 5.3). b) The analog control scheme is very difficult to change and move from one system to another. On the other hand, a microprocessor can easily be reprogrammed, so we may use the same hardware for various control systems. Even more, the microprocessor implementation of control allows one to build a general controller for various robotic systems which may automatically be adjusted to each specific manipulation robot. Such a general controller for robotic systems would be very difficult to implement by analog technique. c) The maintenance of microprocessor systems is much more efficient than the maintenance of systems in analog technique. Besides, the microprocessor technique is more reliable and robust to variations of environment conditions.

It is for these and some other reasons that the microprocessor implementation of control is today exclusively used for robotic systems [1 - 4]. It should be underlined that the higher control levels for robotic systems (strategical control level and/or communication with the operator, and tactical control level) require a digital computer for their implementation. It is quite reasonable to implement the executive control level by microprocessors, too. Thus in the text to follow we shall restrict our attention to microprocessor implementation of control for robotic systems.

It should be mentioned that the hybrid (analog/digital) technique might also be used for control implementation. The local servosystems for joint control might be implemented by analog technique, while global control and higher control levels should be implemented by a digital computer. This option will also be briefly considered here (see Paragraph 5.5). However, this mode cannot be used for a general controller for an arbitrary-type manipulation robot, since it is not simple to adjust analog servosystems from one robotic system to another, in contrast to the case when microprocessor implementation (direct digital control) is used for local controllers, too.

As we have mentioned above, the microprocessor implementation of control offers the possibility to develop a general controller which can easily be adjusted to various types of robotic systems. In Paragraph 5.2 we shall briefly explain the concept of the general controller for an arbitrary type of manipulation robot. Obviously, this controller includes higher control levels: robot language for communication with the operator (for task specification) and tactical control level; it also includes the modules for communication with environment and other systems in the process (intended to synchronize the robot with other systems in the process). We shall briefly explain all these modules that are included in the general controller for robotic system. However, our prime interest in this monograph is the executive control level, so in the next few paragraphs we shall exclusively concentrate on the implementation of executive control level in the scope of general controller. In Paragraph 5.3 we shall consider the numerical complexity of various control laws for various robot structures in order to determine the microprocessor equipment (the number of microprocessors) required for the implementation of executive control level. Then, we shall consider the identification of system parameters (Paragraph 5.4) which is required for the implementation of control. In Paragraph 5.5 we shall consider hybrid and microprocessor implementation of executive control level, and present the experimental results of the control implementation for a specific robotic system.

It should be noted that we shall not discuss in detail various implementations of various control laws [1-4] which were mentioned in Chapters 2 and 3; we shall mainly concentrate on our approaches to control of manipulation robots.

5.2 Concept of General Purpose Controller

In this paragraph we shall briefly describe the general purpose controller for arbitrary type of manipulation robot. We shall briefly describe its main design features, and its main software modules. Our intention is to give the reader an insight of the complexity and structure of such modern concept of the controller for robotic system, so that he can understand advantages of direct digital control and application of the control synthesis described in previous chapters.

Most of the existing control systems are intended for use with a particular robot; for example, the most popular system VAL [5] can be used for control of PUMA family robots only, the AML [6] for IBM System/1 robots etc. Recently, some systems that can be applied to various types of robots have been developed and already announced on the market. However, their adjustment to a particular type of robot is still too complex and tedious job to be efficiently accomplished by a customer. Besides, they usually don't perform compensation of dynamic effects, so that good tracking of fast trajectories cannot be achieved. For these reasons, the development of a new general purpose controller UCS-1 was commenced in the Mihailo Pupin Institute, Belgrade [7].

The system is designed to meet two main advantages over existing controllers: dynamic control ensuring tracking of fast trajectories and easy maintenance and adjustment of the system to nonredundant robots of arbitrary type with up to six degrees of freedom.

The specific goals kept in mind during software design were:

- reduction of the run time computation required for calculation of quantities related to the robot dynamic; in order to meet this goal, which has the principal importance in implementing the system on existing microcomputers, the use of an analytical model of the robot is adopted (see Chapter 1 and Paragraph 5.3);

- hardware transportability, i.e. a possibility of adapting a custommer's system to a particular robot and a particular application without the need for intensive training of the user; for this purpose, an interactive procedure for imposing the parameters necessary for the automatic creation of an analytical model and the selection of control algorithm is developed;

- possibility of adapting the robot operation to the robot environment, especially of synchronizing the robot with the external hardware; this condition is essential for the applicability of the robot in most factory sites; in order to achieve this goal, a set of routines for processing input and output signals and for controlling the order of operation are designed;

- possibility of implementating the system on an inexpensive computer system and possibility of operating the system without utilizing mass memories, so that the probability of system faults in factory conditions is decreased;

- reduction of the human effort necessary for programming the robot task; to this end, a specialized programming language RL is designed; the language supports programming the robot in external coordinates, enables to utilize variables of various types, control structures and user-written subroutines;

- robustness of the system, i.e. protection of the system integrity against an unauthorized use and accidental programming errors.

The controller is implemented by two microcomputers based on powerful 16-bit microprocessors, which enable application of on-line kinematics and dynamic direct digital servosystems including on-line computation of robot dynamics. Also, communication with operator and teaching of the robot by robot program language and teaching box are elaborated.

The complete robot control system with controller UCS-1 is presented in Fig. 5.1. Model UCS-1 ensures interfaces to: a general-purpose computer, video terminal/printer, teaching box, and sensory system. The interfaces to actuators (DC motors or hydraulic) are realized by means of direct analog current or voltage signals from D/A converters.

(Obviously, appropriate amplifiers have to be implemented). Interfaces to a general-purpose computer, videoterminal/printer and teaching box are realized by means of serial lines, and to the sensory system by means of digital and analog lines.

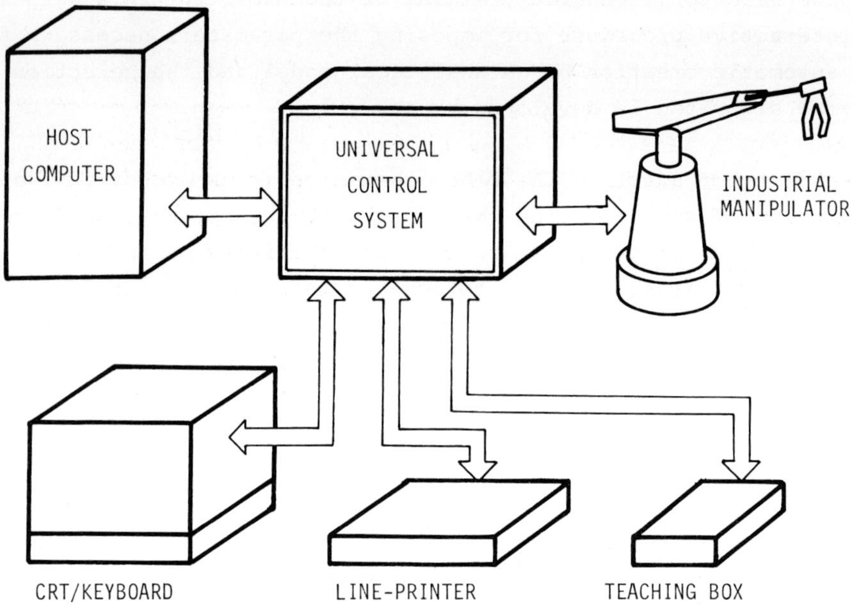

Fig. 5.1. Configuration of general robotic controller

Model UCS-1 contains: 2 microcomputers based on 16-bit microprocessors for processing floating-point operations and calculating various functions, memory (RAM and EPROM), real-time clock, battery back-up system, A/D 12-bit conversion modules, D/A 12-bit conversion modules, digital input/output modules, serial input/output modules, and modules for amplification/attenuation of current and voltage signals.

Controller UCS-1 contains a powerful software support which consists of several interconnected modules:

- software for communication with operator,
- software for initial specification of robot and sensor parameters, i.e. for adjustment of the controller to the particular robot, actuators and sensors system,

- robot programming language,
- program for communication with the teaching box,
- software for on-line robot kinematics, and
- software for direct digital servosystems for tracking trajectories including on-line dynamics.

UCS-1 also contains a number of system routines for servicing discettes/magnetic tape conversion modules, "hardware" - tests, programs for checking the controller's reliability and reactions in emergency situations, etc. The general software organization is presented in Fig. 5.2.

In the text to follow we shall briefly describe some of the modules mentioned above.

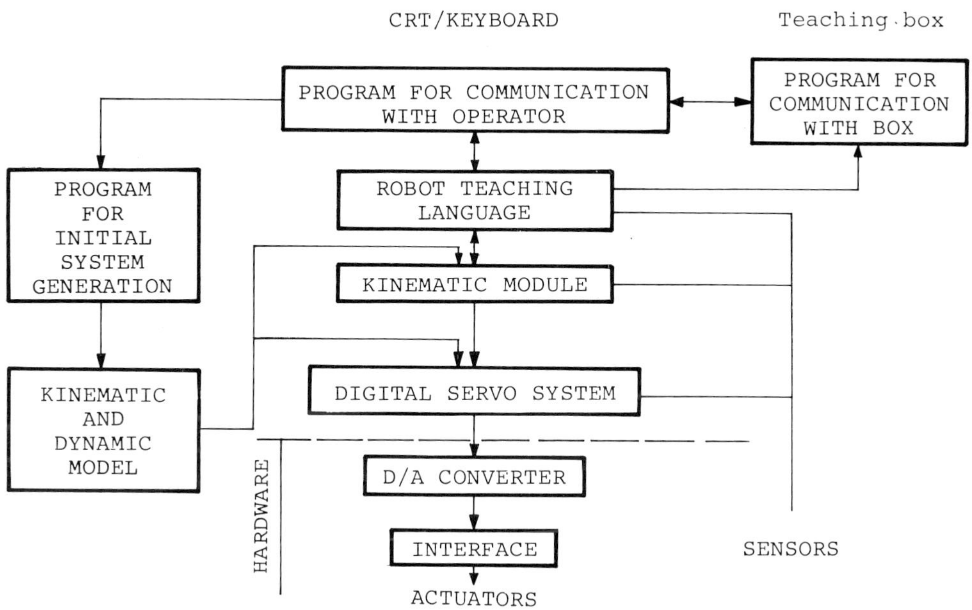

Fig. 5.2. General software organization

1. Software for communication with operator

This module enables communication with the operating system. It allows the operator to call various programs included in the software of the controller, such as: programs for initially setting mechanism parameters, actuators, sensors, programs for teaching the robot, programs of robot tasks, programs for hardware tests, program for adjustment of the system parameters.

2. *Robot programming language*

One of the main objectives stated in design of many robot programming languages was simplicity of use, even for novice programmers unfamiliar with computers and basic programming concepts. We adopted a somewhat different approach; our specific goal was to enable the field engineer, having some experience in programming, to exploit as much as possible the available robot without necessity to learn a new programming language. For this reason we decided to use a PASCAL-like syntax, supposing that the PASCAL is well-known to a majority of system analysis, control and mechanical engineers who are to be the main designers of automated manufacture sites and the main class of users of programming tools in industrial robotics. Besides, the top-down approach in writing programs that is encouraged in PASCAL and clear structure of PASCAL programs were another reason for selecting it as a basis for the RL language.

During the specification of the RL language, we started from a view that the language should facilitate all main phases of robot programming and exploitation; among others these are [8]:

- definition of the robot task;
- writing a robot program which should describe not only the robot behaviour during performing the task but also the procedures for robot teaching, testing, tuning, etc.;
- teaching the robot, i.e. memorizing individual positions and orientations of the robot effector during performing the task, as well as computation of positions, orientations and dimensions of working objects;
- testing the robot program, including eventual reteaching of the positions;
- adjusting the controller parameters in order to meet particular requirements connected to the manipulator mechanical structure and the task to be performed;
- exploitation, which can also include a necessity for occasionally

reteaching particular positions and orientations and adjusting the controller parameters.

The definition of the robot task can be done either in an ad-hoc manner or by using the robot language itself; we decided to encourage the second approach believing that it can speed up the process of writing robot programs and that it can result in more readable programs and thus, programs that are easier to debug, test, tune and expand.

Robot programming language RL is a high level textual language which is implemented in UCS-1 controller in order to simplify communications between the user and the system, i.e. to supply the user with a tool for simply defining particular robot tasks. RL supports the use of program variables, evaluation of logical and arithmetic expressions, use of conditional and repetitive control structures and user-written subroutines. A set of system library subroutines incorporated in the language enables the user to specify manipulator motion, gripper operation, sensor control, synchronization, etc.

Editing, testing and execution of robot language programs are achieved via robot language processor RLP. Robot language processor is a component of UCS-1 software and can be executed in one of the two modes:

- Editor/compiler. In this mode of operation RLP enables interactive creating and editing of a RL program via a terminal keyboard, as well as including sequences of RL instructions generated by the use of the teaching box. During the operation RLP parses input instructions and prepares the RL program for execution. The syntax analysis is performed and diagnostic messages are displayed on the terminal screen immediately after instructions typing.

- Interpreter. In this mode RLP accepts and executes a RL program previously translated into an internal form. RLP permanently monitor the operation of controller and manipulator equipment and informs the user of any hardware or software error detected during the execution.

3. *Teaching box*

Teaching box represents a "manual programming unit" for quick and efficient programming of the robot. This teaching box is implemented by an 8-bit microprocessor in CMOS technology, so that small size and low weight of the box are achieved. These advantages of the teaching box enable programming of the robot in its close surroundings. The teaching box is connected to controller UCS-1 by a long cable and it consists of two alfanumeric display lines, a keyboard for selecting the execution mode, a keyboard for choosing the execution speed and a keyboard for teaching the robot. The keyboard for teaching the robot includes keys for starting and stopping some procedures, for emergency stop of the robot, for memorizing the robot coordinates, defining the coordinate system in which robot coordinates are specified, for defining duration of some actions, for defining the speed of motion, etc.

In defining some control tasks the operator can set some parts of the task using robot language, and some other by using the teaching box. In this way, maximal flexibility in control task specification is provided as well minimum time for teaching the robot.

4. *Software for initial specification of robot parameters*

A program for initial generation of the system (PIGS) provides flexibility of the controller in the sense of its application to an arbitrary configuration of the robot (with up to 6 degrees of freedom), to various types of actuators and sensors. This is why this program is realized as an interactive program with simple communication with the user. It consists of several modules: program for initial specifications of the parameters of the mechanism, actuators, and sensors, program for kinematic model generation, program for dynamic model generation, program for generating models of the actuators and calculation of the digital servosystems parameters.

Using a simple dialogue between the user and the controller, all above mentioned data are imposed by the operator for a specific manipulation robot, and can easily be changed. The user can also specify the types of sensors which are included in the system, together with numerical data on sensors. The controller UCS-1 supports the following types of sensors: binary sensors, multibit sensors, analog sensors, sensors which have to be elaborated by a host-computer (e.g. camera), etc.

The programs for generating kinematic and dynamic models of the robot automatically form the robot models in the machine language of microcomputers. They form analytic models for a specific manipulation type, according to the parameters of the mechanism which have been set by the user [9].

The program for computing the model of the actuators and for the digital servosystems parameters has to synthesize the gains of digital servosystems on the basis of tolerances imposed by the user (Chapter 4).

All these modules do not depend on the type of the robot, so the controller UCS-1 is general and can easily be "adjusted" from one to another robot type.

5. *Program for robot kinematics*

The program for robot kinematics realizes the commands of the robot program language or teaching box which are concerned with the motion of the gripper. This program enables on-line generation of the trajectories of joint coordinates of the robot according to the user's commands, or according to trajectories imposed by the host-computer. This on-line generation of the joint trajectories if external (gripper) trajectories are given, is achieved by off-line prepared kinematic model of the robot. This model has been generated by the PIGS. Kinematic program is designed to enable generation of joint trajectories for various types of external coordinates, taking into account all constraints in the working space of the robot (obstacles) and/or constraints upon the robot joint drives [10]. This program enables implementation of continual and "point-to-point" trajectories with linear and polynomial interpolation.

6. *Program for direct digital servosystems (DDS)*

Controller UCS-1 includes direct digital servosystems for control of DC electromotors or hydraulic actuators in the manipulator joints. Thus, analog servosystems are not needed for robot control. On the basis of data on actuators and mechanism and on the basis of given tolerances which the robot must satisfy, PIGS computes feedback gains for DDS for all joints of the manipulator (Chapter 2). Implementation of DDS requires feedback by position from the shaft-encoder in the manipulator joint and velocity feedback from tachogenerator. The tachogenerator might be omitted, but this would result in worse tracking of fast trajectories. DDS also includes on-line computation of the dynamics of the robot mechanism for global control. In this way very precise tracking of fast trajectories is achieved, since DDS compensates for dynamic effects (inertia and gravity) during the tracking. Thus, the performances of our DDS are improved in comparison with classical solutions of digital or analog servosystems. This matter has been discussed in more detail in Chapter 2. On-line computation of the dynamic model of the mechanism is achieved by a program for automatic generation of the analytical model of the mechanism dynamics on the basis of initially set data on the robot. This problem will be considered in paragraphs to follow.

DDS are designed to withstand variations of parameters of the mechanism and actuators to some extent (variations with respect to initially specified values). However, the user can adjust servo-gains himself through corresponding commands and programs.

7. Software for execution checks

Besides the standard routines for software errors detection, the controller UCS-1 is supplied with a library of routines for continuous check of the controller performance, during programming of the robot and during test execution. These programs detect all irregularities in the execution of the controller and robot and send appropriate messages. The controller is also supplied with special routines for fast responses in emergency situations. In this way reliability of the controller is ensured and the system is protected from all irregularities in the controller execution which could cause demage on the robot or in the working space of the robot. This makes the use of the controller safe and simple, since the controller is protected from eventual irregular handling and programming.

Since our prime interest in this monograph is the executive control level, in the text to follow we shall concentrate on the problem of implementation of the executive control level. It should be underlined that the key problem in the implementation of general purpose controller is accomplishing on-line kinematics [10] and on-line dynamics which is included in direct digital servosystems. The implementation of local and global (non-adaptive and adaptive) control is the main issue of the next paragraphs. Namely, we shall consider the numerical complexity of various control laws in the executive control level in order to determine the equipment necessary for implementing the general purpose controller. It should be understood that the described general purpose controller includes the control synthesis described in previous chapters.

5.3 Numerical Complexity of Control Laws

In this paragraph we shall discuss one of the most important issues related to the real-time applicability of robot control laws. This issue will be referred to as "numerical complexity" of a control algorithm.

Numerical complexity is quantitatively determined by the number of numerical operations required for the implementation of a control law. The mathematical expressions of control algorithms include algebraic operations between real numbers. The numerical complexity can be expressed by the numbers of floating-point multiplications/divisions and additions/subtractions. These numbers determine the real-time applicability of a control law. Our "reference" law-cost microcomputer will be an up-to-date 16-bit micro provided with a floating-point coprocessor, capable of carrying out over 300 floating-point multiplications and over 300 additions/subtractions in 20 ms sampling time.

We shall see in the text to follow that dynamic effects which are taken into account predominantly determine the numerical complexity of a control law. The compensation of Coriolis and centrifugal forces expecially increases the number of floating-point multiplications/additions. We shall use the described numeric-symbolic method for the evaluation of dynamic effects, because of its computational efficiency.

The numerical complexity of dynamic robot models will be considered in Paragraph 5.3.1. The obtained results apply to centralized control laws as well. Various cases are examined, depending on the dynamic effects which are compensated through the control law.

The numerical complexity of decentralized control laws will be considered in Paragraph 5.3.2. We will determine the number of numerical operations required for the evaluation of both local and global components of control law. Here, the global control will be considered as obtained by force measurement or by the real-time computation of dynamic effects.

In the next paragraph, we will consider some of the adaptive control laws described in Chapter 3. First, we will examine the numerical complexity of the centralized adaptive control strategy based on hyperstability theory (Paragraph 3.2.2). Then, the indirect centralized control (Paragraph 3.2.3) will be examined. The self-tuning control strategy will not be considered, because it does not take into account any dynamic effect of a manipulator. Finally, the numerical complexity of the indirect decentralized adaptive control will be determined.

The number of floating-point multiplications/additions will be determined for several robots with 3 and 6 degrees of freedom. The obtained results should give the answer to the question: which dynamic control laws can be applied in real time on typical up-to-date low-cost microcomputers? For the sake of efficiency, the analytical modelling method is applied. This method is also very suitable for multiprocessing applications.

5.3.1. Numerical complexity of dynamic models

Consider a manipulator with n joints and n links powered by electric or hydraulic actuators. Both types of actuators can be presented by the state-space model (2.2.2):

$$S^i:\quad \dot{x}^i = A^i(\theta^i)x^i + b^i(\theta^i)N(u^i) + f^i(\theta^i)P_i \qquad i=1,\dots,n$$

where x^i is an n_i-dimensional state vector of the actuator dirving the joint i,

u^i is the input of the *i*th actuator,

$N(\cdot)$ is the amplitude saturating function defined as

$$N(u^i) = \begin{cases} u_m^i & u^i \geq u_m^i \\ u^i & -u_m^i \leq u^i \leq u_m^i \\ -u_m^i & u^i \leq -u_m^i \end{cases}$$

with $u_m^i > 0$ being a given constant,

P_i is the driving torque/force in joint i,

θ^i is the $\underset{\sim}{\ell}_i$ - dimensional parameter vector of the *i*th actuator, including actuator gain, viscous damping factor, etc.,

A^i is an $n_i \times n_i$ system matrix,

b^i is an $n_i \times n_i$ system matrix,

f^i is an n_i-dimensional load distribution vector.

For example, model matrices describing the dynamics of DC motors are

$$A^i = \begin{bmatrix} 0 & 0 & 0 \\ 0 & -F/J & C_M/J \\ 0 & -C_E/L & -r/L \end{bmatrix}, \quad b^i = \begin{bmatrix} 0 \\ 0 \\ 1/L \end{bmatrix}, \quad f^i = \begin{bmatrix} 0 \\ -1/J \\ 0 \end{bmatrix} \qquad (5.3.1)$$

where F - the viscous damping factor, C_M - actuator gain, C_E - back e.m.f. coefficient, r - rotor resistance, L - rotor inductance, and J - rotor moment of inertia. Here, the state vector is $x^i=[q_i \ \dot{q}_i \ i^i]^T$, with q_i being the ith joint angle and i^i the rotor current. Notice that this model is of order 3. Neglecting inductance, we obtain the second-order model

$$A^i = \begin{bmatrix} 0 & 1 \\ 0 & -C_M C_E/rJ \end{bmatrix}, \quad b^i = \begin{bmatrix} 0 \\ C_m/rJ \end{bmatrix}, \quad f^i = \begin{bmatrix} 0 \\ -1/J \end{bmatrix} \qquad (5.3.2)$$

with state-space vector $x^i = [q_i \ \dot{q}_i]^T$. When a reducer connects the actuator and the joint, it is necessary to include the reduction ratio into θ^i.

Driving torques/forces P_i are arranged by the dynamic equations of motion of manipulator (see Chapter 1)

$$S_M: \quad P_i = \sum_{j=1}^{n} H_{ij}(q, \ d)\ddot{q}_j + h_i(q, \dot{q}, d), \qquad i=1,\dots,n \qquad (5.3.3)$$

where H_{ij} are scalar functions describing the inertial coupling between the degrees of freedom i and j, $q = [q_1 \cdots q_n]^T$ is the vector of joint coordinates, d is an ℓ_i-dimensional vector of the mechanical parameters of manipulator and h_i is a scalar function describing gravitational, centrifugal and Coriolis effects. The matrix form of Eq. (5.3.3) becomes

$$S_M: \quad P = H(q, \ d)\ddot{q} + h(q, \dot{q}, d) \qquad (5.3.4)$$

where $P = [P_1 \cdots P_n]^T$ is the vector of driving torques for revolute joints or forces for sliding ones, H is an $n \times n$ full-rank positive definite inertial matrix and $h = [h_1 \cdots h_n]^T$ is an n vector describing velocity-coupling between links as well as gravitational effects. This vector can be presented in the form (see Eq. (2.2.1)):

$$h = \dot{q}^T C(q, d)\dot{q} + g(q, d) \tag{5.3.5}$$

where C is an n×n×n matrix, and g is an n vector describing gravitational effects. Quadratic form $\dot{q}^T C(q, d)\dot{q}$ represents the n vector $[\dot{q}^T C^1(q, d)\dot{q} \cdots \dot{q}^T C^n(q, d)\dot{q}]^T$ where C^i are n×n matrices.

Actuator model matrices (5.3.2) indicate that their elements are zero except for $A^i(1,2) = 1$ and $A^i(2,2)$, $b^i(2)$ and $f^i(2)$. Let us introduce the constants a_i, b_i and f_i, such that

$$A^i(2,2) = -a_i, \quad b^i(2) = b_i \quad \text{and} \quad f^i(2) = -f_i$$

Now, the actuator model reduces to $\ddot{q}_i = -a_i\dot{q}_i + b_i u_i - f_i P_i$. The matrix form of the system of n actuators becomes

$$S_A: \ \ddot{q} = -\text{diag}(a_i)\dot{q} + \text{diag}(b_i)u - \text{diag}(f_i)P \tag{5.3.6}$$

The entire model that consists of S_M and S_A is

$$S_{(2)}: \ u = \hat{H}(q, \theta)\ddot{q} + \hat{h}(q, \theta) \tag{5.3.7}$$

where θ represents the vector of mechanical and electrical parameters of mechanism and actuators, and

$$\begin{aligned} \hat{H} &= \text{diag}(b_i^{-1}) + \text{diag}(f_i b_i^{-1})H(q, d) \\ \hat{h} &= \text{diag}(a_i b_i^{-1})\dot{q} + \text{diag}(f_i b_i^{-1})h(q, \dot{q}, d) \end{aligned} \tag{5.3.8}$$

Substituting $h = \dot{q}^T C(q, d)\dot{q} + g(q, d)$ into $\hat{h}$, we obtain

$$\hat{h} = \hat{K}(q, \dot{q}, \theta)\dot{q} + \text{diag}(f_i b_i^{-1})g(q, d) \tag{5.3.9}$$

where $\hat{K} = \text{diag}(a_i b_i^{-1}) + \text{diag}(f_i b_i^{-1})\dot{q}^T C(q, d)$. To summarize, the model (5.3.7) - (5.3.9) may be represented by the diagram shown in Fig. 5.3. The state-space representation of model (5.3.7) is

$$S: \ \dot{x} = F(x, \theta)x + B(x, \theta)u + G(x, \theta) \tag{5.3.10}$$

with

$$F = \left[\begin{array}{c|c} 0 & I \\ \hline 0 & -\hat{H}^{-1}(q,\ \theta)\hat{K}(q,\ \dot{q},\ \theta) \end{array}\right] \qquad B = \left[\begin{array}{c} 0 \\ \hline \hat{H}^{-1}(q,\ \theta) \end{array}\right] \tag{5.3.11}$$

$$G = \left[\begin{array}{c} 0 \\ \hline -\mathrm{diag}(f_i b_i^{-1})\hat{H}(q,\ \theta)g(q,\ d) \end{array}\right]$$

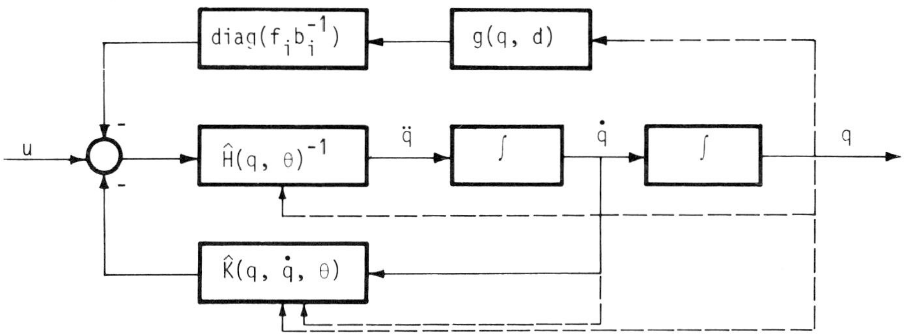

Fig. 5.3. Diagram of the robot model including both the model of mechanical system part and the second-order models of actuators

Under the assumption that the actuators are modelled by third-order systems, the state vector can be taken into the form $x = [q^T\ \dot{q}^T\ p^T]^T$, where q denotes the joint coordinate vector. Vector p is of the form $p = [i^1 \cdots i^n]^T$ (i^j - rotor current of the jth actuator for DC motors, and $p = [p^1 \cdots p^n]^T$ (p^j - differential pressure) for hydraulic actuators. Following the derivation of model (5.3.7), one obtains

$$\begin{aligned} \hat{H}\ddot{q} &= -\hat{K}\dot{q} + \mathrm{diag}(a_{23}^i b_i^{-1})p - \mathrm{diag}(f_i b_i^{-1})g \\ \dot{p} &= -\mathrm{diag}(a_{32}^i)\dot{q} - \mathrm{diag}(a_{33}^i)p + \mathrm{diag}(b_i)u \end{aligned} \tag{5.3.12}$$

with $a_{jk}^i = |A^i(j,k)|$, $b_i = b^i(3)$ and $f_i = f^i(2)$. Matrices $\hat{H}$ and $\hat{K}$ are defined by (5.3.8), (5.3.9) under the condition $a_i = A^i(2,\ 2)$. Now, the robot model may be represented by the diagram shown in Fig. 5.4.

Let us now determine the quantity of floating-point multiplications and additions/subtractions which are to be performed in order to calculate robot model equations. First, we shall suppose that the actuators are modelled by second-order linear time-invariant systems. The input vector, which must be calculated for given q, $\dot{q}$ and $\ddot{q}$, will be denoted by $u_{(2)}$. We see from Eq. (5.3.7) that $u_{(2)}$ can be represented by

$$u_{(2)} = \hat{H}\ddot{q} + \hat{h} \tag{5.3.13}$$

with $\hat{H}$ and $\hat{h}$ determined by (5.3.8). From these expressions, we obtain

$$\begin{aligned} n_M(\hat{H}) &= n_M(H) + n^2, \quad n_A(\hat{H}) = n_A(H) + n \\ n_M(\hat{h}) &= n_M(h) + 2n, \quad n_A(\hat{h}) = n_A(h) + n \end{aligned} \tag{5.3.14}$$

where $n_M(\cdot)$ and $n_A(\cdot)$ are the numbers of floating point multiplications and additions/subtractions, respectively. From (5.3.13), it follows that

$$n_M(u_{(2)}) = n_M(\hat{H}, \hat{h}) + n^2, \; n_A(u_{(2)}) = n_A(\hat{H}, \hat{h}) + n^2 \tag{5.3.15}$$

with $n_M(\hat{H}, \hat{h}) = n_M(\hat{H}) + n_M(\hat{h})$ and $n_A(\hat{H}, \hat{h}) = n_A(\hat{H}) + n_A(\hat{h})$. From (5.3.14) and (5.3.15) we get

$$\begin{aligned} n_M(u_{(2)}) &= n_M(H, h) + 2n^2 + 2n \\ n_A(u_{(2)}) &= n_A(H, h) + n^2 + 2n \end{aligned} \tag{5.3.16}$$

Let us derive the corresponding expressions in the case when the actuators are modelled by third-order systems. From (5.3.12), we see that the input vector $u_{(3)}$ can be expressed as

$$u_{(3)} = \mathrm{diag}(b_i^{-1})\dot{p} + \mathrm{diag}(a_{32}^{i}b_i^{-1})\dot{q} + \mathrm{diag}(a_{33}^{i}a_{23}^{i-1})(\hat{H}\ddot{q}+\hat{h}) \tag{5.3.17}$$

Notice that the computation of $\hat{H}\ddot{q} + \hat{h}$ requires $n_M(u_{(2)})$ multiplications and $n_A(u_{(2)})$ additions, given by (5.3.16). Substituting into (5.3.17), we get

$$n_M(u_{(3)}) = n_M(H, h) + 2n^2 + 5n, \quad n_A(u_{(3)}) = n_A(H, h) + n^2 + 4n \tag{5.3.18}$$

We see that the quantity of numerical operations in equations determining the input vector $u_{(3)}$ depends mostly on the number of multiplications and additions necessary to calculate dynamic model matrices H and h.

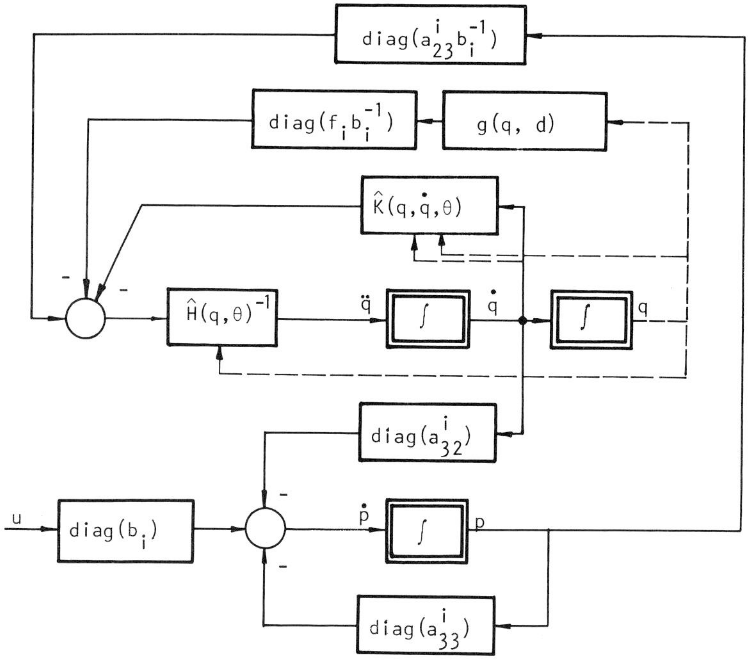

Fig. 5.4. Robot model including both the model of arm and the third-order models of actuators

Depending on the modelling method and the complexity of robot kinematic chain, one obtains different values for n_M (H, h) and n_A(H, h). Let us accept the efficient modelling method described in Chapter 1, and determine n_M and n_A for some typical industrial robots. This method is based on computer-aided generation of analytical expressions for the elements of H, C and g matrices as functions of joint coordinates. To calculate the vector $h = \dot{q}^T C \dot{q} + g$ with a minimal number of numerical operations, we must take into account relations between the elements of C matrices [11]:

$$C^i_{k\ell} = C^i_{\ell k}$$

$$C^i_{k\ell} = -C^k_{i\ell} \quad \text{for} \quad k \geq \ell \quad \text{and} \quad i \geq \ell$$

so that $i,k,\ell \in \{1,\ldots,n\}$. Therefore

$$h_i = \sum_{\substack{k=1\\(k\neq i)}}^{n} C^i_{kk}\dot{q}_k^2 + 2\sum_{\substack{k=1\\(k\neq i)}}^{n}\sum_{\ell=1}^{k-1} C^i_{k\ell}\dot{q}_k\dot{q}_\ell + g_i \tag{5.3.19}$$

with h_i and g_i being the *i*th elements of vectors h and g, respectively. The number of multiplications in (5.3.19) equals $n^2+3n-2(i+1)$, and the number of additions $(n+1)(n+2)/2-i$. For all $i=1,\dots,n$, the total number of multiplications is $n^2(n+1)/2$ and the total number of additions is $n^2(n^2+3)/2$. Now, it follows that

$$\begin{aligned} n_M(h) &= n_M(C) + n_M(g) + \frac{1}{2}n^2(n+1) \\ n_A(h) &= n_A(C) + n_A(g) + \frac{1}{2}n(n^2+3) \end{aligned} \tag{5.3.20}$$

Finally, substituting (5.3.20) into (5.3.16) and (5.3.18), we obtain

$$\begin{aligned} n_M(u_{(2)}) &= n_M(H, C, g) + \frac{1}{2}n(n^2+5n+4) \\ n_A(u_{(2)}) &= n_A(H, C, g) + \frac{1}{2}n(n^2+2n+7) \end{aligned} \tag{5.3.21}$$

as well as

$$\begin{aligned} n_M(u_{(3)}) &= n_M(H, C, g) + \frac{1}{2}n(n^2+5n+10) \\ n_A(u_{(3)}) &= n_A(H, C, g) + \frac{1}{2}n(n^2+2n+11) \end{aligned} \tag{5.3.22}$$

Let us consider the meaning of these relations on several examples. We shall discuss some commercial industrial robots with 3 and 6 degrees of freedom: a cylindrical robot shown in Fig. 5.5 (we will separately consider a version with 3 links CL-3, and a version with 6 links CL-6), a PUMA-like robot shown in Fig. 5.6 (designated by PU-3), a spherical robot shown in Fig. 5.7 (SF-3), and the Stanford manipulator with 6 joints (ST-6) shown in Fig. 5.8. The number of numerical operations necessary for the computation of dynamic model matrices H, C and g, as well as input vectors $u_{(2)}$ and $u_{(3)}$ is given in Table 5.1. This table also contains the computation time obtained on a typical 16-bit microcomputer. Of course, implementation on more powerful microcomputers (such as MOTOROLA 68000 or INTEL 286) would decrease the computation time 2 to 3 times. From this table we also see that the difference between computation times for $u_{(2)}$ and $u_{(3)}$ is minor compared with total times.

In on-line applications, it is necessary to compute $u_{(2)}$ or $u_{(3)}$ at a sampling frequency preferably not lower than 50 Hz, since the resonance

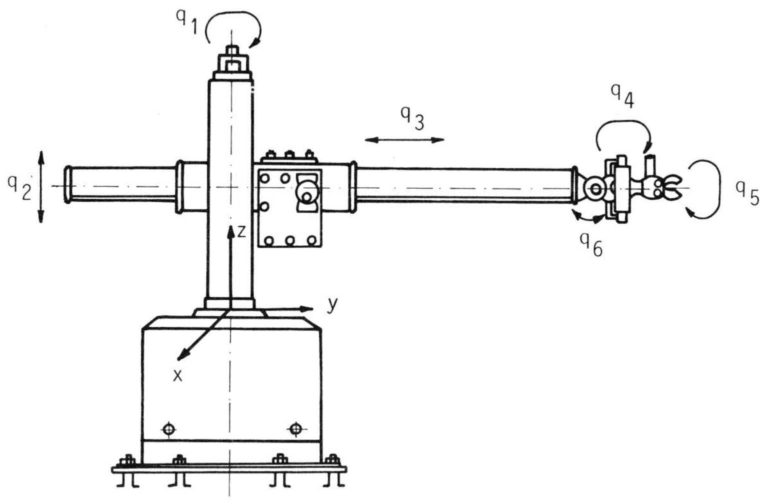

Fig. 5.5. Cylindrical robot (CL)

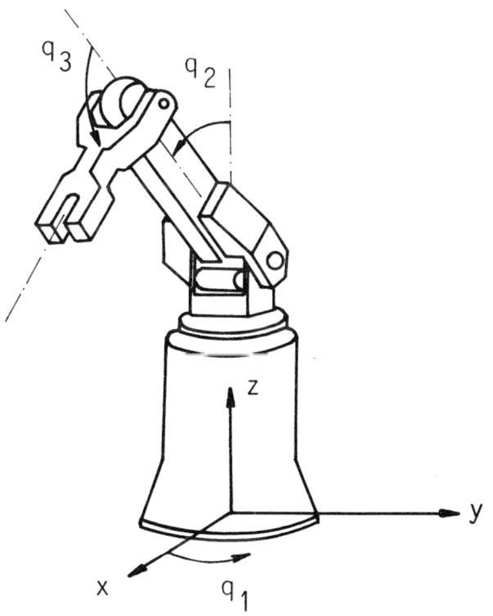

Fig. 5.6. PUMA-like robot (PU)

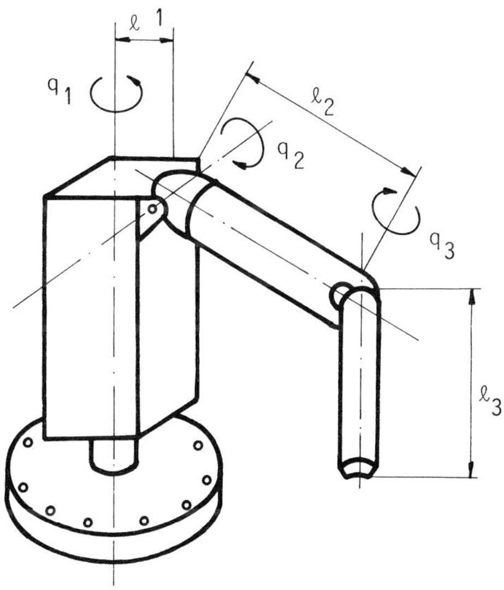

Fig. 5.7. Spherical robot (SF)

frequency of most mechanical manipulators is about 10 Hz. However, the real-time computation (for at most 20 ms) by means of an average 16-bit microcomputer is possible only for 3-degree-of-freedom robots.

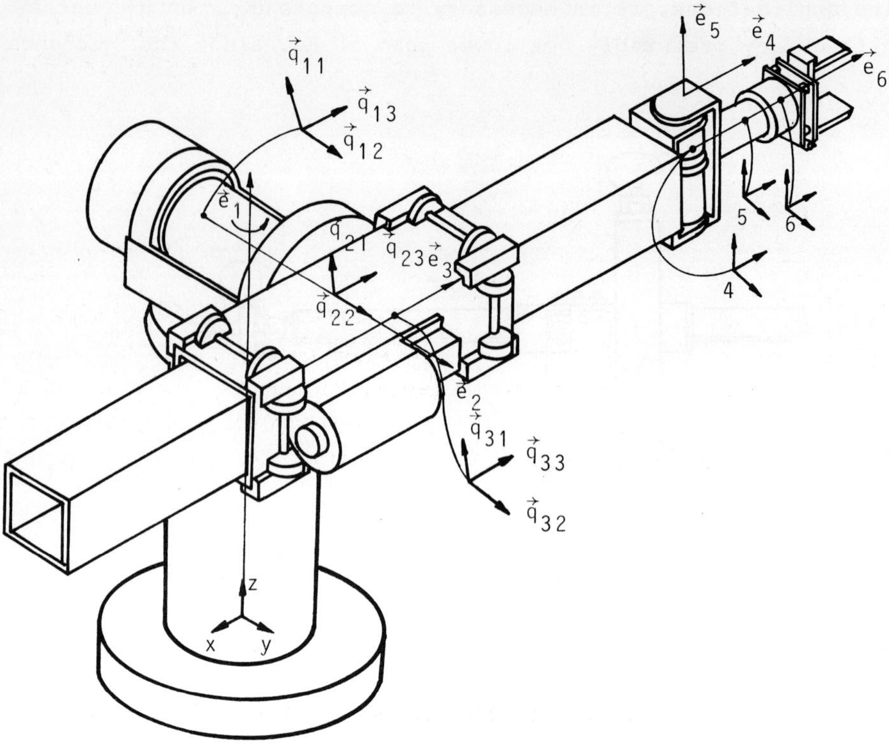

Fig. 5.8. Stanford manipulator (ST)

ROBOT	CL-3	PU-3	SF-3	CL-6	ST-6
$n_M(H, C, g)$	3	49	55	193	372
$n_A(H, C, g)$	3	22	31	80	167
$n_M(u_{(2)})$	45	91	97	403	582
$n_A(u_{(2)})$	36	55	64	245	332
$n_M(u_{(3)})$	54	100	106	421	600
$n_A(u_{(3)})$	42	61	70	257	344
$t(H, c, g)$	0.6	4.0	4.6	12.4	20.8
$t(u_{(2)})$	3.1	6.9	7.4	26.8	37.0
$t(t_{(3)})$	3.7	7.4	8.0	28.0	38.2

Table 5.1. Number of numerical operations and computation time [ms] with microcomputer based on INTEL 8086/8087 at 8 MHz clock

Using more powerful microcomputers, it is possible to reduce the calculation time significantly. However, this makes the price of robot control unit higher.

5.3.2. Numerical complexity of non-adaptive decentralized control laws

Centralized control approach, sometimes called "inverse control structure" or "direct torque method" (see Chapter 2), is implicitly presented in the previous paragraph. For example, both equations (5.3.7) and (5.3.17) may be considered as control laws if the acceleration vector $\ddot{q}$ is computed according to linear feedback structure

$$\ddot{q} = \ddot{q}^{o} + K_p(q-q^{o}) + K_v(\dot{q}-\dot{q}^{o})$$

where K_p and K_v represent $n\times n$ gain matrices, and $q^{o} = q^{o}(t)$ is a nominal trajectory in joint-coordinate-space. In the previous paragraph it was shown that the real-time implementation of (5.3.7) or (5.3.17) would require the use of more powerful 16-bit microprocessors. To develop less expensive controllers, it would be justifiable to consider the numerical complexity of decentralized robot control.

Consider the simplest form of decentralized control law (see Chapter 2)

$$u^{i} = u^{oi} + D_i\Delta x^{i} + k_i^{G}\Delta P_i \tag{5.3.23}$$

where $D_i \in R^{1\times 2}$ includes local gains, $\Delta x^{i} = [\Delta q_i \;\; \Delta\dot{q}_i]^{T}$, and $k_i^{G} = \text{const}$ is the global gain. Suppose that the manipulator joints include torque sensing capability (or force sensing in sliding joints) by means of strain gauges. For example, Stanford manipulator was redesigned to include force sensors [12]. Manipulators like this one can simply be controlled by (5.3.23). Then the number of floating-point operations necessary to realize (5.3.23) becomes

$$n_M = 3n \quad \text{and} \quad n_A = 6n. \tag{5.3.24}$$

In some cases, it is not necessary to include force feedback in all subsystems (actuators). Obviously, the practical stability of the entire system can often be achieved by local feedback loops only. For example, it is not necessary to apply force feedback in joints of a cylindrical robot like the one shown in Fig. 5.5. Global control is usually desirable for the joints in which dynamic coupling effects are significant. Taking only necessary feedback loops into account, the number of floating-point operations in control laws (5.3.23) implemented on robots described in the previous section is shown in Table 5.2. We see that the number of numerical operations is drastically decreased compared to that for centralized control (5.3.7). However, the use of force sensors increases the cost of robot system.

On the other hand, the coupling function ΔP_i can be calculated on-line using robot dynamic equations. For this purpose, it is convenient to apply the mentioned analytical method, which is very efficient regarding the numerical complexity. Unfortunately, the exact computation of driving torques implies the use of high-speed microcomputers. In such a way, the advantages of decentralized control would be lost. Thus, with the decentralized control we will take into account only those dynamic effects which are required to be introduced for a given control task.

	ROBOT TYPE					
CONTROL LAW	M/A	CL-3	PU-3	SF-3	CL-6	ST-6
Eq. (5.3.23)	M A	6 9	9 18	9 18	15 27	15 27
Eq. (5.3.25)	M A	6 9	76 54	87 72	230 147	413 234
Approximate vector P	M A	- -	26 16	30 21	78 48	108 65
Eq. (5.3.26)	M A	6 9	35 31	39 36	93 75	123 82
Approximate vector P	M A	- -	23 12	20 14	18 14	55 33
Eq. (5.3.27)	M A	6 9	32 27	29 29	33 41	70 60
Eq. (5.3.28)	M A	6 9	17 18	13 16	15 27	22 30

Table 5.2. Number of floating-point multiplications (M) and additions/subtractions (A) for different control laws

Let us first systematically suppose that the dynamic model is computed exactly. Then, the decentralized control law (5.3.23) becomes

$$u^i = u^{oi} + D_i \Delta x^i + k_i^G \left(\sum_{i=1}^{n} H_{ij}(x, \ d)\ddot{q}_j^o + h_i(x, \ d) - P_i^o \right) \qquad (5.3.25)$$

Notice that (5.3.25) includes all dynamic effects: inertial, gravitational, centrifugal and Coriolis' effects. The numerical complexity of (5.3.25) is presented in Table 5.2. This table shows that is is possible to implement the control law (5.3.25) "on-line" even for 6-degree-of-freedom robots having relatively simple kinematic chains.

As the next version of decentralized control law we shall discuss the control structure including inertial and gravitational effects

$$u^i = u^{oi} + D_i x^i + k_i^G \left(\sum_{j=1}^{n} H_{ij}(x, \ d)\ddot{q}_j^o + g_i(x, \ d) - P_i^o \right) \qquad (5.3.26)$$

Here, centrifugal and Coriolis' effects are omitted in the expressions for global control component. Using the proposed analytical method for computing inertial and gravitational forces, we obtain the number of floating-point multiplications and additions given in Table 5.2. The numerical complexity of approximated driving torque vector is also given in this table. Compared to the numerical complexity of the centralized control law, we see that the number of floating-point operations is reduced by about 40%.

A further simplification of global control leads to the elimination of cross-inertial components. Here, only diagonal elements of inertia matrix are taken into account. Then, we have

$$u^i = u^{oi} + D_i x^i + k_i^G (H_{ii}(x, \ d) q_i^o + g_i(x, \ d) - P_i^o) \qquad (5.3.27)$$

with P_i^o precalculated according to the expression $H_{ii}(x^o, \ d)\ddot{q}_i^o + g_i(x^o, \ d)$. The results for both control vector and approximated driving torques are given in Table 5.2. We see that even for the most complex robot structures the computing time is shorter than 10 ms.

Finally, if we take into account gravitational forces only, we obtain

$$u^i = u^{oi} + D_i \Delta x^i + k_i^G (g_i(x, \ d) - P_i^o) \qquad (5.3.28)$$

The number of operations is now very small and offers the possibility for implementation on 8-bit microcomputers. The dependence of the number of floating-point multiplications on global control complexity is presented in Fig. 5.9. The computing times measured on INTEL 8086/8087 are given in Table 5.3. Cosine and sine functions of joint coordinates are included, too.

CONTROL LAW	ROBOT TYPE				
	CL-3	PU-3	SF-3	CL-6	ST-6
Eq. (5.3.24)	0.9	2.25	2.25	3.4	3.4
Eq. (5.3.28)	0.6	3.0	2.4	3.4	3.9
Eq. (5.3.27)	0.6	3.5	3.4	4.7	6.9
Eq. (5.3.26)	0.6	3.8	4.1	8.3	9.8
Eq. (3.5.25)	0.6	6.25	7.3	16.3	26.8

Table 5.3. Computing times ms necessary for implementation of various decentralized control laws (INTEL 8086/8087 at 8 MHz)

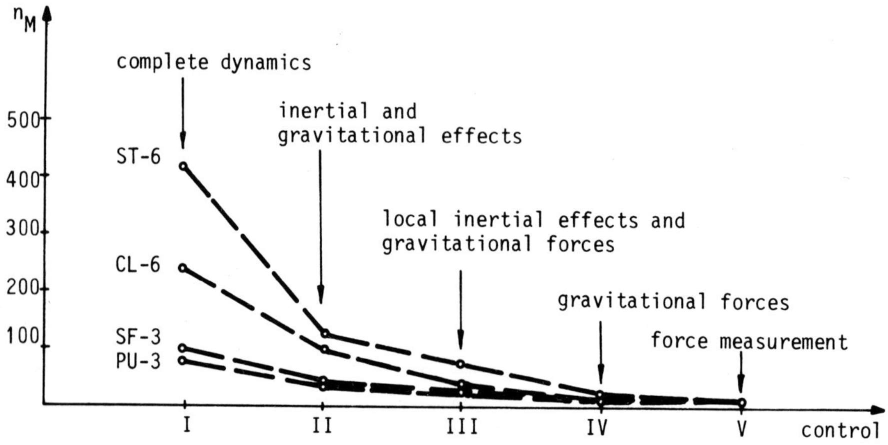

Fig. 5.9. Number of floating-point multiplications versus control law (I - Eq. (5.3.25), II - Eq. (5.3.26), III - Eq. (5.3.27), IV - Eq. (5.3.28) and V - Eq. (5.3.24))

Finally, let us point out the following results. It is shown that for the well-known Stanford arm, it is necessary to implement from 15 to 400 floating-point multiplications depending on the dynamic effects which are compensated. This corresponds to the range of 3.4 to 27 ms on a typical 16-bit microprocessor (for instance, INTEL 8086/8087). Having accepted 20 ms as a maximal sampling period for most robots, we obtain that it is necessary to use 2 processors in a parallel manner. However, if we neglect centrifugal and Coriolis effects, the computational time drops to 10 ms. If the exact nonlinear stability analysis justifies the neglect (as is often the case, except for high-precision/ /high-speed tracking tasks), then the controller may be realized by a single processor unit. Obviously, by the use of powerful microcomputers (which are 2 to 3 times faster), one obtains the possibility to achieve real time evan if entire dynamics is included.

5.3.3. Numerical complexity of adaptive control laws

We described in Paragraph 3.2.2 the centralized reference-model following control based on hyperstability theory. Now, we will consider its on-line applicability on low-cost microcomputers. The numerical complexity of the control structure shown in Fig. 3.3 can be obtained by adding the number of floating-point multiplications/additions for reference model, non-adaptive and adaptive control terms.

To evaluate the reference model (3.2.38):

$$S_M: \dot{x}^o = A_M x^o + B_M u^o$$

using the second-order subsystems, it is necessary

$$n_M(u^o) = 3n, \qquad n_A(u^o) = 2n \tag{5.3.29}$$

multiplications and additions/subtractions, respectively. For the computation of the non-adaptive control term (3.2.49):

$$u = [K_p(\xi, \bar{\theta}) + \mathcal{L}_p(\xi, \theta)]\xi + [K_u(\xi, \theta) + \mathcal{L}_u(\xi, e)]u^o$$

it is necessary to form the dynamic model matrices. Thus, it is convenient to apply the numeric-symbolic method, described in Chapter 1. But, the numbers of floating-point operations are obviously the same as those obtained in Paragraph 5.3.1. These numbers will be denoted by

$$n_M = n_M(u_{NA}) \quad \text{and} \quad n_A = n_A(u_{NA}) \tag{5.3.30}$$

where u_{NA} denotes the non-adaptive component:

$$u_{NA} = K_p(\xi, \bar{\theta}) + K_u(\xi, \bar{\theta})u^o. \tag{5.3.31}$$

Let us consider now the numerical complexity of the adaptive component

$$u_A = \mathcal{L}_p(\xi, v) + \mathcal{L}_u(\xi, v)u^o \tag{5.3.32}$$

using the adaptation mechanism (3.2.60). Taking

$$n_M(||v||^{-1}) = n, \qquad n_A(||v||^{-1}) = n-1$$

we get

$$\begin{aligned} n_M(\mathcal{L}_p) &= 2n^2 + n + 1, \qquad n_A(\mathcal{L}_p) = n-1 \\ n_M(\mathcal{L}_u) &= n^2 + n + 1, \qquad n_A(\mathcal{L}_u) = n-1 \end{aligned} \tag{5.3.33}$$

The overall number of operations is

$$\begin{aligned} n_M(\mathcal{L}_p, \mathcal{L}_u) &= 3n^2 + n + 3 \\ n_A(\mathcal{L}_p, \mathcal{L}_u) &= n. \end{aligned} \tag{5.3.34}$$

Notice that $||v||$ includes one square-root. Applying table look-up method, we see that it is necessary to perform at least one multiplication and one addition.

For the complete centralized control law we have

$$n_M(u) = n_M(u^o) + n_M(u_{NA}) + n_M(u_A)$$

$$n_A(u) = n_A(u^o) + n_A(u_{NA}) + n_A(u_A)$$

or

$$\begin{aligned} n_M(u) &= n_M(u_{NA}) + 3n^2 + 4n + 3 \\ n_A(u) &= n_A(u_{NA}) + 3n. \end{aligned} \tag{5.3.35}$$

The numbers of numerical operations for the robot structures shown in Figs. 5.5 - 5.8 are given in Table 5.4. We see that the considered adaptive control law is on-line applicable only for robots with up to 3 degrees of freedom. For a typical 6 degree-of-freedom robot (Stanford arm) it is necessary to use no less than 3 parallel processors.

Number of operations	ROBOT TYPE				
	CL-3	PU-3	SF-3	CL-6	ST-6
$n_M(u_{NA})$	45	91	97	403	582
$n_A(u_{NA})$	36	55	64	245	332
$n_M(u)$	87	133	139	538	717
$n_A(u)$	45	64	73	263	350

Table 5.4. Number of floating-point multiplications/additions for the computation of non-adaptive and adaptive control law

Let us now discuss the class of centralized indirect adaptive control. As described in Paragraph 3.2.3, the indirect control consists of two blocks: parameter estimation block and variable gain block. Using the numeric-symbolic method we shall determine the number of floating-point operations.

The control law (3.2.67):

$$u = D_a\ \phi(q,\ \dot{q},\ \ddot{q}^o + D_p(q-q^o) + D_v(\dot{q}-\dot{q}^o),\ \theta_k)\theta_d$$

is practically the same as the centralized nonadaptive control law (Paragraph 5.3.1). The number of floating-point multiplications and additions will be denoted as

$$n_M = n_M(u), \qquad n_A = n_A(u) \tag{5.3.36}$$

and is given in Table 5.1. The norm of the generalized error $||v||$ requires, at least,

$$n_M(||v||) = n, \quad n_A(||v||) = n-1. \tag{5.3.37}$$

The estimation algorithm (3.2.70)

$$\bar{\theta}_d(t_{k+1}) = \bar{\theta}_d(t_k) + \phi^T(t_k')[\phi(t_k')\phi^T(t_k')]^{-1}D_a^{-1}v(t_k')$$

includes the function ϕ, i.e. the dynamic robot model. Thus

$$n_M(\phi) = n_M(u), \qquad n_A(\phi) = n_A(u). \tag{5.3.38}$$

As $D_a^{-1}v$ can be included into the generalized error v, we shall assume that it does not increase the number of operaitons. Further,

$$n_M(\phi\phi^T) = n^2 n_d, \qquad n_A(\phi\phi^T) = n^2 n_d \tag{5.3.39}$$

where n_d is the number of dynamic parameters. The computation of $(\phi\phi^T)^{-1}v^*$ (with $v^* = D_a^{-1}v(t_k')$) requires

$$\begin{aligned} n_M((\cdot)^{-1}v^*) &= \frac{1}{2}(n^2-1)(n+2) \\ n_A((\cdot)^{-1}v^*) &= \frac{1}{2}(n^2-1)n, \end{aligned} \tag{5.3.40}$$

and $\phi^T[(\phi\phi)^{-1}v^*]$ requires

$$n_M(\phi^T(\cdot)) = n n_d, \qquad n_A(\phi^T(\cdot)) = n n_d \tag{5.3.41}$$

additional operations. Summing the expressions (5.3.38) - (5.3.41), we get

$$\begin{aligned} n_M(\bar{\theta}_d) &= n_M(u) + n(n+1)n_d + \frac{1}{2}(n^2-1)(n+2) \\ n_A(\bar{\theta}_d) &= n_A(u) + n(n+1)n_d + \frac{1}{2}(n^2-1)n \end{aligned} \tag{5.3.42}$$

From (5.3.36) and (5.3.37), we obtain

$$n_M = 2n_M(u) + n(n+1)n_d + \frac{1}{2}(n^3+2n^2+n-2)$$

$$n_A = 2n_A(u) + n(n+1)n_d + \frac{1}{2}(n^3+n-2)$$

The number of dynamic parameters n_d depends on the number of parameters which describe the dynamics of an isolated link. A link may be described by at least one parameter (mass or moment of inertia), and by at most 10 parameters (mass and 9 terms of inertia tensor). It is often enough to include 4 parameters (mass and the main moments of inertia), or even 3 parameters (mass, longitudinal and transversal moment of inertia). For the last case, we get

$$n_d = 3n. \tag{5.3.44}$$

Then, (5.3.43) becomes

$$\begin{aligned} n_M &= 2n_M(u) + \frac{1}{2}(7n^3+8n^2+n-2) \\ n_A &= 2n_A(u) + \frac{1}{2}(7n^3+6n^2+n-2) \end{aligned} \tag{5.3.45}$$

Taking $n_M(u)$ and $n_A(u)$ from Table 5.1, we obtain the number of floating-point multiplications/additions given in Table 5.5. We see that this control law can be implemented "on-line" for the simplest robot structures only.

	CL-3	PU-3	SF-3	CL-6	ST-6
n_M	221	325	462	1708	2066
n_A	194	232	250	1356	1530

Table 5.5. Number floating-point operations in centralized indirect adaptive control law

Finally, let us consider the numerical complexity of the decentralized indirect adaptive control, described in Paragraph 3.3.3. The local adaptive controllers

$$\Delta u_i^L = [D_{i(min)} + \frac{\Delta\theta_p^e}{|\Delta\theta_p|_{(max)}}(D_{i(max)} - D_{i(min)})](\xi^i - \xi^{io})$$

require

$$n_M(u) = 4n, \qquad n_A(u) = 6n \tag{5.3.46}$$

numerical operations. The number of operations for the estimator (3.3.34) equals

$$\begin{aligned} n_M(\Delta\theta_p^e) &= n_M(S_p) + 2n+1, \\ n_M(\Delta\theta_p^e) &= n_A(S_p) + 3n \end{aligned} \tag{5.3.47}$$

where $n_M(S_p)$ and $n_A(S_p)$ denote the number of operations necessary for the calculation of $S_p(\cdot)$. As shown in [13], we can write approximately

$$n_M(S_p) \approx n_M(P)/n \qquad \text{and} \qquad n_A(S_p) = n_A(P)/n;$$

thus

$$n_M(\Delta\theta^e) = \frac{n_M(P)}{n} + 2n+1$$
$$n_A(\Delta\theta^e) = \frac{n_A(P)}{n} + 3n \qquad (5.3.48)$$

where (P) indicates the computation of the complete system dynamics. The estimator (3.3.34), (3.3.36) requires

$$n_M = \frac{n_M(P)}{n} + 3n+1$$
$$n_A = \frac{n_A(P)}{n} + 3n. \qquad (5.3.49)$$

These results are also presented in Table 5.6. The estimation with and without force measurement for several typical industrial robots is analyzed. This table includes the computation of sensitivity functions when all of the dynamic effects are incorporated, and the case when the centrifugal and Coriolis effects are neglected. The results show that the number of multiplications is smaller than 150 in all cases. Thus, the real time is easily achievable with our reference microcomputer.

Effects included	M/A	CL-3	PU-3	SF-3	CL-6	ST-6
Inertial, Coriolis gravitational (3.3.35)	n_M	24	41	46	103	130
	n_A	31	40	45	87	96
Inertial gravitational (3.3.35)	n_M	23	28	29	57	59
	n_A	30	32	34	66	97
Inertial, Coriolis gravitational (3.3.36)	n_M	27	44	49	109	136
	n_A	31	40	45	87	96
Inertial gravitational (3.3.36)	n_M	26	31	32	63	66
	n_A	30	32	34	66	97

Table 5.6. Number of floating-point multiplications/additions in the decentralized indirect adaptive control law

5.4 Parameter Identification

In this paragraph we shall present some practical results devoted to the identification of parameters of robotic actuators. Actually, the idea is to estimate the parameters of robot subsystems, which include not only actuators but also reducers and transmissions. Such driving units are supposed to be integrated with robotic arms. Thus, we can measure joint coordinates and velocities and actuator inputs. In order to estimate the parameters of the *i*th subsystem, we shall suppose that we excite only the *i*th actuator input, producing the motion of the *i*th robot degree of freedom. Further, we shall suppose that the robot subsystems include local analog servoloops. The introduction of analog servoloops is convinient from the standpoint of identification process convergence. However, once the parameters of actuators together with analog servoloops are identified, it is simple to determine parameters of actuators without servoloops if we want to apply direct digital control.

Mathematical model of a servosystem can usually be presented as (see Paragraph 2.4)

$$u^i = D^i(y^i - y^{io}) \tag{5.4.1}$$

where $y^i \in R^{m_i}$ represents the output vector of subsystem S^i, $y^{io} \in R^{m_i}$ is the input vector for the servosystem, and $D^i \in R^{1\times m_i}$ is a gain matrix. The input y^{io} will be denoted by $\bar{u}^i$. Thus, the input vector of the system becomes $\bar{u} = [\bar{u}^{1T} \cdots \bar{u}^{nT}]^T$. For single-input servosystems we can present y^{io} as $y^{io} = [v_i \; 0 \cdots 0]^T$. The corresponding output vector is $y^i = [q_i \,|\, y_r^{iT}]^T$. Now, the model (5.4.1) becomes

$$u^i = [d_i \,|\, d_r^{iT}] \begin{bmatrix} q_i - v_i \\ \hline y_r^i \end{bmatrix} \tag{5.4.2}$$

where $d_i \in R$ is the position gain, and $d_r^i \in R^{m_i - 1}$ is the gain vector corresponding to the remaining coordinates y_r^i. Thus, we obtain

$$u^i = d_i(q_i - v_i) + d_r^{iT} y_r^i. \tag{5.4.3}$$

We see that the actuator input u^i and subsystem input v_i are related by (5.4.3). In the case when there are no analog servoregulators, we have $u^i = v_i$.

Let us introduce the model of mechanical part of the robotic system (i.e. the model of the ith joint dynamics) into the model of actuator. This issue has been already considered in Paragraph 2.4. However, for the sake clarity let us briefly repeat this consideration.

Let us consider now the model of the mechanical arm, when only the ith joint is moved. Starting from the model

$$P = H(q, d)\ddot{q} + \dot{q}^T C(q, d)\dot{q} + g(q, d)$$

we can extract the differential equation describing the dynamics of the ith degree of freedom:

$$P_i = H_r^i(q, d)\ddot{q} + \dot{q}^T C^i(q, d)\dot{q} + g_i(q, d) \tag{5.4.4}$$

where $H_r^i(q, d)$ represents the ith row of the inertial matrix $H(q, d)$, $C^i(q, d)$ is an n×n matrix of Coriolis and centrifugal effects, and $g_i(q, d)$ is the gravitational vector assigned to the ith degree of freedom. When changing only the ith joint coordinate, the joint velocity vector becomes $\dot{q} = [0 \cdots 0\ \dot{q}_i\ 0 \cdots 0]^T$, and the joint acceleration vector becomes $\ddot{q} = [0 \cdots 0\ \ddot{q}_i\ 0 \cdots 0]^T$. Model (5.4.4) reduces now to

$$P_i = H_{ii}(q, d)\ddot{q}_i + C_{ii}^i(q, d)\dot{q}_i^2 + g_i(q, d) \tag{5.4.5}$$

where $H_{ii}(q, d)$ represents the (i, i) element of inertial matrix, and $C_{ii}^i(q, d)$ is the ith diagonal element of $C^i(q, d)$. But, from Chapter 1 we know that

$$C_{ii}^i(q, d) = 0$$

for $\forall i \in \{1,\ldots,n\}$. Thus, centrifugal and Coriolis effects do not influence the system dynamics when only one joint coordinate is changed.

In the case when q_j = const for $j \neq i$, Eq. (6.2.16) becomes

$$P_i = H_{ii}(q_i, d)\ddot{q}_i + g_i(q_i, d) \tag{5.4.6}$$

It is obvious that H_{ii} is actually independent of q_i, i.e.

$$H_{ii}(q_i, d) = \bar{H}_{ii} = \text{const.}$$

Thus, we obtain

$$P_i = \bar{H}_{ii}\ddot{q}_i + g_i \tag{5.4.7}$$

Now, let us join together models (5.4.7) and (2.4.2). For this purpose, let us introduce $1\times n_i$-dimensional transformation matrix T^i which maps $\dot{x}^i$ into $\ddot{q}_i$:

$$\ddot{q}_i = T^i\dot{x}^i.$$

Now, the entire model becomes*)

$$\dot{x}^i = A^i x^i + b^i u^i + \bar{H}_{ii} f^i T^i \dot{x}^i + f^i g_i \tag{5.4.8}$$

Notice that $f^i T^i$ is a 3 3 matrix. It should be pointed out that g_i depends on q_i, i.e. x^i. We can use a linear approximation of $f^i g_i$ (see (2.4.5)) and take $f^i g_i = f^i \bar{h}^i \hat{T}^i x^i$ and joint $f^i \bar{h}^i \hat{T}^i$ to A^i. On the other hand, the gravity term g_i can also be approximated by a constant, when the estimation is performed about a single point in the work space. It follows now that

$$\dot{x}^i = W^i A^i x^i + W^i b^i u^i + W^i f^i g_i \tag{5.4.9}$$

where $W^i \in R^{n_i \times n_i}$ is given by

$$W^i = (I - \bar{H}_{ii} f^i T^i)^{-1} \tag{5.4.10}$$

Let us include now the local analog servosystems (5.4.3). We suppose that $q_i \in x^i$ and $y_r^i \in x^i$. Then, we can determine a vector $d^i \in R^{n_i}$ such that

$$d_i q_i + d_r^{iT} y_r^i \equiv d^i x^i$$

where d_i and d_r^i are defined by (5.4.3). Substituting into (5.4.3), we obtain the servosystem model

$$u^i = d^i x^i - d_i v_i. \tag{5.4.11}$$

Including (5.4.11) into (5.4.9), we get

$$\dot{x}^i = W^i(A^i + b^i d^i)x^i - W^i b^i d_i v_i + W^i f^i g_i \tag{5.4.12}$$

---

*) We assume that the load acting upon the actuator is equal to the ith joint driving torque, $M_i = P_i$.

Thus, we obtain the model (5.4.12) when the system includes analog feedback gains, and (5.4.9) when the system includes only actuators and transmission. Both models encompass the inertial effects of manipulator arm. These models incorporate some parameters which can be treated as well known, and some which are not exactly known. For example, in the case of DC motor drives the following parameters are exactly known: inductance and resistance of rotor coil, torque gain and back E.M.F. constant, etc. On the other hand, viscous and static frictions are temperature dependent and variable. Translating (5.4.9) and (5.4.12) into discrete time domain and scalar form, we obtain

$$\dot{q}_i(k) = \alpha_i(T, \theta^i)\dot{q}_i(k-1) + \beta_i(T, \theta^i)v_i + e_i(k) \tag{5.4.13}$$

and

$$\dot{q}_i(k) = \alpha_i(T, \theta^i)\dot{q}_i(k-1) + \beta_i(T, \theta^i)\cdot \cdot (v_i - q_i(k-1)) + e_i(k) \tag{5.4.14}$$

Here, $e_i(k)$ represents the error due to finite sampling time, coupling influence, etc. The coefficients α_i and β_i should be estimated using a method for parameter identification [14, 15, 16]. The estimation procedure should be simple enough to be implementable on microcomputers in real time. In the text to follow we will use the least squares method. The results will be obtained on a real robot, i.e. without using a simulation technique.

Let us consider the evaluation of the system (5.4.13) and (5.4.14) over N sampling intervals (N>>2) and denote the observation vector by $Y_N = [\dot{q}_i(k+1) \cdots \dot{q}_i(k+N)]^T$, and the estimated vector of parameters according to N measurements as $\hat{\theta}_N = [\alpha_i^N \ \beta_i^N]$. The estimation after N+1 measurements $\hat{\theta}_{N+1}$ can be obtained using the estimates $\hat{\theta}_N$ and a term proportional to the difference between the prediction

$$x_{N+1}^T\hat{\theta}_N = [\dot{q}_i(k+N) \quad u^i(k+N)]^T\hat{\theta}_N$$

in case of model (5.4.13), or

$$x_{N+1}^T\hat{\theta}_N = [\dot{q}_i(k+1) \quad -q_i(k+N) + v_i(k+N)]^T\hat{\theta}_N$$

for the model (5.4.14), and the measured output $\dot{q}(k+N+1)$:

$$\hat{\theta}_{N+1} = \hat{\theta}_N + K_N(\dot{q}(k+N+1) - X^T_{N+1}\hat{\theta}_N) \qquad (5.4.15)$$

The matrix K_N according to the least squares method is

$$K_N = P_N X_{N+1}(1+X_{N+1}P_N X_{N+1})^{-1} \qquad (5.4.16)$$

with

$$P_{N+1} = P_N - P_N X_{N+1}(1+X^T_{N+1}P_N X_{N+1})^{-1}X^T_{N+1}P_N \qquad (5.4.17)$$

where P_N is a square matrix of dimension corresponding to the numbers of unknown parameters. Let us consider the following example:

Estimation of cylindrical robot parameters

Let us consider an example with a 3-degree-of-freedom cylindrical robot, shown in Fig. 5.5. This robot is powered by DC motors. The rotor shafts are connected to reducers producing either revolute or sliding motion at the outputs. For example, joint 3 is a sliding one (see Fig. 5.10). The actuators are controlled by servosystems which can be modelled by (5.4.11). Here, the state vector of the *i*th subsystem is $x^i = [q_i \; \dot{q}_i]^T$. The servosystems include position and velocity feedback loops. The gain vector can be presented as $d^i = [-a_{pi} \; a_{vi}]$, and $d_i = - a_{pi}$. It holds

$$u^i = -a_{pi}(q_i - v_i) + a_{vi}\dot{q}_i \qquad (5.4.18)$$

The measured input-output diagrams of the realized servoregulators are presented in Fig. 5.11. Position gains a_{pi} and velocity gains a_{vi} are thus obtained.

Control of the cylindrical robot is realized by an experimental microcomputer system. Thus, the measurements of robot coordinates and velocities are realized by using A/D converters. The estimation algorithm (5.4.15) - (5.4.17) is implemented in a high-level programming language. First, the estimation of parameters of the subsystem $S^i(i=1)$ is realized. The control $v_i(t)$, joint coordinate $q_i(t)$ and velocity $\dot{q}_i(t)$ during $t\in[0, 5.8s]$ are presented in Fig. 5.12. These data are obtained by measurements combined with computer graphics. The estimation process is illustrated in Fig. 5.13. The curves are normalized to values printed over them. For example, the joint coordinate shown in Fig. 5.12 is normalized to 0.5414 rad. The time is normalized to 5.8 s. We obtain

$\alpha_1 = 0.82$ and $\beta_1 = 0.40$ from Fig. 5.13. The estimation process of parameter β_i during the first 2 seconds is presented in Fig. 5.14. We see

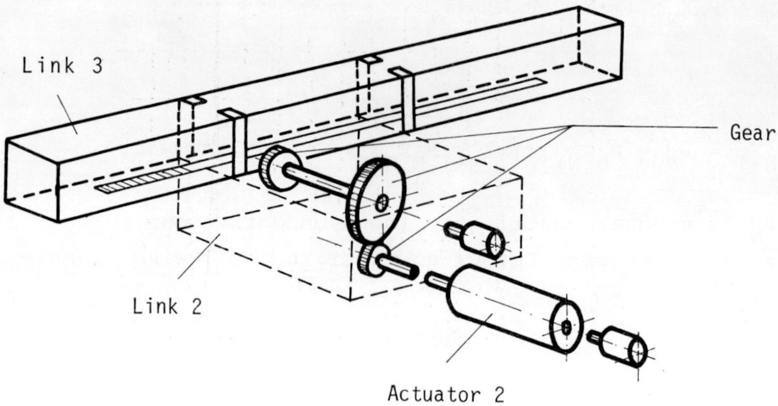

Fig. 5.10. Motor and reducer between the links 2 and 3

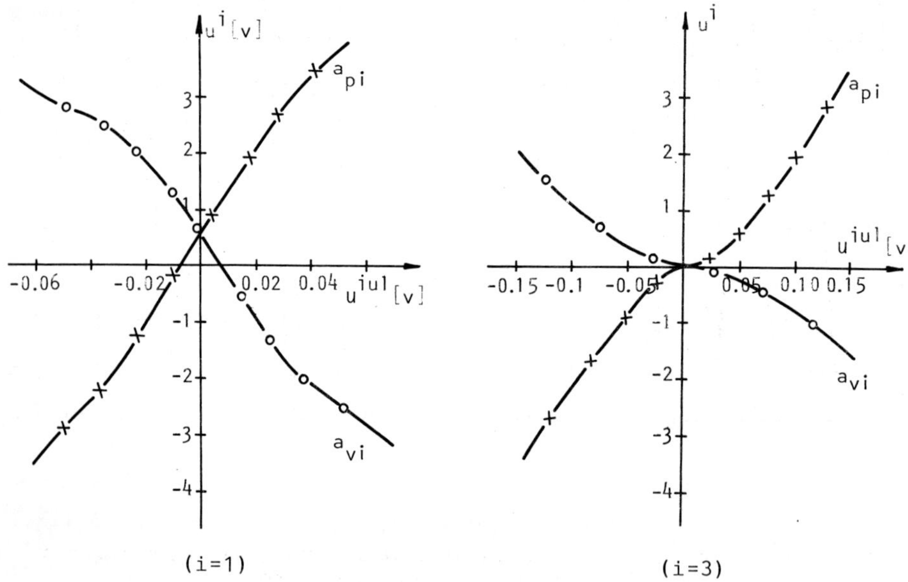

Fig. 5.11. Input-output characteristics of servosystems S^1 and S^3 (a_{pi} is obtained under the conditions $q_i=0$, $\dot{q}_i=0$ and $u^i=v_i$, and a_{vi} under the conditions $q_i=v_i=0$ and $u^i=\dot{q}_i$)

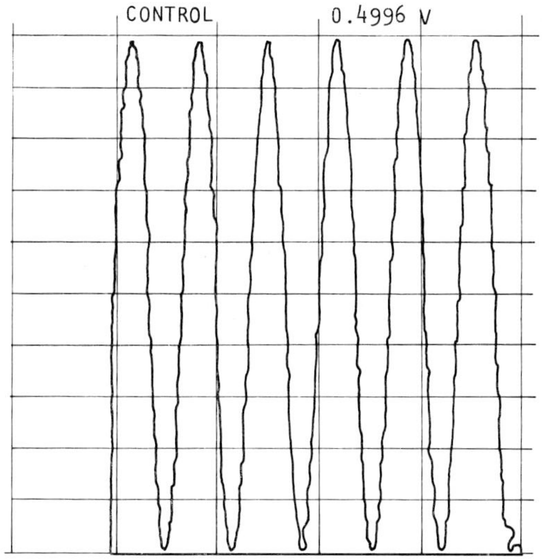

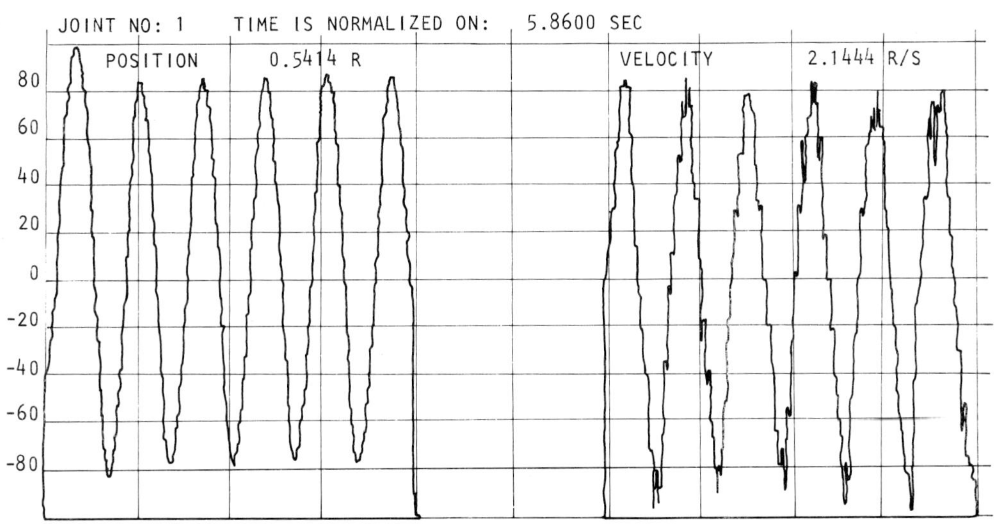

Fig. 5.12. Control $v_i(t)$, joint coordinate $q_i(t)$ and velocity $\dot{q}_i(t)$ for the first degree of freedom

that the estimation process converges after a few seconds. During this time only the first joint coordinate is changed. The dynamic model (5.4.7) now becomes $P_1 = H_{11}\ddot{q}_1$. Combining this model with the servo-system model and the model of the DC - motor/reducer, we obtain that $\bar{a}_i = a_i + a_{vi}b_i$ and $\bar{b}_i = a_{pi}b_i$ are of the following forms:

$$\bar{a}_i = (1-\alpha_i)(1+f_iH_{ii})/T, \quad \bar{b}_i = \beta_i(1+f_iH_{ii})/T$$

where a_i and b_i are the parameters of subsystem models without servo-systems, T is a sampling interval. The parameter $f_i = 1/J_{ri}N_i^2$ can be considered as known because the rotor moment of inertia and the reduction ratio N_i are known. For the first degree of freedom we have $J_{ri} = 0.000056\ kgm^2$, $N_i = 31.17$ and $H_{ii} = 0.76\ kgm^2$. Thus, we obtain $\bar{a}_i = 134.7$ and $\bar{b}_i = 299.4$. Using the gain coefficients from Fig. 5.11, we obtain the parameters of subsystems without analog servoloops.

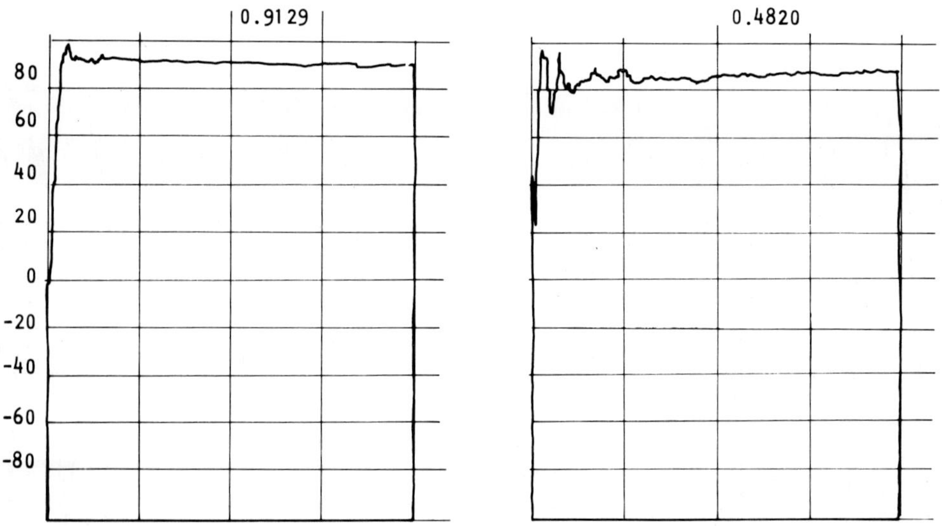

Fig. 5.13. Estimation process of α_1 and β_1

Finally, we will present the results of the estimation of parameters of subsystem 3. The input is excited by triangular voltage signals (2 Hz), as shown in Fig. 5.15. Joint coordinate q_3 and velocity $\dot{q}_3$ are presented in Fig. 5.15. The estimates α_3 and β_3 given in Fig. 5.17.

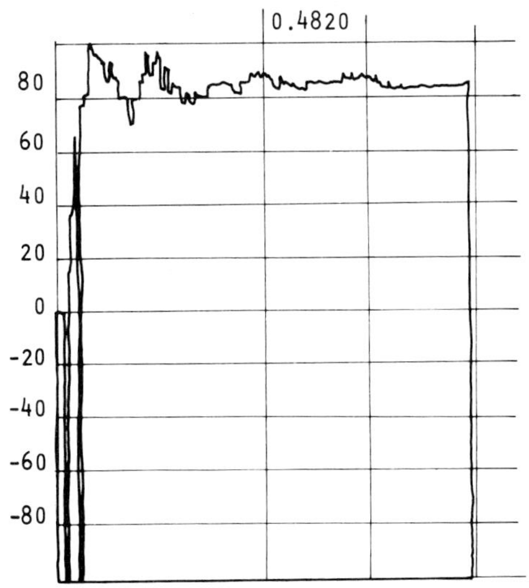

Fig. 5.14. Estimation of β_1 during t=2 s

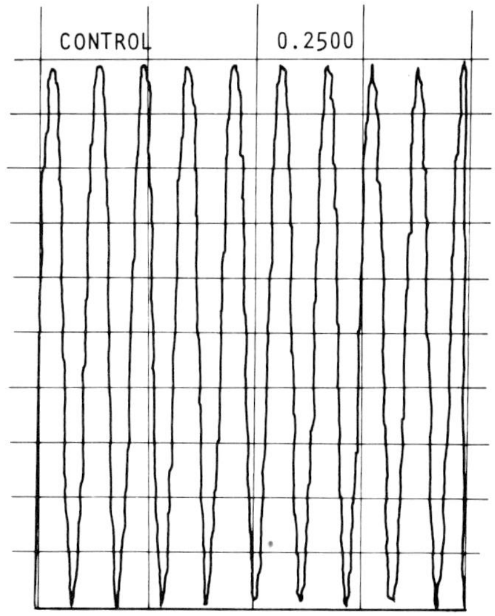

Fig. 5.15. Control $v_3(t)$

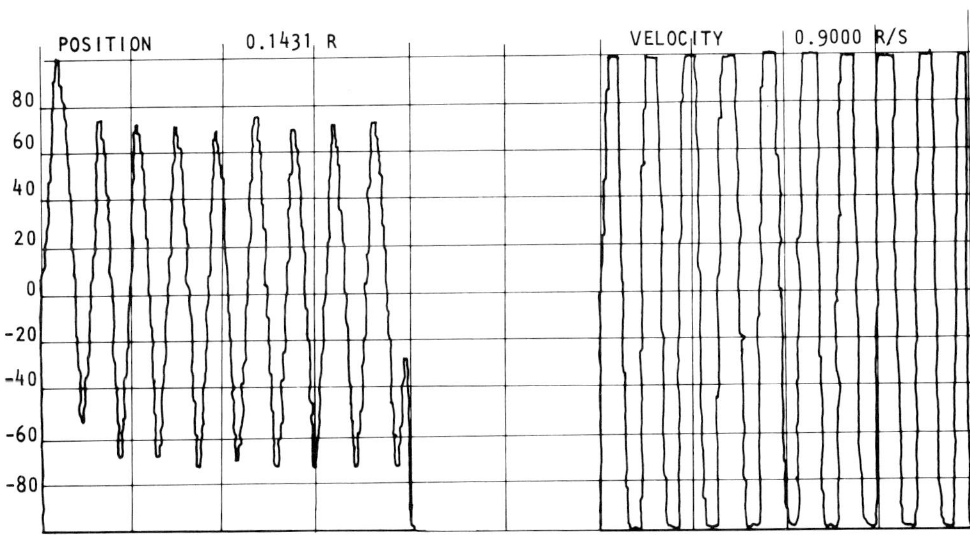

Fig. 5.16. Joint coordinate $q_3(t)$ and velocity $\dot{q}_3(t)$

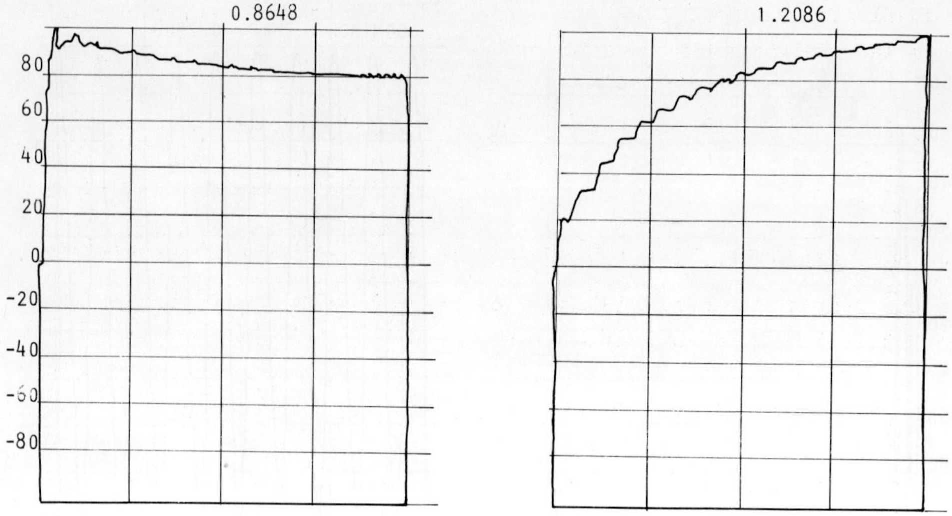

Fig. 5.17. Estimation of α_3 and β_3

5.5 Microprocessor Implementation of Non-Adaptive Control Laws

In this paragraph we shall present the implementation of non-adaptive control for a particular robotic system. The synthesis of non-adaptive control has been explained in Chapters 2 and 4. In Paragraph 5.3 we have considered the numerical complexity of various non-adaptive control laws (for various manipulation structures). This numerical complexity determines the microprocessor equipment that has to be applied in order to implement a corresponding control law. We shall present the experimental results of control implementation by various techniques: analog, microprocessor (direct digital control) and hybrid to permit a comparison to be made.

It should be remembered that the microprocessor implementation of control requires discrete time models of the system, i.e. the model that takes into account the sampling period required by microprocessors for control computation (Paragraph 4.2.11). The control parameters should be synthesized taking into account the sampling period. These problems have already been considered in [17]. Here, we shall briefly list various control forms in the discrete time domain.

The model of actuators (2.4.3) in the discrete time domain can be written as (assuming that $M_i = P_i$):

$$S^i: \quad x^i(k+1) = A_D^i x^i(k) + b_D^i N(u^i(k)) + f_D^i P_i(k), \quad \forall i \in I$$

$$k=0,1,2,\ldots \qquad (5.5.1)$$

where $A_D^i = \exp(\bar{A}^i T)$, $b_D^i = \int_0^T \bar{b}^i e^{\bar{A}^i \tau} d\tau$, $f_D^i = \int_0^T \bar{f}^i e^{\bar{A}^i \tau} d\tau$, $x^i(k) = x^i(kT)$, T is the sampling period and the model of the mechanical part of the robot (2.2.1) can be approximated by:

$$P_i(k) = H_i(q(k))\ddot{q}(k) + h_i(q(k), \dot{q}(k)), \ \forall i \in I \quad k=0,1,2,\ldots \qquad (5.5.2)$$

The microprocessor (kinematic module) computes the nominal trajectories $q^o(k)$, $k=0,1,2,\ldots,L$ of the joint angles (displacements) in order to perform the motion specified by the operator. The task of the control system is to ensure tracking of the imposed nominal trajectories.

As already explained in previous chapters, the control is taken in one of the following forms [18, 19]:

- nominal centralized control and local control

$$u^i(k) = u^{io}(k) + \Delta u_i^L(k) \qquad (5.5.3)$$

- local (decentralized) nominal control and local control

$$u^i(k) = \bar{u}^{io}(k) + \Delta u_i^L(k) \qquad (5.5.4)$$

- local (decentralized) nominal control and local control and global control

$$u^i(k) = \bar{u}^{io}(k) + \Delta u_i^L(k) + \Delta u_i^G(k) \qquad (5.5.5)$$

The nominal centralized programmed control is synthesized to satisfy:

$$x^{io}(k+1) = A_D^i x^{io}(k) + b_D^i u^{io}(k) + f_D^i P_i^o(k), \ \forall i \in I \quad k=0,1,\ldots,L \qquad (5.5.6)$$

where $x^{io}(k) = (q_i^o(k), \ \dot{q}_i^o(k))^T$; and P_i^o is the nominal driving torque acting upon the ith d.o.f., which is calculated from:

$$P_i^o(k) = H_i(q^o(k))\ddot{q}^o(k) + h_i(q^o(k), \dot{q}^o(k)), \ \forall i \in I, \ k=0,1,\ldots,L \tag{5.5.7}$$

The control satisfying (5.5.6) can be calculated using the minimum inverse of vector b_D^i [20]:

$$u^{io}(k) = (b_D^{iT} b_D^i)^{-1} b_D^{iT}(x^{io}(k+1) - A_D^i x^{io}(k) - f_D^i P_i^o(k)), \quad \forall i \in I \quad k=0,1,2,\ldots,L \tag{5.5.8}$$

The local control (2.4.23) in the discrete time domain can be taken as:

$$\Delta u_i^L(k) = \hat{D}_D^{iT} \Delta x^i(k) + \hat{Q}_D^i \sum_{j=0}^{k} \Delta x^i(j), \ \forall i \in I \tag{5.5.9}$$

where $\hat{D}_D^i \in R^{1 \times n_i}$ is the feedback gain vector and $\hat{Q}_D^i$ is the integral feedback gain, synthesized in the discrete time domain.

The local nominal control is synthesized to satisfy:

$$x^{io}(k+1) = A_D^i x^{io}(k) + b_D^i \bar{u}^{io}(k), \quad \forall i \in I, \quad k=0,1,2,\ldots,L \tag{5.5.10}$$

Similarly to (5.5.8), we can obtain the local nominal control as:

$$\bar{u}^{io}(k) = (b_D^{iT} b_D^i)^{-1} b_D^{iT}(x^{io}(k+1) - A_D^i x^{io}(k)) \tag{5.5.11}$$

The global control ((2.6.1) when the local nominal control is applied) can be taken as:

$$\Delta u_i^G(k) = -K_i^G(x^i(k))\tilde{P}_i(k) \tag{5.5.12}$$

where $K_i^G: R^{n_i} \to R^1$ is the global gain, $\tilde{P}_i(k)$ is the on-line computed driving torque around the ith joint. On-line computation of the driving torques using various approximate models (and local control) has been considered in Paragraph 5.3.

In addition to these control laws (which have already been discussed in previous chapters), we shall consider the so-called on-line computation of nominal control in the form:

$$u^{io}(k, x(k)) = (b_D^{iT} b_D^i)^{-1} b_D^{iT}(x^{io}(k+1) - A_D^i x^i(k) - f_D^i \tilde{P}_i(x(k)) \tag{5.5.13}$$

In the text to follow we shall describe microprocessor implementation of the dynamic control for the manipulator UMS-2. We shall present the results of control synthesis which have been obtained using our software package. We shall also present realizations of tracking the nominal trajectories using various control laws and the three above mentioned versions of control implementation: (Ia) when servosystems are implemented by analog techniques, (Ib) when microprocessors are added to improve tracking, and (Ic) when only microprocessors are used for the control implementation.

The manipulation robot UMS-2 is shown in Fig. 5.5. It has five degrees of freedom: two linear and three rotational*). Parameters of the manipulator are presented in Table 5.7 (lengths, masses and moments of inertia of the links). D.C. motors are used as actuators and their paramaters are given in Table 5.8. These parameters have been identified as described in Paragraph 5.4.

The dynamic model of this manipulator is extremely simple so it has been easy to achieve on-line calculation of dynamics. We have used "DIGITAL" microprocessors of PDP 11-03 type. We have found out that it is possible to implement control by two microprocessors only, because of the simplicity of the dynamic model. The block scheme of the three versions of control implementation is presented in Fig. 5.18 [21].

The first microprocessor is used for the synthesis of nominal trajectories and communication with the operator or with higher control levels. In the latter case additional microprocessors or a computer have to be used for the implementation of decision-making levels. The first microprocessor calculates the nominal trajectories in accordance with the operator's requirements. It can calculate nominal trajectories either on-line or off-line and memorize them. It sends nominal angles and velocities at each sampling interval to the second microprocessor.

The executive control level is implemented by the second microprocessor. This microprocessor calculates the nominal programmed control either on-line or off-line using the nominal coordinates and velocities obtained from the first microprocessor. It also calculates local control (5.5.9) and it may calculate global control (5.5.12) by on-line calculation of the driving torques and forces.

We shall not discuss in detail the problems of programming of both

*) Actually, the manipulator of Fig. 5.5 has six d.o.f. but the 6th d.o.f. was locked in our experiments.

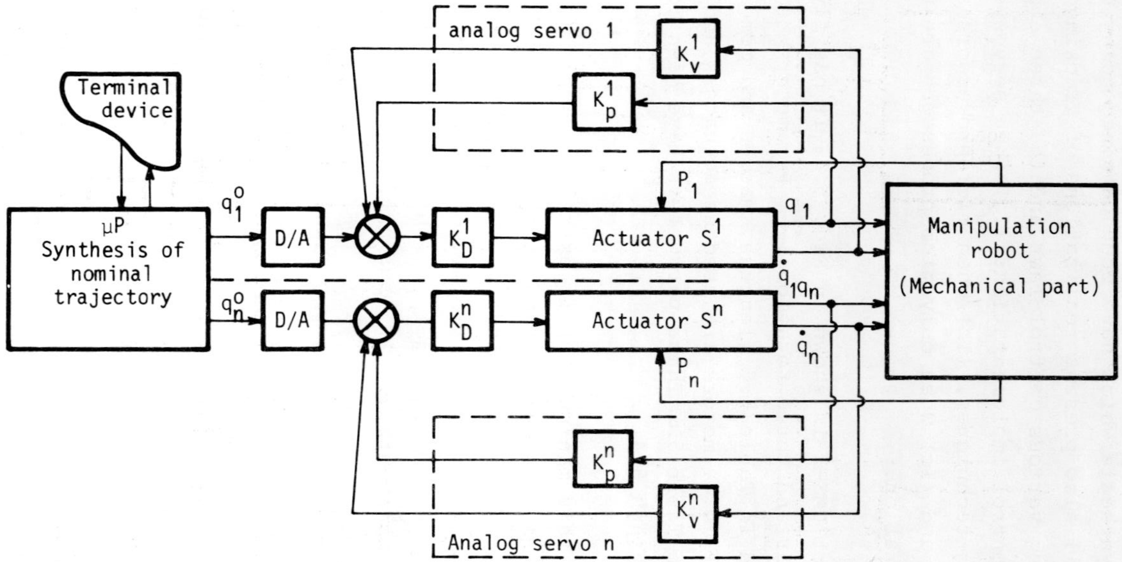

Fig. 5.18. Control scheme (a) analog version

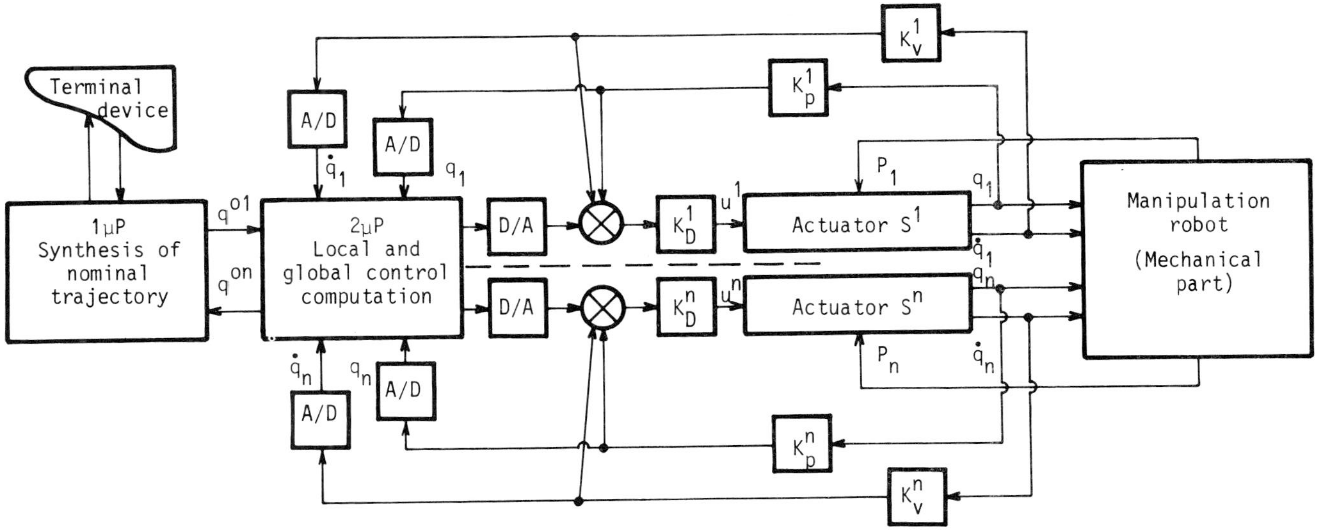

Fig. 5.18(b) Hybrid version

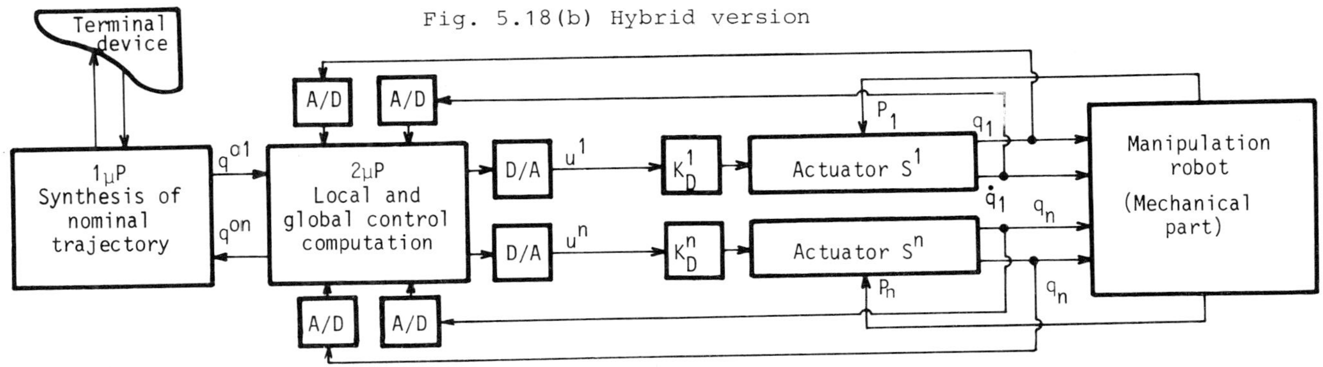

Fig. 5.18(c) Direct digital control

microprocessors. Using our software package (Chapter 4) we synthesize local and global gains for various forms of control implementation. We first synthesize analog gains to stabilize each servosystem (5.4.1). The gains are given in Table 5.9 (Ia). This is the case when analog servomechanisms (5.4.1) are implemented (as presented in Fig. 5.18a). If we want to improve the tracking of nominal trajectories we can add local discrete servoloops (5.5.9) implemented by the second microprocessor and we can add nominal programmed control calculated off-line (5.5.8). The additional discrete feedback gains are also synthesized using our software package and are given in Table 5.10 (Ib). We have chosen the sampling period T = 0.02 s which enables on-line calculation of local and global control (if it is necessary).

Links	1	2	3	4,5
Mass	10.0	7.0	4.15	0.5
Lengths	0.38	0.02	0.45	0.05
J_{xi} (kgm^2)	-	-	-	0.001
J_{yi} (kgm^2)	-	-	-	0.001
J_{zi} (kgm^2)	0.029	0.055	0.318	0.0015

Table 5.7. Parameters of the mechanism

i	$J^i[kgm^2]$	$C_E^i[V/rad/s]$	$C_M^i[Nm/mA]$	$R[\Omega]$	$F_D^i[Nm/rad/s]$
1	0.03	1.43	1.50	1.6	0.006
2	205.3	120.3	125.4	1.6	40.3
3	74.	72.2	75.5	1.6	1.
4,5	0.0325	0.325	1.36	3.32	2.86

Table 5.8. Parameters of the actuators (J^i - rotor moment of inertia; C_E^i - electromotor coefficient; C_M^i - moment coefficient, R^i - resistance of the rotor circuit, F_D^i - viscous friction coefficient; gear speed and moment ratios are taken into account)

As we have already stated, in this specific case the mathematical model of manipulator dynamics is very simple, so we can achieve on-line calculation of the driving torques by one microprocessor only. Actually, we have calculated on-line driving torques only for the first three degrees of freedom of the manipulator (minimal configuration of the manipulator), since the influence of the gripper dynamics is negligible. So we have introduced global control in the first three degrees of freedom of the manipulator. We have calculated global gains using our software package. The dynamic model of this manipulator is given in Appendix 2.B. The number of multiplications and additions which have to be realized for local and global control is given in Paragraph 5.3.

	1	2	3	4,5
Positional gain [V/rad]	975.	15000.	3037.	50.
Speed gain [V/rad/s]	59.	934.	462.	-

Table 5.9. Analog feedback gains

	1	2	3	4
Positional gain	7.1	3.47	12.2	1.3
Speed gain	0.086	0.064	0.18	-

Table 5.10. Local discrete gain for T = 0.02 s with analog feedback

Fig. 5.19 shows experimental results of tracking the nominal trajectories with various control laws. In order to compare various control laws, we have implemented trackings of the same nominal trajectory with various laws under equal conditions. We have made initial condition errors in the first and third degree of freedom of the manipulator, i.e. we have started the tracking of the nominal trajectories from the initial position of the manipulator which deviates from the nominal initial position by $\Delta q_1(0) = -0.1$ rad and $\Delta q_3(0) = -0.1$ m, where Δq_i denotes deviation of the position (angle) q_i of the ith degree of freedom from its nominal value q_i^o ($\Delta q_i(t) = q_i(t) - q_i^o(t)$). In the remaining degrees of freedom tracking starts from the nominal positions.

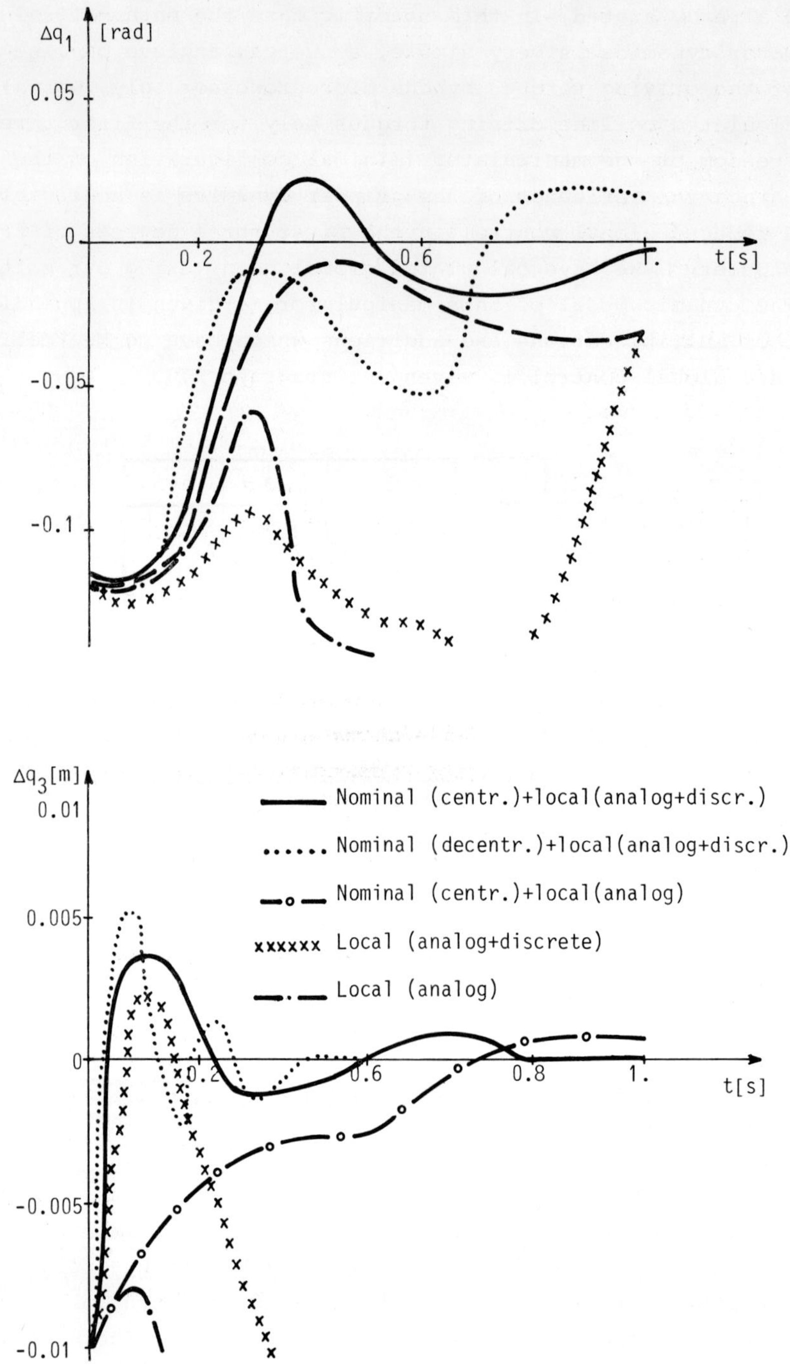

Fig. 5.19. Implementation of tracking of trajectories with discrete and analog control - experimental results

First, the tracking has been realized by analog servosystems only, when the microprocessor sends directly nominal trajectories $q_i^o(k)$ through D/A converters. From Fig. 5.19 it is obvious that the tracking is very poor. In the figures the tracking for the first and third degrees of freedom only is presented, since these two degrees of freedom are coupled. The tracking is very poor since servosystems are not compensated for either dynamics of actuators and mechanism or acceleration along the nominal trajectory.

Then, we have considered the tracking when nominal programmed control $u^{oi}(k)$ is the output from the microprocessor. In this case the computer calculates off-line nominal programmed control and sends it through D/A converters. Obviously the tracking is much better (see Fig. 5.19). In order to improve the tracking we have introduced additional discrete local feedback loops (Fig. 5.18b). In this case, the second microprocessor on-line calculates local control $\Delta u_i^L(k)$, adds it to the nominal control $u^{oi}(k)$ and sends it through D/A converters. The results of tracking are also shown in Fig. 5.19.

We have also considered tracking with decentralized nominal control $\bar{u}^{oi}(k)$ (5.5.11), when the nominal control is calculated using the approximate decoupled model of the manipulator. The results are also presented in Fig. 5.19. We also present the results of tracking when only local control (analog (5.4.1) and discrete (5.5.9)) is implemented.

The next step in the implementation of various control laws has been the omission of the analog servomechanism (Fig. 5.18c). We have implemented discrete controllers (Ic) only to track the same nominal trajectories under the same conditions as before. We have chosen the same sampling period T = 0.02 s. Now, discrete local gains are given in Table 5.11 (when analog servomechanisms are omitted). The results of tracking the nominal trajectory are given in Fig. 5.20. We have implemented both on-line (5.5.13) and off-line (5.5.8) calculation of nominal programmed control. We have realized the tracking with off-line calculated programmed control u^{oi} with local discrete dontrol (5.5.9). Then, we have realized the tracking with nominal control plus local plus global control. Obviously the robot performance is much better in the second case. However, a similar performance can be achieved by on-line calculation of the nominal control (5.5.13). In this case the microprocessor again calculates on-line driving torques in the manipulator joints and then using these torques, the control which should lead the mani-

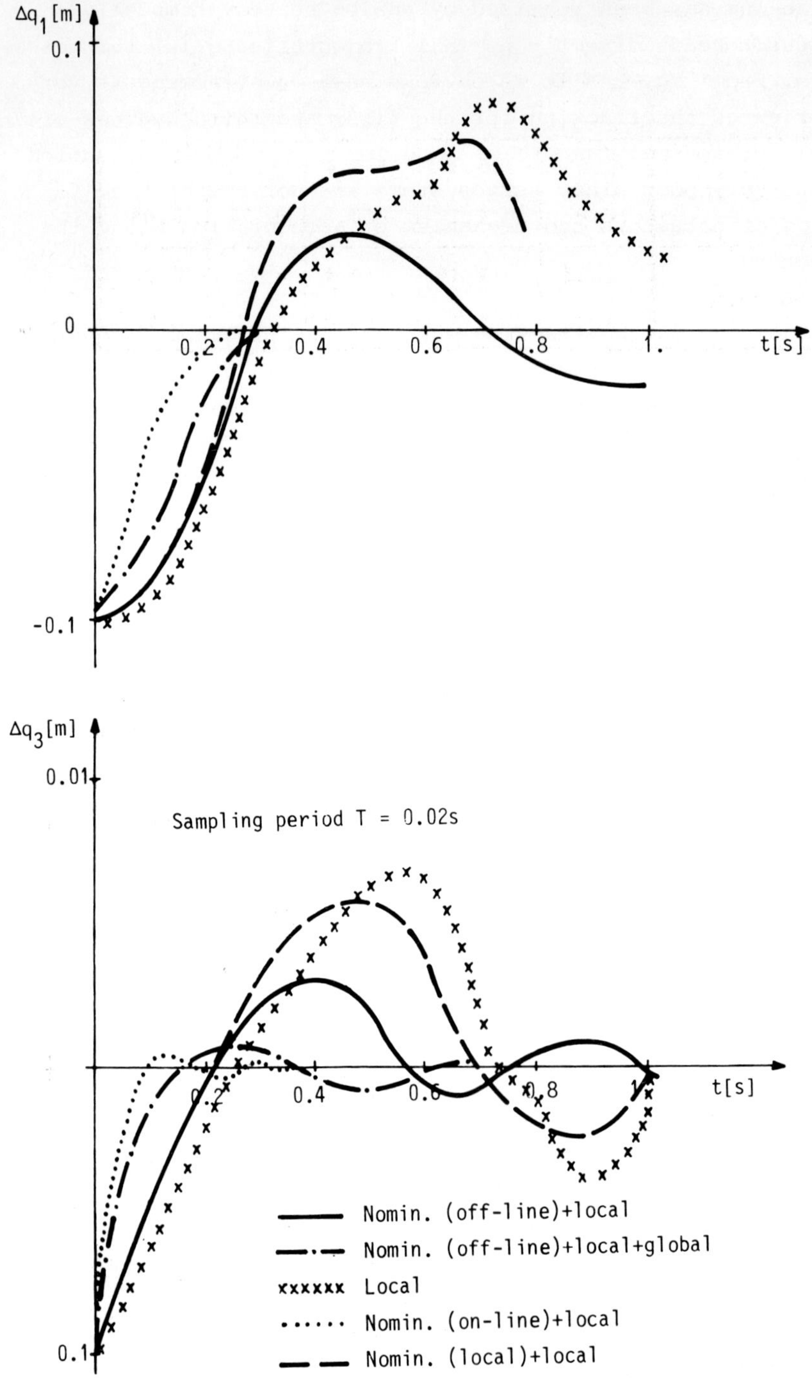

Fig. 5.20. Implementation of tracking of trajectory with discrete control - experimental results

pulator along the nominal trajectory.

	1	2	3	4, 5	T_D[s]
Positional gain	20.07	356.3	355.4	1.3	0.02
	10.0	4.1	100.	0.37	0.06
Speed gain	1.53	7.49	10.4	-	0.02
	4.03	0.28	10.0	-	0.06

Table 5.11. Local discrete gains (when analog servomechanisms are omitted)

We have also considered the case when the nominal control is synthesized using the decentralized model of the robot (3.5.11). If the third degree of freedom is fixed, the tracking of the nominal trajectory for first degree of freedom is the same as in the centralized case (Fig. 5.20). However, if the third degree of freedom is also moving, the tracking is rather poor, since this decentralized control does not take care about the coupling which exists between the first and the third degree of freedom. The results of tracking when only local discrete control (5.5.9) is implemented are also presented. The tracking is poor.

In Fig. 5.21 control signals that are sent through D/A converters for various control laws are presented.

In order to stress the importance of the choice of the sampling period, we have implemented all the above mentioned control laws with the sampling period T = 0.06 s. Local discrete gains (when analog servomechanisms were omitted) are also given in Table 5.11. The experimental results of tracking the same nominal trajectory with various control laws are presented in Fig. 5.22. It can be seen that the robot performance is worse than in the case when the sampling period was shorter. This results from the fact that 0.06 s is too long a period for the manipulator dynamics. The gains have to be rather low in order to keep stability of the robot, and the settling time is thus too long. The control signals are shown in Fig. 5.23.

In order to check the validity of our software package for control synthesis (which we used for the calculation of local and global gains),

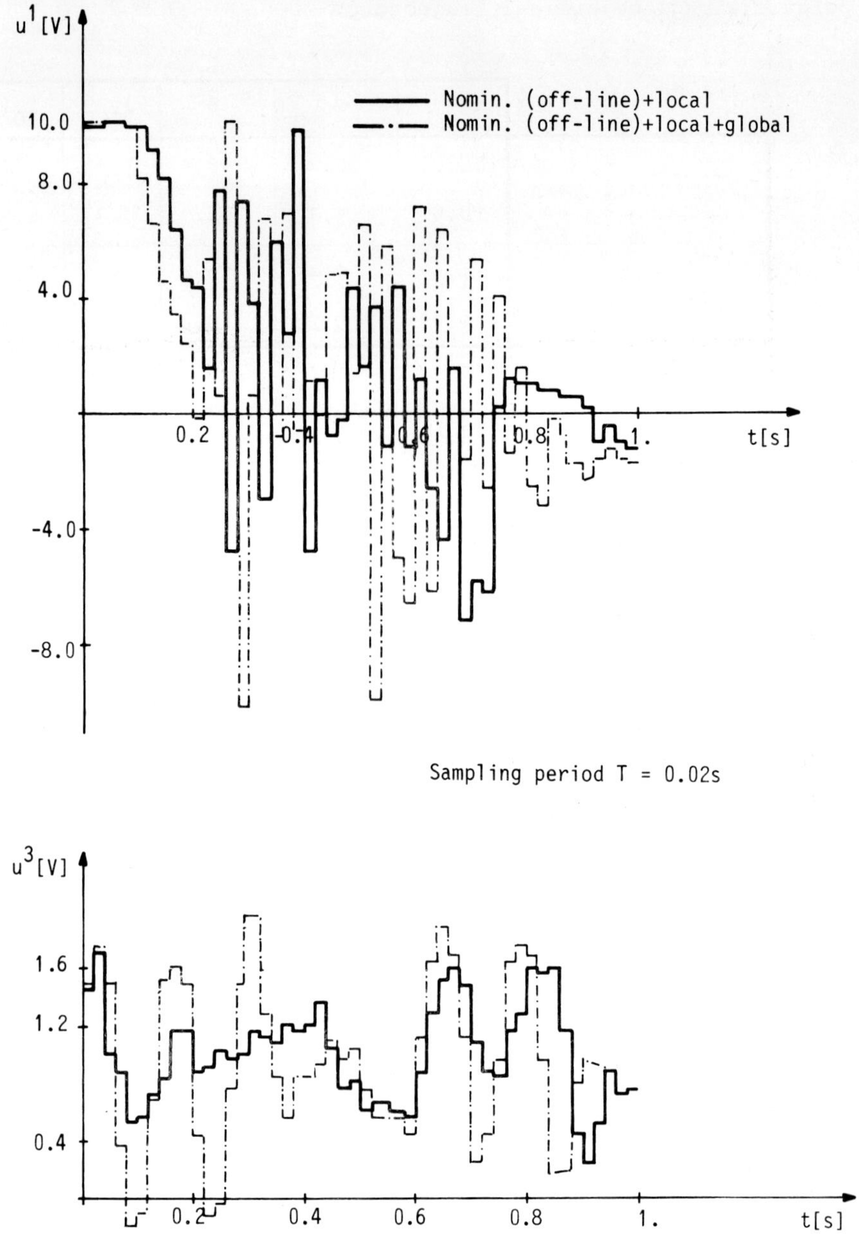

Fig. 5.21. Control signals during tracking of trajectory for UMS-2 - experimental results

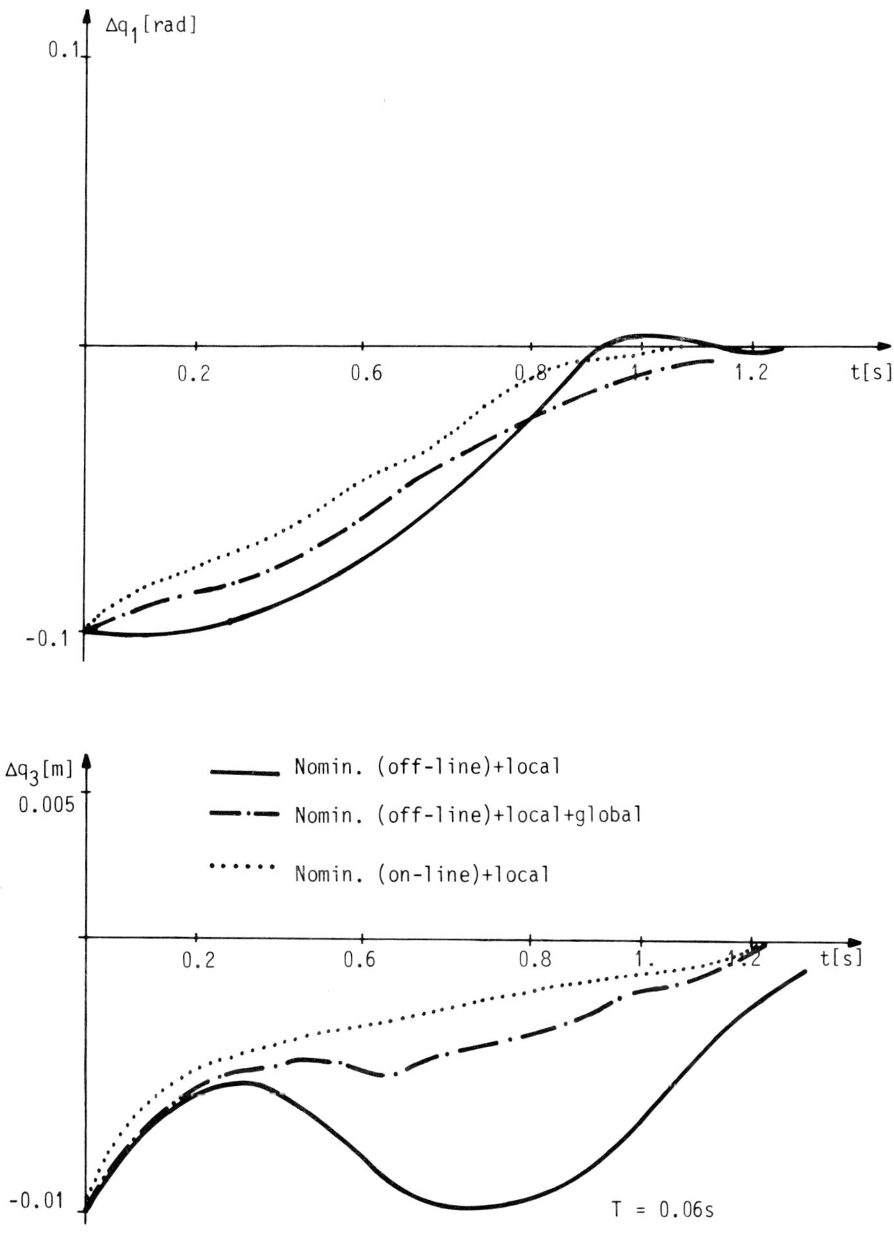

Fig. 5.22. Implementation of tracking of trajectory with discrete control - experimental results

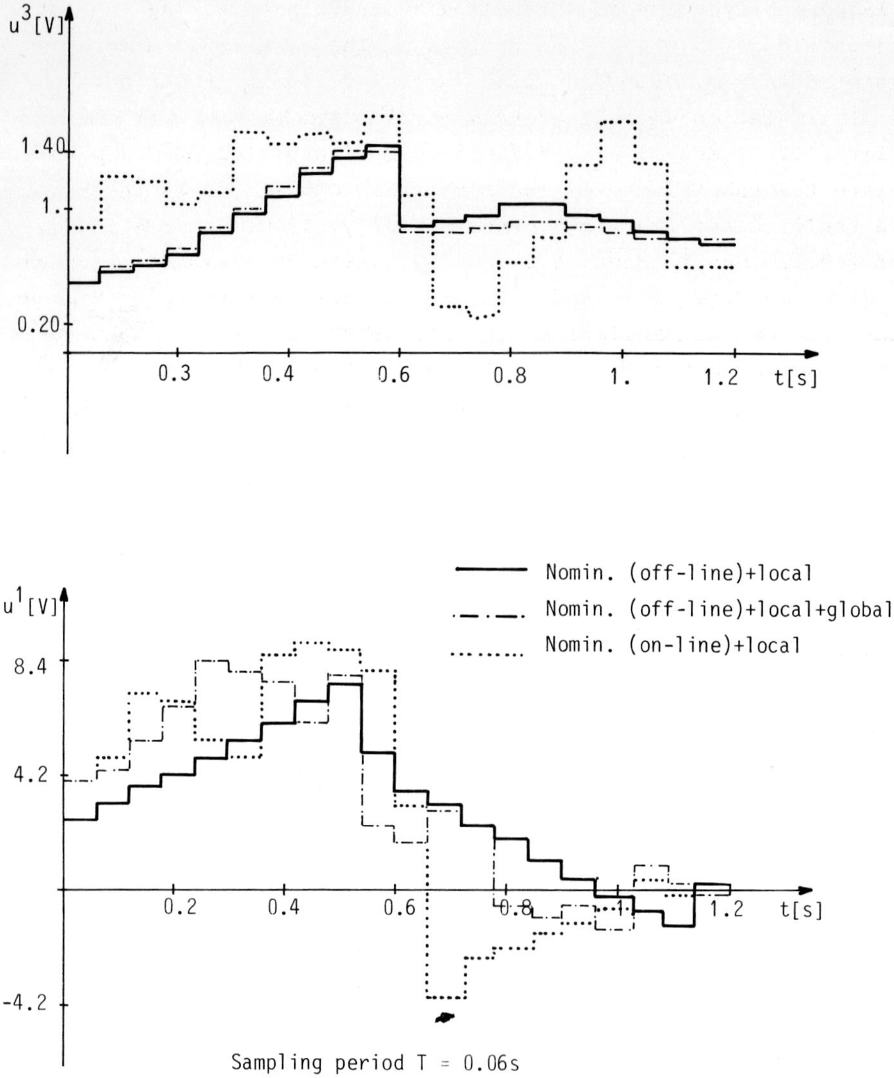

Fig. 5.23. Control signals during tracking of trajectory for UMS-2 - experimental results

we simulated the dynamics of the manipulator during tracking the nominal trajectories with various control laws. The results of simulation are presented in Fig. 5.24 and 5.25. In Fig. 5.24 are presented the results of simulation when discrete control is synthesized with sampling period T = 0.02 s, and in Fig. 5.25 the sampling period is T = 0.06 s. We simulate the robot dynamics under the same conditions as those under which we realized the trackings experimentally. If we compare the results of simulations in Figs. 5.24 and 5.25 with the results of experiment, given in Figs. 5.20 and 5.22, we can see that there is a good agreement between the simulation and real behaviour of the robot. This means that the mathematical model of the robot dynamics used in the package is well stated and that the parameters of the robot are well identified. This is the reason why the feedback gains synthesized using the software package were appropriate for the control implementation.

The results of implementation of various control laws show that a better performance is achieved with the control which takes care about the dynamics of the robot. However, such control is much more complex to be implemented since on-line calculation of the robot dynamics has to be achieved. If we have to ensure precise tracking of a desired trajectory it is necessary to take care about the robot dynamics through either nominal or global control.

If we compare various implementations of the proposed control laws it is obvious that the best results are achieved by analog discrete implementation of the local controllers. However, such implementation is the most complex. Discrete implementation of the controllers is the simplest from the "hardware point of view" (simple microprocessors are necessary for the calculation of trajectory). Discrete implementation is convenient from the point of view of possible changes and robot evaluation, as explained in Paragraph 5.2.

References

[1] Albus S.J., Barbera J.A. and Fitzgerald L.M., "Programming a Hierarchical Robot Control System", Proc. of the 12th. ISIR, pp. 505--517, Paris, June 1982.

[2] Megahed S. and Renaud M., "Minimization of the Computation Time Necessary for the Dynamic Control of Robot Manipulators", Proc. of the. 12th. ISIR, pp. 469-478, Paris, June 1982.

[3] Luh Y.S.J., Walker W.M., Paul C.P.R., "On-line Computational Scheme for Mechanical Manipulators", ASME Trans. J. Dynamic Syst.,

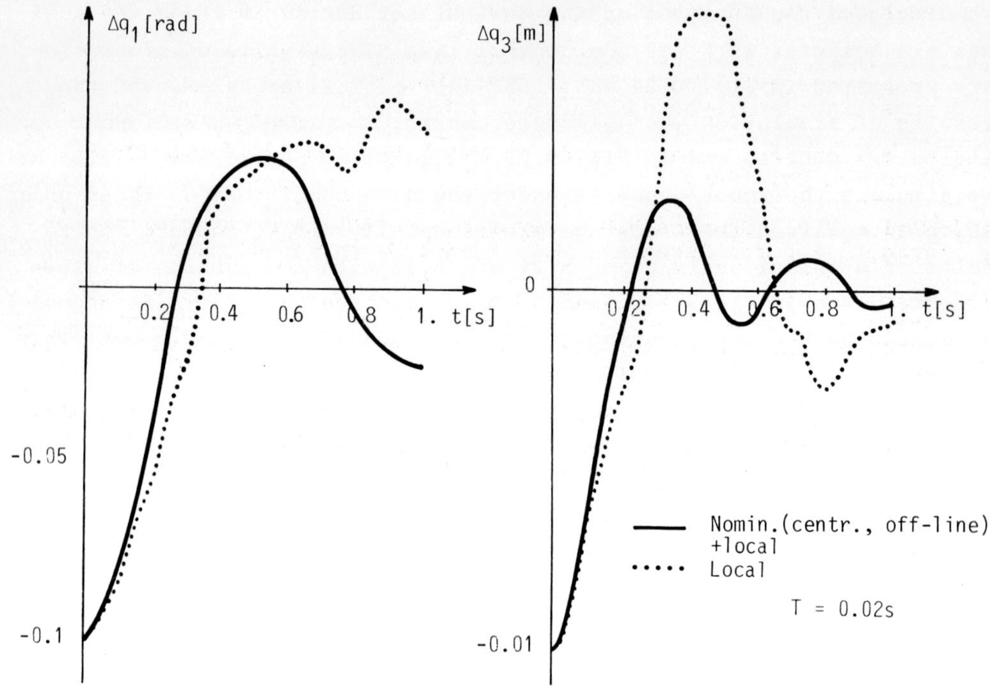

Fig. 5.24. Simulation of tracking of trajectory with discrete control

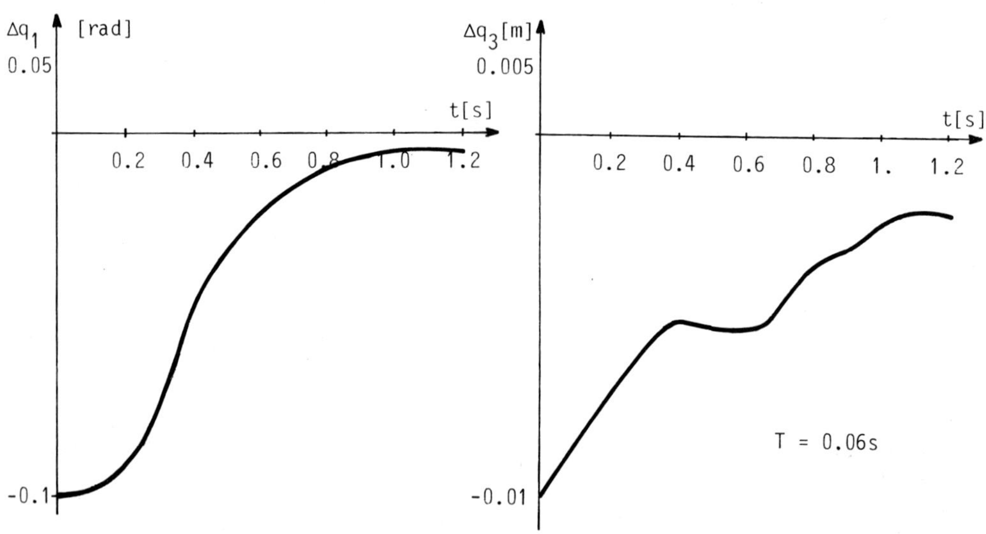

Fig. 5.25. Simulation of tracking with discrete control

Measurement and Contr., Vol. 102, No. 2, pp. 69-76, June, 1980.

[4] Luh Y.S.Y., Lin S.C., "Scheduling of Parallel Computation for a Computer-Controlled Mechanical Manipulator", IEEE Trans. on Systems, Man, and Cybernetics, Vol. SMC-12, No. 2, March/April, 1982.

[5] Unimation Inc. "User's Guide to VAL", Version 12, Unimation Inc., Danbury, CT, 1980.

[6] Taylor R.H., Summers P.D., Meyer J.M., "AML: A Manufacturing Language", Robotic Research, Vol. 1, No. 3, pp. 19-41, 1982.

[7] Vukobratović K.M., Kirćanski M.N., Stokić M.D., Kirćanski V.M., Karan B., "General Purpose Controller for Industrial Manipulators", Proc. of the Second Yugoslav-Soviet Symposium on Applied Robotics, Belgrade, pp. 1-15, 1984.

[8] Karan B., Vukobratović K.M., Kirćanski M.N., Stokić M.D., "A Software System for Teaching and Commanding the Industrial Robots", Proc. of the First Symposium on Robot Control, Barcelona, 1985.

[9] Vukobratović K.M., Kirćanski M.N., "Computer Assisted Generation of Robot Dynamic Models in Analytical Form", Acta Applicandae Mathematicae, No. 2, pp. 49-70, 1985.

[10] Vukobratović K.M., Kirćanski V.M., Scientific Fundamentals of Robotics, 3, Kinematics and Trajectory Planning, Monograph Springer-Verlag, Berlin, 1985.

[11] Vukobratović K.M., Kirćanski M.N., Real-Time Dynamics of Manipulation Robots, Scientific Fundations of Robotics, Springer-Verlag, 1984.

[12] Luh Y.S.J., Fisher W.D. and Paul R.P.C., "Joint Torque Control By a Direct Feedback for Industrial Robots", IEEE Trans. on Automatic Control, Vol. AC-28, No. 2, 1983.

[13] Vukobratović K.M., Kirćanski M.N., "Computer Assisted Sensitivity Model Generation in Manipulation Robots Dynamics", Journal of Mechanisms and Machine Theory, No. 1, 1984.

[14] Vukobratović K.M., Kirćanski M.N., "Decentralized Adaptive Control for a Class of Large-Scale Mechanical Systems", III Symp. on Large Scale Systems: Theory, Applications and Impacts, Warsaw, 1983.

[15] Mahmoud M., Singh M., Large Scale Systems Modelling, International Series on Systems and Control, Vol. 3, Perg. Press, 1981.

[16] Hassan M., Singh M., "A two Level Costate Prediction Algorithm for Nonlinear Systems", Automatica - The IFAC Journal, 6, 1977.

[17] Vukobratović K.M., Stokić M.D., Scientific Fundamentals of Robotics 2, Control of Manipulation Robots: Theory and Application, Monograph, Springer-Verlag, Berlin, 1982, also in Russian, Nauka, 1985.

[18] Vukobratović K.M., Stokić M.D., "Is Dynamic Control Needed in Robotic Systems, and if so to What Extent?", International Journal of Robotic Research, Vol. 102, June, 1982.

[19] Vukobratović K.M., Stokić M.D., "One Engineering Concept of Dynamic Control of Manipulators", Journal of Dynamic Systems, Measurement and Control, Trans. of the ASME, Vol. 103, No. 2, pp. 108 - 118, 1981.

[20] Stokić M.D., Timotijević M., "Algorithms of Dynamic Control of Manipulation Robots and its Microprocessor Implementation", The VI IFToMM Congres, New Delhi, 1983.

[21] Vukobratović K.M., Stokić M.D., "Microprocessor Implementation of Dynamic Control for Manipulation Robots", Digital Systems, (to be published), 1985.

[22] Stokić M.D., Vukobratović K.M., Hristić S.D., "Implementation of Force Feedback in Control of Manipulation Robots", International Journal of Robotic Research, (to be published), 1985.

Subject Index